Protected Cultivation of Horticultural Crops

AF765185

NIPA® GENX ELECTRONIC RESOURCES & SOLUTIONS P. LTD.
New Delhi-110 034

www.nipaers.com
Browse, Search, Read & Buy...
Print Books
eChapters
eBooks
Publishers
Forthcoming
Subject Catalogues
New Titles
Online Resources on:
✓ Current Affairs
✓ Reasoning, Logic & Aptitude
✓ Competitive Examinations
✓ English Language Lab
✓ Writing & Pronunciation Tools
✓ Personality Development
✓ Online Programmes for Professionals
✓ Interview Preparation
Pay Using
PhonePe
Google Pay
paytm
PayPal
UPI
CC Avenue

Protected Cultivation of Horticultural Crops

D.K. Singh
Professor
Department of Vegetable Science
G.B. Pant University of Agriculture & Technology
Pantnagar – 263 145, Uttarakhand

&

K.V. Peter
Director
World Noni Research Foundation
Chennai – 600 096, Tamil Nadu

NIPA® GENX ELECTRONIC RESOURCES & SOLUTIONS P. LTD.
New Delhi-110 034

NIPA® GENX ELECTRONIC
RESOURCES & SOLUTIONS P. LTD.

101,103, Vikas Surya Plaza, CU Block
L.S.C. Market, Pitam Pura, New Delhi-110 034
Ph : +91 11 27341616, 27341717, 27341718
E-mail: newindiapublishingagency@gmail.com
web: www.nipabooks.com

For customer assistance, please contact
Phone: + 91-11-27 34 17 17
Fax: + 91-11- 27 34 16 16
E-Mail: feedbacks@nipabooks.com

ISBN: 978-81-19254-15-6

Composed and Designed by NIPA®.

DEVOTION

Times Magazine rated Prof. M S Swaminathan as one among the icons of Indian History-Mohandas Karamchand Gandhi and Ravindranath Tagore. The Mahatma dreampt of a hunger free India.Tagore called hunger a national shame. It is the legendary Monkombu Sambasivan Swaminathan who made hunger a thing of the past by the Ever Green Revolution. We devote this book as a token of gratitude and love to him.

It is a divine destiny that the historical Food Security Bill which makes access to food a fundamental right is piloted by Professor Kuruppassery Varkey Thomas, the Union Minister of State (Food, Civil Supplies, Public Distribution). I join my late parents Varkey and Rosa to facilitate him. We devote this book to Prof. K.V. Thomas who goes to history by accomplishing the dreams and actions of Mahatma Gandhiji, Tagore and Swaminathan.

डॉ. पीतम चन्द्र
निदेशक

Dr. Pitam Chandra
Director

भाकृअनुप
ICAR

केन्द्रीय कृषि अभियांत्रिकी संस्थान
(भारतीय कृषि अनुसंधान परिषद)
नबी बाग, बैरसिया रोड, भोपाल - 462 038 (म.प्र.), भारत

Central Institute of Agricultural Engineering
(Indian Council of Agricultural Research)
Nabi Bagh, Berasia Road, Bhopal - 462 038 (M.P.), India

Off.: 91-755-2737191, 2521000, 2521001; Fax: 2734016; Res.: 2521203; e-mail : <director@ciae.res.in>, <pc1952@gmail.com>

Dated: 27th July 2013

Foreword

India is on its course to become the most populous country in the world, even surpassing China. Availability and accessibility of food and nutrition to this increasing population continues to be a daunting task for agricultural planners, researchers, farmers and others involved in the production consumption chain. Climate change and resource constraints are posing serious constraints on the profitability and sustainability of food production technologies.

Productivities of vegetables, fruits, ornamental plants, spices and aromatic plants are lower in India on account of vagaries of weather and dwindling availability of space, water, fertilizer, labour and energy. Efficient technologies for horticultural production in terms of space, labour and other inputs are emerging in a big way to meet the requirements of the expanding consumer base. Protected cultivation in the form of greenhouse, cloches, row covers, mulching, kitchen gardening, indoor gardening, terrace gardening, vertical farming, hydroponics, aeroponics etc. is drawings the attention of young farmers for more productive and dignified agriculture. High density planting, use of root stocks, pruning and training integrated pest and disease management, use of good agricultural practices, harvesting and post harvest technologies and establishment of food supply chains are all components of successful protected cultivation technology value chain.

The ICAR initiated efforts to conduct R&D in the area of protected cultivation in the early 80s. The All India Coordinated Research Project on Plastics in Agriculture and the Plasticulture Development Centre Scheme of the National Committee on the use of plastics in agriculture were the early structured programmes on development of protected cultivation in India. These schemes have over the three decades, resulted into appropriate technologies on greenhouses, low tunnels, plastic mulches, integrated pest, disease and nutrient management and appropriate machinery. The centre for protected cultivation technology at IARI, New Delhi has created awareness on the feasibility and economics of protected cultivation of horticultural crops in the country. Needless to say that, at present, several thousand hectare area in the country has been brought under protected cultivation.

The present book edited by Dr. Dinesh Singh, Professor, G.B. Pant University of Agriculture and Technology and Dr. K.V. Peter, Director, World Noni Research Foundation, Chennai is an excellent reference material being made available at the appropriate time when many state governments are promoting hi-tech agriculture. I congratulate the scientists who have contributed to this references book and the New India Publishing Agency, New Delhi to make this book available to the stakeholders.

Pitam Chandra

Preface

Productivity of vegetables, fruits, ornamentals, spices and aromatic plants are lower in India, being the above crops left to vagaries of weather; space,water, fertilizers, labour and energy being limiting factors. Space, water and energy saving horticulture are emerging in a big way in urban and peri-urban areas. Protected cultivation is known as vertical horticulture, terrace gardening, indoor gardening, kitchen gardening and with innovation has turned to hydroponics and aeroponics. High density planting, use of root stocks, pruning and training, integrated pest and disease management, use of good agricultural practices, harvest and post harvest technologies and above all establishment of cold chains and scientific retailing where producer farmers get at least 60% of customers price are all components of protected cultivation. The All India Co-ordinated Research Project on Plastics in Agriculture concentrates on structures in Protected Cultivation Houses. Water and nutrients are saved and efficiently used by drip fertigation. Mulching is done to save water and manage weeds in the polyhouses/net houses/rain shelters. Integrated Pest, Disease and Nutrient management technologies are used to produce clean fruits and vegetables. Agro-machineries are available for precision farming, prepararion of nursery beds, grafting and planting, regulating light and heat inside and waste disposal at the end. Self sufficiency in organic manures are assured by composting of the left over plant biomass.

The Centre for Protected Cultivation Technology (CPCT), IARI, New Delhi has demonstrated the economics and feasibility of protected cultivation of horticultural crops.

The present book comes up in the most appropriate time when many State Governments are promoting hi-tech agriculture. We congratulate the scientists who contributed the chapters and New India Publishing Agency, New Delhi for publishing the book.

Editors

Acknowledgement

We acknowledge the Vice-Chancellor, G B Pant University of Agriculture and Technology, Pantnagar for the support and encouragement extended to the first editor (Dr D K Singh).

Dr K V Peter acknowledges Prof. P I Peter, Chairman Noni Biotech, Chennai for the fullest support and Dr Kirti Singh, Chairperson, World Noni Research Foundation for guidance. No words are sufficient to express gratitude to both of them.

Mrs. Dinesh Kumar Singh and Ms Vimala Peter need special mention for their patience and endurance. We acknowledge all the contributors for putting their best to the chapters.

We also acknowledge The Staff of NIPA for excellent printing of the book.

Anvar K Peter, Ajay K Peter and Cynara Justin provided needed encouragements.

Editors

Contents

Contributors

A.A. Patil
Hi-Tech Horticulture Project
University of Agricultural Sciences
Dharwad-580 005
Karnataka

A.S. Sidhu
Division of Vegetable Crops
Indian Institute of Horticultural Research
Hessaraghatta, Bangalore – 560 089
Karnataka

A.K. Nair
Division of vegetable Crops
Indian Institute of Horticultural Research
Hessaraghatta, Bangalore – 560 089
Karnataka

A.K. Sadanandan
Proj. Coordinator Spices (Rtd.)
IISR, Calicut – 673 012, Kerala

Ajay Bhardwaj
Department of Olericulture
College of Horticulture
Kerala Agricultural University
Kerala

Balraj Singh
National Research Centre on Seed Spices
Ajmer -305206
Rajasthan

Brahma Singh
Former Director
FRL (DIHAR) Leh
Directorate of Life Sciences, DRDO
New Delhi

Deepa H. Dwivedi
BBA University
Vidya Vihar, Raebarelly Road
Lucknow – 226 025, UP

D.K. Singh
Department of Vegetable Science
G.B. Pant University of Agriculture & Technology
Pantnagar – 263 145
Uttarakhand

Ganga Joshi
G.B. Pant University of Agriculture & Technology
Pantnagar – 263 145, Uttarakhand

Ganesh B. Kadam
Directorate of Floricultural Research
New Delhi – 110 012

H.P. Bhagya
Division of Vegetable Crops
Indian Institute of Horticultural Research
Hessaraghatta, Bangalore – 560 089
Karnataka

I.M. Mannikeri
Hi-Tech Horticulture Project
University of Agricultural Sciences
Dharwad – 580 005
Karnataka

J.K. Chinnu
Division of Vegetable Crops
Indian Institute of Horticultural Research
Hessaraghatta Lake P.O.
Bangalore – 560 089
Karnataka

Jyoti Bhaskar
Communication Centre
Kerala Agricultural University
P.O. Mannuthy – 680 651, Kerala

K.V. Peter
Director
World Noni Research Foundation,
12 Rajiv Gandhi Salai
Sreenivasanagar P O Perungudi
Chennai – 600 096

M.S. Biradar
Hi-Tech Horticulture Project
University of Agricultural Sciences
Dharwad – 580 005, Karnataka

M. Prabhakar
Division of Vegetable Crops
Indian Institute of Horticultural Research
Hessaraghatta Lake P.O.
Bangalore– 560 089, Karnataka

P.S. Naik
Indian Institute of Vegetable Research
Post Bag. No.01, Shahanshahpur (Jakhini)
Varanasi, UP

P. Soman
Head, Agronomy and Training & Extension
Jain irrigation Systems Ltd.
Jalgaon, Maharashtra

P.K. Singh
G.B. Pant University of Agriculture & Technology
Pantnagar – 263 145, Uttarakhand

Prabhat Kumar
HNBGU
(A Central University)
Srinagar, Garhwal, Uttarakhand

Ranjan Srivastava
Department of Horticulture
G.B. Pant University of Agriculture & Technology
Pantnagar – 263 145
Uttarakhand

Ramesh Kumar
Directorate of Floricultural Research
New Delhi – 110 012

S.N.S. Chaurasia
Indian Institute of Vegetable Research
Post Bag. No.01, Shahanshahpur (Jakhini)
Varanasi, UP

S.S. Hebbar
Division of Vegetable Crops
Indian Institute of Horticultural Research
Hessaraghatta, Bangalore – 560 089
Karnataka

Sanjai K. Dwivedi
Defence Research and Development Organisation (DRDO)
Delhi – 110 054

S.S. Hebbar
Division of Vegetable Crops
Indian Institute of Horticultural Research
Hessaraghatta Lake P.O.
Bangalore – 560 089
Karnataka

S.S. Sindhu
Centre for Protected Cultivation Technology (CPCT)
IARI, Pusa
New Delhi – 110 012

Satish Chand
Department of Horticulture
G.B. Pant University of Agriculture & Technology
Pantnagar – 263 145
Uttarakhand

S.M. Mantur
Hi-Tech Horticulture Project
University of Agricultural Sciences
Dharwad – 580 005
Karnataka

T. Pradeepkumar
Department of Olericulture
College of Horticulture
Kerala Agricultural University, Kerala

Introduction

As per Wittwer and Castilla(1995) protected cultivation, which enables some control of wind velocity, moisture, temperature, light intensity and atmospheric composition, has contributed and will continue to contribute much to a better understanding of growth factor requirements and inputs for improving crop productivity in open fields. Structures for crop protection began in the early part of the Roman Empire to accommodate the Emporer Tiberius Caesar (14-37AD).It was not until the late 15th to 18th centuries that the precursors of greenhouses appeared, primarily in England, Holland, France, Japan and China. Paroda (2013) in his address at the First National Seminar on "Advances in Protected Cultivation" dealt in detail the scenario of protected cultivation, human resource development, components of protected cultivation, urban-horticulture, government support, protected cultivation clusters, strategies for accelerating production-policy, research, development and cluster-.To quote him "Since ancient times, agriculture is an outdoor or open field production of crops. Open field production is climate and weather dependent. In fact, growth and development of crops under a particular set of climate parameters define geographical location, productivity and production period of different crops. The magnitude of impact of climate and weather on agricultural productivity and quality of produce is appreciated by farmers and the scientific community, including horticulturists. Abiotic and biotic environments govern crop production potential and quality of products. Among the major constraints in production of horticultural crops are temperature (hot or cold), sunlight duration and quality, water deficiencies or excesses, atmospheric moisture (relative humidity), weeds, deficiency of nutrients, heavy winds, carbon dioxide and host of diseases and insect pests. There are ecological optima for obtaining production potential of each of the crops. Deviation from these conditions results in yield losses partially and sometimes totally. However, near optimal climatic conditions could be created by controlling the climate with the help of greenhouse using different protected structures/methods/devices and such cultivation under controlled environmental conditions is termed as protected cultivation".

The Scenario

Besides, sporadic efforts made by Defence Research and Development Organization (DRDO) in Ladakh, protected cultivation technology in India for commercial production is hardly three decades old, whereas in developed countries, namely Japan, Holland, Russia, UK, China and others, it is about two century old.

China started protected cultivation in 1990's and today the area under protected cultivation in China is more than 2.5 m ha and 90 per cent area is under vegetables. In China, low-cost protected technology, viz. plastic mulches, plastic low tunnels and walk-in- tunnels are being used on 80 per cent of the total area under protected cultivation and perhaps this is the basic reason that today China is the largest producer of vegetables in the world(400 million tones/year).

In recent years, Israel is one country which has taken big advantage of this technology by producing quality vegetables, flowers, fruits, etc. in water deficit desert area for meeting not only its small domestic demand but also the huge export demands. Recently, India established collaboration with Israel for demonstration of this technology at the Center for Protected Cultivation Technology (CPCT), Indian Agricultural Research Institute, New Delhi, which has paid good dividents.

Two Centers of Excellence for Vegetables and Fruit Crop Nurseries established in Haryana are functioning very well at Karnal and Sirsa, respectively in collaboration with Israel. Also Indo- Israel collaborative projects on protected cultivation in other states are doing well. The major challenges for the adoption of the protected cultivation are the lack of awareness, cost-effectiveness of the methods and lack of market linkages. If these aspects are properly addressed by the industrial players and the government, then protected cultivation industry will showcase remarkable growth in the next 5 years.

India, at present, is the second largest producer of vegetables in the world. The area under all forms of protected cultivation is around 25,000 ha. Indo-American Hybrid Seeds (India), Bangalore, is the pioneer in India to make use of greenhouse technology since 1965 for commercial productions of flower seeds

Protected Cultivation

Protected cultivation is one of the most promising areas of agriculture in the current context. It is an upcoming and alternative production system involving high-tech and intensive practices mainly for urban and export demands of horticultural and ornamental crops for food, nutrition and economic security.

Burgeoning population, fragmentation of land holdings, depletion and erosion of natural resources are all adversely affecting agricultural productivity. The protected cultivation offers several advantages to grow high-value crops with improved quality even under unfavourable and marginal environments. It has the potential of fulfilling the requirements of small growers as it can increase the yield manifold per unit area. The crops can also be grown round the year, including off-season with increased profitability. The technology has already been adopted in many parts of the country. Protected cultivation is one area where Government initiatives have been far more forthcoming though not as complimentary for human resource development and technical support to the farmers adopting protected cultivation. Protected cultivation or greenhouse cultivation or controlled environment agriculture is the most contemporary approach to produce mainly horticultural crops which are highly productive, conservative of water and land, and also protective to the environment. This involves manipulation of plant growth conditions to yield higher and quality produce during on and off-season or round the year. Off-season production of horticultural crops is one such area the importance of which can hardly be emphasized as it satisfies almost all stakeholders.

Human Resource Development

The protected cultivation is a new emerging field and there is dire need to train/ get exposure to graduates/postgraduates and senior scientists to oversee the works being done at Institutes of repute in advanced countries such as Israel and Holland. This trained manpower can help in strengthening R&D programes on protected cultivation in the country. Education and training of extension agents, growers and end-users on use of complete technology for protected cultivation of high-value vegetables and flowers, including IPM, post-harvest handling, on-farm value-addition, packaging, etc. and marketing are needed. Adoption of improved package of practices in holistic way, by using technologies such as design of structures and drip fertigation system, modern nursery raising and complete production technology are extremely important and need to be adopted for enhancing production and raising income. Issues like judicious use of pesticides and safe inputs, crop health, workers' safety, healthy packaging interventions and projection of GAP issues need to be addressed across all the stakeholders. The same should be duly inculcated through training/ HRD programes for addressing relevant environment safety and bio-safety issues. Developing software based databases on production protocols would facilitate knowledge transfer across different working partners. For group mobilization for harnessing the full potential of technology, timely technical advice should be provided.

Components of Protected Cultivation

Protected cultivation has two major components of technology. One is the infrastructure involving frames, cladding materials, irrigation system, tools, implements and other engineering inputs. These inputs ensure optimal light, air temperature, water and plant growth requirements. This optimal aspect of climatic parameters involves simple to most advanced engineering inputs such as automation to regulate several parameters such as ventilation which is one of the most important components in a successful greenhouse production. In the absence of proper ventilation, greenhouses and their plants can become prone to problems like high temperatures and development of pathogens, etc. Air circulation in protected structures is a must which is done by providing ventilation devices. Importance of cladding materials in protected cultivation can hardly be overemphasized. Their quality and cost are important besides certification. Micro-irrigation and fertigation involve a lot of science and technology, demanding research for continuous improvement. It is the engineering aspect of the protected structure which provides plants optimal conditions to grow normally.

Another aspect of protected cultivation is crop production technology which involves development of high-yielding varieties and hybrids of crops suitable for protected cultivation. This aspect has not yet been given due attention in India. Besides, development of varieties, the production protocol for protected cultivation is different from open field production. Stresses-biotic and abiotic- are different, good agricultural practices(GAP) are different and the integrated pest, water, nutrient, weed management, pollination, training of crops, harvesting practices, etc. are crop-specific and different than the ones followed in open field conditions. In nut shell, protected cultivation being a specialized farming, has engineering and production components, demanding full package of practices of its own.

Urban Horticulture

The technology of protected farming mainly of vegetables and flowers is the most suited to promote much needed peri-urban and urban agriculture being useful to both small-scale and large-scale farmers. The bulging metros would need this technology if they want to have fresh vegetables, flowers, fruits and mushrooms as transportation of these from distant places mainly rural areas would not render them as fresh produce. Moreover, post-harvest losses can be reduced significantly when production is located close to consumers especially of perishables. We have to think of developing multistoried vertical farms in each locality to meet the requirement of fresh produces. Mushroom production in some cities in the country is already a part of urban agriculture and so is the case with ornamental plant nurseries.

The Technology

Protected cultivation technologies cover climate control or high-tech greenhouses, poly/net-houses, naturally-ventilated green/polyhouses, hydroponics, aeroponics, plasticulture, drip irrigation, fertigation, mulching, integrated greenhouse pest management, lowcost protected structures like nethouses insect-proof and shade houses, hydroponics, aeroponics and vertical farms. At present, in India, we are not in a position to provide power to manipulate and regulate climate in protected structures with the help of sensors and nano materials. In research and developmental programes, we actually accord priority to them but for this purpose we have to outweigh the returns to justify high cost involved. For example, use of high-tech or climate controlled greenhouses for propagation of horticultural crops to multiply disease free planting material will go a long way to improve the production and productivity of these crops. The single use of this high-tech horticulture by establishing climate controlled disease-free nursery production centers can double production of horticultural crops. As stated above, low-cost protected structures such as cloches, low tunnels, mulch, crop covers, and rain shelters provide modification of plant growing environment. At the same time, it gives consumers safe vegetables at affordable prices and protects the environment.

Returns

It is estimated that annual returns per unit area from protected cultivation could be 10-100 times than those of open field cultivation. Protected cultivation ensures reduced use of water and land and it is becoming extremely important under current scenario of climate change. Protected cultivation is friendly to marginal farmers, to the environment, to unemployed youths, to meet requirements of fresh produces like vegetables, fruits and flowers round the year, urban and peri-urban farming and to one and all interested in this farming. The country's vision on protected cultivation should be to target the potentialities and convert them into reality with the help of stakeholders, including private sector through public private joint programes.

Government Support

The Government has taken an initiative paving the way for future growth of the protected cultivation in India. Protected cultivation is fast emerging as one of the progressive areas in agriculture through various central and state Government schemes such as National Horticulture Mission, *Rastriya Krishi Vikas Yojana* and National Horticulture Board. The subsidies are available on different components of protected cultivation such as Hi-tech nursery development, fertigation, mulching material, etc. A few growers are taking advantage of it. There is need to educate the farmers to create awareness on the issue that protected cultivation as a whole, is a potential farming system to augment their income.

Haryana Kisan Ayog has constituted several Working Groups including the one on "Development of Protected Cultivation in Haryana". The group submitted its report highlighting the importance of trained manpower, area and crop specific affordable protected structures, facilities and expertises for maintenance of structures. The group has come up, besides highlighting the necessity and potential of protected cultivation in Haryana, with specific recommendation worth emulating by each state in the country, i.e. the establishment of School of Protected Cultivation at

Chaudhary Charan Singh Haryana Agricultural University, Hisar, and a diploma course on Maintenance and Repair of Protected Structure/Greenhouses, Micro irrigation Systems and Related Aspects at each Industrial Training Institute. The proposed school has to train farmers also periodically with the help of *Krishi Vigyan Kendras*. The step taken by Government of Haryana to promote protected cultivation seems to be the first of its kind in the country.

Protected Cultivation in clusters

Undertaking protected cultivation in clusters avoids and minimizes a host of problems faced by a farmer undertaking this farming in isolation. Such an approach would help creation of electronic auction market mechanism linking growers to export houses/agencies besides getting empowered to get remunerative returns.

Strategies for Accelerating Production

Policy

- Government support needs to be extended for self-fabrication module of temporary low-cost structures like insect-proof net houses, shade net houses, walk-in-tunnels, and self fabricated low-cost protected structures for production of vegetables and flowers.
- There is need for regular uninterrupted power supply to maintain optimal growing conditions within protected structures.
- To provide better returns to the farmers, there is need to provide support with proper market linkages/ market intelligence and cool chain

Research

- Quality seeds/nursery/other planting materials suitable for protected structures should be ensured
- The most suitable crop sequences need to be developed for protected structures and seasons based on research
- Self-help groups with scientific support should be promoted for multiplication of quality planting materials for protected cultivation
- Selection and promotion of suitable design of protected structures need to be undertaken for different regions of the country for cultivation of high-quality long-duration vegetables and flowers.

Development

- Large-scale motivation and training of educated unemployed youths in the field of protected cultivation.

- Major Technical Institutes in the country should offer a diploma course on fabrication of protected structures, repair and their maintenance
- Large-scale promotion of low pressure drip irrigation system should be ensured for low-cost and small-scale protected cultivation in various parts of the country
- Practical demonstration unit at each state level needs to be established for ensuring demonstrations of all low-cost protected structures along with production and pest management strategies

Cluster

- Government should support and promote protected cultivation using cluster approach especially in peri-urban areas.
- Development of input hubs should be undertaken on priority for protected cultivation in multi-locations in PPP mode in various regions of the country.
- Electronic auctioning of produces through cluster representatives should be initiated.
- All the protected cultivation clusters should be mandatorily clubbed with rain water harvesting infrastructures and facilities

The states and union territories which have adopted protected cultivation in the country have increased from 9 states in 2007 to 30 states and union territories in 2012. The phase-wise implementation of protected cultivation has bolstered the adoption of protected cultivation across the country.

Combined efforts of the national boards such as NHM, NHB and RKVY have created awareness and are providing financial assistance to the farmers to adopt protected farming methods for agricultural cultivation. The states that have consistently expanded the area under protected cultivation for the period of 2007 to 2012 are Andhra Pradesh, Gujarat, Maharashtra, Haryana, Punjab, Tamil Nadu and West Bengal. In the coming years, it is expected that the states shall actively engage themselves in promoting protected cultivation in the states through various subsidy schemes, incentive programmes and human resource development for protected cultivation. With the rising involvement of the government national boards and the respective state agricultural agencies, the area under protected cultivation is expected to rise at about 84.2 per cent for the period 2013 to 2017"-Paroda (2013) concluded.

References

Paroda, R.S.2013. Strategies for protected cultivation. Inaugural Address delivered at the First National Seminar on 'Advances in protected cultivation' Indian Society for Protected Cultivation, CPCT, IARI, New Delhi

Sylvan H. Wittwer and Nicolas Castilla 1995.Protected Cultivation of Horticultural Crops Worldwide. *Hort.Technology* **5**(1):6-22

Pioneering Polyhouse Vegetable Breeding In India

D.K. Singh

Pant Polyhouse Tomato -2

India's population is increasing and simultaneously the demand of vegetables is also on the increase. They are potent source against mal nutrition. It is estimated that total requirement of vegetables will rise to 250 million tonne by 2050. To meet the demand, there is a high prerequisite of growing vegetables under protected condition as it implies 3-5 fold increase in their production.

But very little efforts have been made to exploit vast potential of vegetable production. Therefore, it is important to promote protected cultivation by breeding suitable varieties or hybrids of vegetables for protected cultivation by breeding suitable varieties or hybrids of vegetable for protected cultivation.

Pant Polyhouse Hybrid Tomato -1

Keeping in view the above facts, the concept of polyhouse vegetable breeding programme was started by Dr. D.K. Singh, Senior Vegetable Breeder

Pant Parthenocarpic Cucumber -2

during 2002 in cucumber and tomato. A large number of genepool/germplasm lines were collected from different parts of country and abroad. These accessions were evaluated for the search of *Pc* and *F* gene in cucumber and *sp*, *hp* and *og* gene in tomato; out of which some of good genotypes having these gene were isolated and exploited with specific objectives suitable for ideotype breeding for protected cultivation. Finally one hybrid PPBTH-1 and one pureline PPBT-2 in tomato and PCUCP-2 and PCUCP-3 in parthenocarpic cucumber were developed. These purelines and hybrid were tested in multilocational trails at Tarai, mid hills and higher hills under polyhouse condition. These identified genotypes were released by Uttarakhand State Variety Release Committee during May 2011, for commercial cultivation at farmers' polyhouse. Their main characters/potentials are as follows:

Pant Parthenocarpic Cucumber-2 (PCUCP-2) having yield potential 1755 q/ha and Pant Pathenocarpic Cucumber-3 (PCUCP-3) having yield potential 1605 q/ha and tomato, Pant Polyhouse hubrid tomato-1 (PPBTH-1) with yield potential of 1900 q/ha and avarage fruit weight of 130-140 g and Pant Polyhouse Tomato-2 (PPBT-2) having yield of 1500 q/h with average fruit weight 100-105 g. Besides Uttarakhand, these varieties were tested in Punjab and Karnataka.

Pant Parthenocarpic Cucumber -3

1

Protected Cultivation of Horticulture Crops – Present Scenario and Future Prospects

D.K. Singh and K.V. Peter

Abstract

Demand for horticultural commodities is increasing over a period of time. There is increasing pressure on available land for growing of such crops. The production of quality fruits, vegetables and other horticultural crops can be enhanced either by increasing productivity, use of improved cultivars/hybrids and effective production system, bringing additional land under horticultural crops or by promotion of protected cultivation which supports production in adverse climatic conditions and off seasons. Among horticultural crops, production of vegetables, flowers, medicinal and aromatic plants under protected condition is the best alternative to use the land and other resources more efficiently. Greenhouse cultivation is useful and is in practice in all kinds of climatic conditions, may it be severe and extreme cold (Canada and Russia), intolerant summer (Israel, UAE and Kuwait), excessive rains (Phillippines) and in moderate climate in several other countries. Nursery production of crops under protected conditions is highly economically viable.

The area under protected cultivation was only 0.23 per cent of the total area under horticultural crops in India in 2012. To boost up this, a number of varieties are bred for cultivation especially under polyhouse conditions.

Greenhouse for vegetable production encompasses glasshouse, polyhouse, insect-proof net house, low tunnel polyhouse and zero energy polyhouse. Protected structures are of different kinds, viz. open-ventilated; closed polyhouse with fan and pad system; shade net house; sloped roof, rain shelter etc. Micro-irrigation has proved its efficacy in saving water, improving the yield and also is responsible for reducing diseases which develop in rather moist conditions. Protected structures appeared as the best means and are being used by many nursery growers for multiplication of healthy planting materials and are becoming popular especially in the hilly regions. Use of techniques like soilless culture, production of high value and low volume flowers and vegetables, hybrid seed of important vegetable crops and grafting of promising varieties on desirable root stocks can be made more efficiently under protected environment. Seed yield of many crops can be 3-4 times more under cover compared to their open field cultivation along with ensured quality.

Protected cultivation technologies in Himalayan high altitudes of India for the welfare of the soldiers and the society are a landmark in cultivation of horticultural crops in non-conventional areas. The protected production system has the potential to contribute greatly for achieving high productivity and stability to production in the present scenario of climate change. Extension of growing/harvesting period including off season is another desirable feature associated with protected cultivation.

Keywords: protected cultivation, horticultural crops, off-season production, non–conventional area.

India is the second largest producer of fruits and vegetables in the world; however, as per the Medical Council of India, the production is much less than the requirement if balanced diet are to be provided to every individual. The blooming targeted production can be achieved by different ways, e.g., bringing additional area under horticultural crops, using hybrid seeds/ good planting materials and use of improved agro-techniques. Another potential approach is perfection and promotion of protected cultivation technology. However, the protected cultivation technologies are still in its preliminary stage in India and Concentrated efforts are required from all concerned agencies to bring it at par with the global standards. In view of the fragmented landholdings of the farmers in India, contract farming/ cooperative/consolidated farming needs to be encouraged so that modern technologies of cultivation can be adopted for increasing productivity and quality. With adoption of precision farming practices (polyhouse and micro irrigation along with water source augmentation structures), socio-economic status of small and marginal farmers can be enhanced with generation of gainful employment. Area under protected cultivation and especially low cost greenhouses can be increased manifold in peri-urban areas. Production of high value vegetables during off-season vis-à-vis production of high value cut flowers in protected structures would benefit growers due to high price of the produce in metropolitan market.

The consumption pattern in India is shifting towards fruits and vegetables hence; the demand for horticultural commodities is increasing over a period of time. In spite of a few distinct advantages, the horticultural produce is not providing the farmers with adequate returns, due to absence of necessary infrastructure, technology adoption and marketing. The development of horticulture depends crucially on three factors (i) Natural favourable conditions; (ii) Infrastructure and institutions; and (iii) Incentives for horticultural producers. Therefore, there is a need to adopt new technological interventions supplemented with institutional support for growth of the sector and remunerative returns to the stake-holders.

Why protected cultivation?

Protected cultivation means some level of control over plant microclimate to alleviate one or more of abiotic stresses for optimum plant growth which can be achieved in green houses, poly house, net house, poly-tunnels, cold frames, etc. Usually under open conditions, plants experience short cropping season but under protected conditions, the environmental factors are controlled or altered to a desirable extent to provide large growing period of the crop. In protected conditions, the production and availability of crop can be taken successfully all the year round for many crops, because protected production technology is a specialized form of cultivation of vegetables, flowers, medicinal and aromatic plants besides some fruits where environmental and edaphic conditions are adjusted as per the crops requirements. The protected structures are designed as per the climatic requirements of the crop so that optimum growth and yield could be realized. In the changing scenario of increasing population, rapid urbanization and industrialization, the natural resources particularly land and water are shrinking. Consequently, there is increasing pressure on available land for crop production and that too with effect of climate change. The huge gap between present production and future requirements necessitates for high production and productivity of horticultural crops for ensuring availability. The production of quality fruits, vegetables and other horticultural crops can be enhanced either by increasing productivity, use of improved cultivars/hybrids and effective production systems, bringing additional land under horticultural crops or by promotion of protected cultivation which can support production in adverse climatic conditions. Among horticultural crops, production of vegetables, flowers, medicinal and aromatic plants under protected condition has proved the best alternative to use the land and other resources more efficiently. The protected cultivation particularly in vegetables and flower crops has made it possible for year round production and availability of quality produces both for domestic use and export purposes. Yield level of horticultural produce under these structures can be several times higher than those of open field conditions. The quality of produce is also superior and input use efficiencies are usually higher under such structures. More than 90 per cent of Indian farmers have land holdings with an average farm size of about 1.57 hectare comprising nearly 55 per cent of the arable land. These farmers are resource-poor but are capable to sustain an intense and efficient agro-production system if provided with critical inputs like, technology, marketing and credit. The most important aspect is to ensure greater participation of small and marginal farmers and convert their inherent weaknesses into opportunities. In order to realize the full benefit of the protected cultivation technology, appropriate package of practices for suitable varieties at various locations need to be promoted.

Advantages of protected cultivation

- The yield may be 10-12 times higher than that of outdoor cultivation depending upon the type of greenhouse, type of crop and environmental control facilities.
- Possibility of better quality produce is higher under greenhouse cultivation.
- Ideally suited for vegetables and flower crops.
- Extended period and off-season production of vegetable and flower crops.
- Disease-free and superior quality transplants can be produced for round the year planting.
- Efficient utilization of chemicals and pesticides to control pests and diseases.
- Water requirement of crops is very limited and easy to control.
- Maintenance of stock plants, cultivating grafted plant-lets and micro propagated plantlets.
- Hardening of tissue cultured plants.
- Most useful in monitoring and controlling the instability of various ecological factors.
- Modern techniques of Hydroponics (Soil less culture), Aeroponics and Nutrient film techniques are possible only under greenhouse cultivation.

Adoption of appropriate technologies

Protected cultivation has the potential to enable farmers to achieve higher productivity, quality and maximize monetary returns. The major research and development issues include development of user friendly technology and reduced cost.

The field experiments help in examining the operational weaknesses and identification of appropriate structure or design for target areas and devise remedial measures for early adoption of technologies. Adoption of technologies helps in the development of industry and service sector for establishment of structures and marketing of final and quality produces to the terminal and the local market. Greenhouse production technology of horticultural crops emphasizes the need for having appropriate organic intervention for enhanced productivity and economy. Some of the advantages adopting protected cultivation for production of horticultural crops are as follows:

1. Higher productivity and better quality of produce.
2. Cultivation of cash crops under protective cover provides safeguard against adverse weather and enhances life span of the crops.
3. Higher productivity on sustainable basis through adoption of precision farming practices like precision planting/seeding, fertigation, and micro-irrigation, weed management, pest and disease management etc.
4. Efficient inputs use and reduction in cost of production.
5. Conservation of scarce resources i.e. water,space,energy,nutrients.
6. Farmers can advance/delay and segregate production of cash crops as in accordance with market demand.
7. Safeguard against impact of biotic and abiotic stresses.

Greenhouse Technology in the current framework

Since time immemorial, man has learnt how to grow plants under natural environmental conditions. In some of the temperate regions where the climatic conditions are extremely adverse and no crops can be grown, man has developed methods of growing some high value crops continuously by providing protection from the excessive cold, which is called as Greenhouse Technology. Greenhouse technology is the technique of growing crops by providing favorable environment/ growing conditions to the plants. It is used to protect the crops from adverse climatic conditions and to provide optimum conditions of soil, light, temperature, humidity carbon dioxide, etc. for the best growth of plants to achieve maximum yield coupled with quality.

Greenhouses are framed or inflated structures covered with transparent or translucent material large enough to grow crops under partial or fully controlled environmental conditions to get optimum growth and productivity. A greenhouse being a covered structure protects the plants from high wind velocity, excess evaporation, unusual precipitation, and temperature extremes. The CO_2 released by the plants during night is also trapped inside the greenhouse which is subsequently taken up by the plants itself during day time in the presence of light, and thus enhances photosynthesis. In temperate regions where temperature goes beyond -20°C, it is difficult to grow crops but by using greenhouse effect, it is made possible to grow vegetables. An ideal micro climate around the plants is created in this production system, where the environmental conditions are so modified that one can grow any plant in any place at any time by providing suitable environmental conditions with minimum labour.

Universal scenario of protected cultivation

The greenhouse technology is more than 200years old and Europeans were the pioneers in this field. Later, with the advent of plastics during World War II; a new phase in the greenhouse technology emerged. At present nearly 90 per cent of the new greenhouses are being constructed by utilizing ultra violet (UV) stabilized polythene sheets as the glazing material. There are more than 50 countries now in the world where cultivation of crops is undertaken on a commercial scale under cover. The Netherlands alone has around 9,000 greenhouse enterprises that operate over 10,000 hectares of greenhouses and employ about 150,000 workers, efficiently producing US$4.5 billion worth of fruits, vegetables, plants, and flowers, some 80 per cent of which are exported. One of the largest greenhouse complexes in the world is in Almeria, Spain, where greenhouses cover almost 50,000 acres (200 km^2) and is sometimes referred to as the 'sea of plastics'. In Spain, simple tunnel type greenhouses are generally used without any elaborate environmental control equipments mostly using UV stabilized polyethylene film as cladding material. Nonetheless, the greatest expanse of protected cropping occurs in Asia, especially in China, South Korea, Chinese Taipei and Japan. Extensive use is made of plastic greenhouses to modify winter conditions, achieve early and longer production seasons and to provide protection from adverse environmental conditions such as rain. Crops that are produced are mostly vegetables such as tomatoes and capsicums; flower crops such as roses, gerbera, lilium, carnations and cymbidium orchids; and fruit crops as diverse as bananas, strawberry and table grapes.

The United States of America has a total area of about 4000 ha under greenhouses mostly used for floriculture with a turnover of more than US $ 2.8 billion per annum and the area under greenhouses is expected to go up considerably, if the cost of transportation of vegetables from neighboring countries continues to rise. The area under greenhouses in Italy is 18,500 ha and is used mostly for growing vegetable crops like watermelon, capsicum, strawberries, beans, cucumbers and tomatoes. In Canada, the greenhouse industry caters both to the flower and off-season vegetable markets. The main vegetable crops grown in Canadian greenhouses are tomato, cucumbers and capsicum. Hydroponically grown greenhouse vegetables in Canada find greater preference with the consumers and could be priced as much as twice the regular greenhouse produce. The Netherlands is the traditional exporter of greenhouse grown flowers and vegetables all over the world.

With about 89,600 ha under cover, the Dutch greenhouse industry is probably the most advanced in the world. Dutch greenhouse industry however relies heavily on glass framed greenhouses, to cope up with very cloudy conditions prevalent all the year round. A very strong research and development component has kept the Dutch industry in the forefront. The development of greenhouses in Gulf countries is primarily due to extremity in the prevailing climatic conditions. Israel is a leading exporter of cut flowers and has wide range of crops under greenhouses (15,000 ha) and Turkey has an area of 10,000 ha under cover for cultivation of cut flowers and vegetables. In Saudi Arabia, cucumbers and tomatoes are the most important crops contributing more than 94 per cent of the total production. Egypt has about 1000 ha greenhouses consisting mainly of plastic covered tunnel type structures. Arrangements for natural ventilation are made for regulation of temperature and humidity conditions. The main crops grown in these greenhouses are tomatoes, cucumbers, peppers, melons and nursery plant material.

In Asia, China and Japan are the largest users of greenhouses. The development of greenhouse technology in China is faster than in any other country in the world. With a modest beginning in late seventies, the area under greenhouses in China has increased to 48,000 ha in recent years. Out of this, 11,000 ha is under fruits like grapes, cherry, Japanese persimmon, fig, loquat, lemon and mango. Most of the greenhouses in China are unheated and use straw mats to improve the heat retention characteristics. Japan has more than 40,000 ha under greenhouse cultivation of which nearly 7500 ha is devoted to only fruit orchards. Greenhouses in Japan are used to grow wide range of vegetables and flowers with a considerable share of vegetable demand being met from greenhouse production. Even a country like South Korea has more than 21,000 ha under greenhouses for production of flowers and fruits.

Thus, greenhouses permit crop production in areas where winters are severe and extremely cold as in Canada and Russia, and also permit production even in areas where summers are extremely intolerable as in Israel, UAE, and Kuwait. Greenhouses in Philippines make it possible to grow crops in spite of excessive rains and also in moderate climate of several other countries. Thus, in essence, greenhouse cultivation is being practiced and is possible in all types of climatic conditions.

Scenario of protected cultivation in India

In India, the technology is still in its nascent stage. The area under greenhouse cultivation in India, as reported by National Horticulture Mission, Government of India was about 2000 hectare including 500 ha net house, shade nets and 1500 ha green house in the year 2005 which has grown up to 80795.81 hectare by the end of 2012 (Table 1). The area under protected cultivation was 0.23 per cent of the total area under the horticulture cultivation in India in the year 2012. The major crops grown in the protected cultivation are tomato, capsicum, cucumber, melons, rose, gerbera, carnation, lilium and chrysanthemum. With the concerted efforts of the Central and State governments, protected cultivation is gaining popularity in India. The states and union territories which have adopted protected cultivation in the country have increased from 9 states in the year 2007 to 30 states and union territories in 2012. The phase wise implementation of protected cultivation has bolstered the adoption of protected cultivation across the country. There exists a vast scope for further expansion of greenhouse technology in India. The greenhouses which existed in 1960s were used for commercial purposes but the impetus and the much required thrust came only in the mid 1980's with the emergence of the industries manufacturing UV-stabilized low-density polyethylene (LDPE) and the development of the indigenous technology for low-cost greenhouses. Production of brinjal, capsicum, tomato, and cucurbits is taken in the summer months on a large scale, whereas the green leafy vegetables are being grown in the long frozen winter months when the average temperature reaches up to - 40°C. Underground greenhouses and soil trenches are also being used on a large scale in these remote areas. In India, the use of greenhouse technology started during 1980's for research activities. This may be because of our emphasis, so far had been on achieving self-sufficiency in food grain production. In recent years in view of the globalization of international market and tremendous boost and fillip that is being given for export of agricultural produce, there is a spurt in the demand for greenhouse technology. The National Committee on the use of Plastics in Agriculture (NCPA-1982) has recommended location specific trials of greenhouse technology for adoption in various regions of the country.

Greenhouses are being built in the Ladakh region for extending the growing season of vegetables from 3 to 8 months. In the North-East, greenhouses are being constructed essentially as rain shelters to permit off-season vegetable production. In the Northern plains, seedlings of vegetables and flowers are being raised in the greenhouses either for capturing the early markets or to improve the quality of the seedlings. The increase in yield of various crops under protected cultivation is reported up to 200-320% in capsicum, 370% in brinjal, 441% in chilli, 252-348% in tomato and upto the tune of 1300% in cucumber.

Propagation of hard-to-root tree species is also very encouraging. Several commercial floriculture ventures are coming up mainly in Maharashtra, Tamil Nadu and Karnataka states and in other parts of country as well to meet the demands of both domestic and export markets. The commercial utilization of greenhouses started from 1988 onwards and now with the introduction of Government's liberalization policies and developmental initiatives (Table 1), several corporate houses have entered to set up 100 per cent export oriented units. In just four years, since implementation of the new policies in 1991, 103 projects with foreign investment of more than Rs.800 million have been approved to be set up in the country at an estimated cost of more than Rs.10 thousand million around Pune, Bengaluru, Hyderabad and Delhi. The area under climatically controlled greenhouses of these projects is around 300 ha. out of which many have already commenced exports and have received very encouraging results in terms of the acceptance of the quality in major markets abroad and the price obtained. Presently, the total protected structures including green house, polyhouse, net house cover about 2000 ha area in the country.

Table 1: Area (in ha.) covered under Green House during 2005-06 to 2011-12 under National Horticulture Mission

States	Hi-tech Green House (GH)	Naturally Ventilated GH
Andhra Pradesh	5.39	20
Bihar	NA	4.94
Chhattisgarh	1.72	2.64
Delhi	1.07	7.05
Goa	0.01	0.55
Gujarat	8.15	77.12
Haryana	1.74	13.87
Jharkhand	16.3	3.8
Karnataka	46.02	131.91
Kerala	2.43	10.9
Madhya Pradesh	2.05	14.96
Maharashtra	42.3	153.04
Odisha	1.68	4.79
Puducherry	NA	6.2
Punjab	13.26	17.49
Rajasthan	10.0	21.03
Tamilnadu	116.57	44.46
Uttar Pradesh	0.25	18.68
West Bengal	5.39	16.27
Andaman & Nicobar Islands	0.1	2.48

Polyhouse Vegetable Breeding

Population is increasing and simultaneously the demand of vegetables is also on the increase. They are potent source of balanced nutrition. It is estimated that total requirement of vegetables will rise to 250 million tones by 2030. To meet the demand, there is a high perquisite of growing vegetables under protected condition as it implies 3-5 fold increase in their production. But very little efforts are made to exploit vast potential of vegetable production. To boost polyhouse vegetable breeding, a strong research support is required (Singh, 2013). It is important to promote protected cultivation by breeding suitable varieties or hybrids of vegetables for protected cultivation.

Keeping in view the above facts, the concept of polyhouse vegetable breeding programme was started by Dr. D. K. Singh, Senior Vegetable Breeder during 2002 in cucumber and tomato at G.B.Pant University of Agriculture and Technology, Pantnagar,Udham Singh Nagar, Uttarakhand.

A large number of genepool/germplasm lines were collected from different parts of country and abroad. These accessions were evaluated for the search of the *Pc* and *F* gene in cucumber and *sp*, *hp* and *og* gene in tomato; out of which some of good genotypes having these genes were isolated and exploited with specific objectives suitable for ideotype breeding for protected cultivation (Singh, 2013). One hybrid PPBTH-1 and one pureline PPBT-2 in Tomato and PCUCP-2 and PCUCP-3 in parthenocarpic cucumber were developed. The pureline and hybrid were tested in multilocational trials at Tarai, mid hills and higher hills under polyhouse condition. These identified genotypes were released by Uttarakhand State Variety Release Committee during May 2011, for commercial cultivation at farmers' polyhouses. Their main characters/potentials are as follows:

A) Pant Polyhouse Hybrid Tomato-1 (PPHT-1)

The variety has 7-8 fruits per cluster, single fruit weight 130-140 g , average yield potential of 1900 q/ha. This cultivar has better keeping quality because of its thick pericarp (1.0-1.25 cm).

B) Pant Polyhouse Tomato-2 (PPBT-2)

The variety has 5-6 fruits per cluster, single fruit weight 100-105 g with average yield potential of 1500 q/ha. This cultivar has better storage quality because of its thick pericarp (0.9-1.0 cm).

C) Pant Parthenocarpic Cucumber-2 (PPC-2)

Plant bears only female flowers (gynoecious), 551 numbers per plant of having single fruit weight 630 g with average yield potential of 1755 q/ha. Plant produces seedless fruits (Pathenocarpic in nature).

D) Pant Parthenocarpic Cucumber-3 (PPC-3)

Plant bears only female flowers (gynoecious), 465 number per plant of having single fruit weight 415 g with average yield potential of 1605 q/ha. Plant produces seedless fruits (Pathenocarpic in nature).

Pant Parthenocarpic Cucumber-3 (PPC-3), Pant Parthenocarpic Cucumber-2 (PPC-2), Pant Polyhouse Hybrid Tomato -2 (PPHT-2), Pant Polyhouse Tomato - 2 (PPBT-2), bell-pepper hybrids (Swarna and Natasha) were the best for commercial cultivation under greenhouse condition in mid-hills to high hills of Uttarakhand in an experiment conducted by Singh *et al.*, (2013).

One promising advanced line in cherry tomato namely Pant Polyhouse Cherry Tomato-1 is in pipeline for release. This variety is early flowering, heavy fruiting (120-150 fruits/cluster), 8 per cent TSS content and 30-40 mg (in 100g tomato flesh) vitamin C. This genotype accounts 10-13 fruit cluster per plant, having about 10g average fruit, fruit yield 750 q/ha.

Design and inputs for protected structures

The main principle involved in protected agriculture is green house effect for heating cultivated space using sun rays and ventilation for cooling and air (CO_2) circulation. The cultivated area is isolated and covered with plastic film or glass which is transparent to incoming short wave infrared radiation from soil and plants. As a result, protected space maintains higher temperature than the ambient. The tapped heat is gradually dissipated through conduction, convection/ventilation and radiation. By incorporating heat sinks, reduced ventilation and application of radiation shields, either the rate of cooling is slowed down or the supplemented heat source is placed inside. The term greenhouse means different things to different people. A greenhouse is a structure formed of glass, with a heating (or cooling) system that is used year-round, but especially in winter. A greenhouse is generally covered with a transparent material like polythene or glass but in the present scenario, normally polycarbonate sheets are used because of their longevity (up to 12 year), durability, and transparency. The greenhouse covered with simple plastic sheet is termed as polyhouse. Depending upon the cladding material and its transparency, major fraction of sunlight is absorbed by the growing crops and other objects. These objects in greenhouse, in turn, emit long wave thermal radiations for which cladding material has lower transparency. With the result, solar energy is trapped which raises the temperature inside the greenhouse. The greenhouse generally reflects back 43 per cent of the net solar radiation incident upon it allowing transmittance of the photosynthetically active solar radiation in the range of 400 -700 nm wavelength. In summer season, above-stated phenomenon, ventilation, exhaust fans and cooling pads are required to maintain the temperature well inside the structure below 35°C.

Design of polycarbonate and fibre-reinforce polyester (FRP) sheet-clad greenhouses are found suitable for production of indeterminate hybrids of tomato, capsicum, chilli, brinjal, gynocious hybrid of cucumber and zucchini during summer from 2,400-3,600 m altitude and leafy vegetables, i.e., Swiss chard, beet leaf, kale, lettuce, etc. during winter. Above 3,600 m altitude, selection of the crops suitable to grow in the greenhouse is changed as cabbage, cauliflower, radish, turnip and short duration varieties of tomato. The greenhouse made of locally available material except cladding material is found suitable to raise early nursery and round the- year production of leafy vegetables. However, the trench greenhouse is the most suitable to produce off-season potato, vegetable nursery and certain leafy vegetables during March-May. The early protected nursery especially of cucurbitaceous crops, can be produced in seasonal greenhouse as they provide better scope for hardening of the seedlings before transplantation. Moreover, such types are also suitable for flower crops.

Greenhouse for high value low volume crops particularly flowers and vegetable crops production encompasses glasshouse, polyhouse, insect-proof net house, low tunnel polyhouse and zero energy polyhouses. Protected structures are of different kinds, viz. open-ventilated; closed polyhouse with fan and cooling-pad system; shade net house; sloped roof, rain shelter etc. Greenhouse structures are of different kinds based on shape (lean to type, evan span type, ridge and furrow type etc.), utility (temperature and humidity controlled), construction (wooden, pipe or truss framed), and covering material (glass, fibreglass, plastic-film). Plastic film covering materials are of different types like acrylic, polycarbonate, fibre glass reinforced polyester, polyethylene film and polyvinyl chloride film. Plastic glazed greenhouses have many advantages over glasshouses, the main being cost. (Montero *et al.*, 2005). Plastic also is adapted to various greenhouse designs, usually resistant to breakage, lightweight and fairly easy to use. Most greenhouse crops grow the best in light whose wavelengths range from 400 to 700 nm and hence the glazing materials should be highly transparent. The different types of greenhouses are as follows:

(i) Hi-tech or climate controlled greenhouse

This type of greenhouse is constructed to achieve higher degree of climate control to enhance the cultivation period of crops. Evaporative cooling and exhaust fans are used to maintain required temperature inside greenhouse spaces as and when needed. The climate is usually accomplished through automatic system. Such types of greenhouses are used for growing high value blemish free crops for long period. These greenhouses are being used mainly in European countries, Morocco, Israel, USA and Japan. These are computer controlled for environmental, edaphic and nutritional factors. It is regularly monitored through computer and regulated over a longer period of time. These greenhouses are mostly used for long duration cultivation of tomato and sweet pepper besides a variety of flowers like rose, gerbera, orchids, anthurium, etc. In most of the countries, the greenhouses are made of plastic except the glass houses as in The Netherlands. Glasshouses and rigid plastic houses are longer life structures, and are therefore mostly located in cold regions where these structures can be used for crop production throughout the year. Asian continent has maximum area under greenhouses and China leads amongst countries followed by Italy, Spain and Japan.

(ii) Semi-climate controlled greenhouses

In this kind of greenhouses, the structural frame is made of Galvanized Iron pipes, like the climate controlled greenhouse but only exhaust fans with evaporative cooling pads are provided to maintain a favourable temperature and humidity during summer. These types of greenhouses are suitable for flower and vegetable cultivation during mild winter and mild summer only. They are also being operated for nursery raising of different horticultural crops, hardening of tissue cultured plants etc. The basic cost of erection of these greenhouses is half to that of climate controlled greenhouses and are used in Turkey, Spain, USA, Morocco, Israel and Japan.

(iii) Naturally ventilated or low cost greenhouses

These are simple greenhouses with low initial investment. The frame may be of Galvanized Iron pipes, wooden logs or steel pipes but no heating and cooling systems are provided for the structure. The top of the greenhouses is covered with plastic but the side walls have insect proof nets from ground to a height of 1.5 to 2 m. with or without manually rollable plastic cover. The central height is 6-8m. for better air circulation and effective cooling.

The initial cost of these greenhouses is less than half to that of semi-climate controlled greenhouses. These are mainly used for production of tomato, cucumber, muskmelon, gerbera, lilium, carnation, etc. and are being used extensively in China where the area during the last two decades has increased rapidly, followed by Spain and South Korea. Other countries using such greenhouses are Japan, Israel, Turkey, Spain, Italy, etc (Table 2).

Table 2: Major greenhouse vegetable production countries of the world (>500 ha only)

Country	Ha ('000)	Country	Ha ('000)
China	81.0	Argentina	2.2
Spain	70.4	Chile	2.1
South Korea	47.0	Jordan	2.0
Japan	36.0	Belgium	1.6
Turkey	33.5	Russia	1.4
Italy	25.0	Germany	1.4
Morocco	16.5	Australia	1.3
France	10.0	Tunisia	1.3
Poland	5.2	Romania	1.3
Hungary	5.4	Egypt	1.2
Algeria	5.0	Canada	1.2
Greece	5.0	Bulgaria	1.1
Netherlands	4.6	Libya	1.0
Columbia	1.2	Serbia/ Montengr	1.0
Mexico	4.3	Lebanon	1.0
Israel	4.0	Brazil	1.0
Iran	4.0	United Arab Emir.	0.8
Palestine	3.3	India	0.7
Syria	3.1	New Zealand	0.7
Ukraine	2.7	United kingdom	0.7
Ecuador	2.7	USA	0.7
Portugal	1.5	Moldova	0.5

Source: Hickman 2011

(iv) Plastic low tunnels

Plastic low tunnels or low covers are simple and low cost structures for off-season production of vegetables in open field. These are flexible transparent coverings installed over single or multiple rows of vegetables crops during winter season to enhance plant growth as the micro-climate around the plants gets congenial. The soil temperature gets raised and the plants are also protected from hails, cold wind injury, frost etc. Plastic low tunnels or row covers originated during the 16th and 17th centuries with the use of jars, cloches and cold frames placed over the plants during the cold period. Low tunnels can be supported over the plants using hoops of wire. A transparent white plastic is stretched over the hoops and the sides are secured by placing it in the soil. The plastic of the tunnel can be vented during the cropping season as the temperature increases within the tunnel environment. Since construction and dismantling of these facilities occur frequently, therefore, statistical data are less reliable than that of permanent structures, like greenhouses, glasshouses, walk-in-tunnels, etc. although use of these protected structures started in West and North European countries and Japan, but presently, China has the largest area under plastic low tunnels followed by Japan and European countries (Table 3). In India, although the area under plastic low tunnel cultivation of horticultural crops is not more than 1000 hectares yet these are highly suitable and profitable for off-season growing of vegetables in peri-urban areas of Northern plains, cold dessert areas and other areas of hill states like Jammu & Kashmir, Himachal Pradesh and Uttarakhand. The production technology for off-season cultivation of cucurbits under plastic low tunnels has already been standardized under Indo-Israel Project at Indian Agricultural Research Institute, New Delhi for commercial cultivation of vegetable crops . Moreover, with the introduction and availability of biodegradable plastics for making low tunnels, there may be a big breakthrough in this technology in India as it will be more eco-friendly and environmentally sustainable. Moreover, the technology has already proved to be a boon for regions like Ladakh and Auli. The major schemes funded by Government of India on Plasticulture applications are given in Table 4.

Table 3: Area (in ha.) under plastic low tunnels for protected cultivation of horticultural crops in different countries

Country	During 1990-91	During 1999-2000
Africa	-	80,000
Egypt	15,000	50,000
USA	5,000	15,000
China	20,000	85,000
Japan	53,624	56,500
Italy	100,000	24,000
Spain	100,000	17,500
France	50,000	16,000
Israel	2,000	5,000
India	100	1,000
World	-	882,000

Source: International Committee for Plastic in Agriculture, 2000.

Table 4: Major Govt. of India schemes on Plasticulture applications

S. No.	Name of the Scheme	Plasticulture component
1.	National Mission on Micro Irrigation (NMMI)	Drip & Sprinkler Irrigation
2.	National Horticulture Mission (NHM)	Greenhouse, Shadenet House, Mulching, Low Tunnel, Community Pond
3.	Horticulture Mission for North Eastern & Himalayan States (HMNH)	Greenhouse, Shadenet House, Mulching, Low Tunnel, Community Pond, Hail Protection nets
4.	Rashtriya Krishi Vikas Yojana (RKVY)	Drip & Sprinkler Irrigation, Greenhouse, Shadenet house, Hail Protection nets
5.	NABARD – Rural Infrastructure Development Fund (RIDF)	Drip & Sprinkler Irrigation (Project Based)
6.	Integrated Scheme on Oilseeds, Pulses, Oilpalm & Maize (ISOPOM)	Drip & Sprinkler Irrigation
7.	National Food Security Mission (NFSM)	Drip & Sprinkler Irrigation
8.	National Horticulture Board (NHB)	Drip & Sprinkler Irrigation, Greenhouse, Shade-net House, Mulching, Low Tunnel, (Project based)
9.	Technology Mission on Cotton	Drip Irrigation

(v) Floating mulch or floating row cover

Floating mulches are also the temporary and low cost protected structures made without any mechanical support of wire hoops, which mainly provide protection against adverse temperature conditions for out of season cultivation of several horticultural crops, especially tender warm season vegetables to advance maturity, improve quality and extend the growing season. Floating mulch or row covers are made of spun bounded or non-woven fabrics which can be used to cover either single row or multiple of rows.

The floating covers have minimal effect on temperature and light transmission. For self-pollinating vegetable crops and leafy vegetables, the floating covers can be kept on for the entire duration of the crop. However, when the temperature rises under the covers during sunny days these need to be removed. Europe has maximum area under floating mulch (Table 5). These structures can also be used in India for out of season production of cucurbits in peri-urban area to catch early markets.

Table 5: World Area under floating mulch (ha)

Country	During 1990-91	During 1999-2000
U. K.	9000	12000
France	8000	10000
Italy	3000	10000
USA	1500	5000
China	1000	8000
Japan	4000	6000
Egypt	-	5000
World	-	86000

Source: International Committee for Plastics in Agriculture, 2000.

(vi) Use of plastic mulches

Mulching is the practice of covering soil around plants with an organic or synthetic material to make conditions more favourable for plant growth, development and crop production. By proper selection of plastic mulch composition, colour and thickness, it is possible to precisely control the soil environment. Black plastic mulch is mostly used during summer for soil moisture conservation and weed control, whereas the transparent mulch of 30-40 micron is largely used during winter to increase the soil temperature and micro-climate around the plants for protection against cold. Coloured plastic mulches are used for reduction of insect infestation. Yellow colour mulch is mainly used in several countries like Israel, USA, European countries for reduction of virus incidence in several vegetable crops like tomato, chilli, sweet pepper, okra etc. Plastic mulches have many advantages, *viz.* (i) Early crop growth (ii) Check losses through evaporation (iii) Effective weed control (iv) Reduce nutrient loss (v) Reduce soil compaction (vi) Cleaner crop produce (vii) Checking insect incidence on crop (viii) Provide natural carbon-dioxide enrichment to crop plants.

Soil-less cultivation

Soil system under protected environment requires frequent replacing of greenhouse soils due to problems associated with soil degradation, over-fertilizing or soil borne diseases. This led to development of soil-less cultivation where local materials like rock wool, peat moss, coir, sand, perlite, coconut fiber etc. can be combined to form a substrate or medium to support the plants to realize their full potential. Soil-less culture methods may be categorized as either solid or liquid-medium systems. Sterilized, soil-less growing media mostly comprising coco-peat, perlite and vermiculite in the ratio of 3:1:1 with addition of hydrogel, biological agents and organic amendments not only helps in raising high density seedlings but also evades soil-borne diseases and some insect-pests. Seedlings are watered when necessary and liquid fertilizer is applied based on leaf colour and growth analysis. Liquid systems also called 'hydroponics' are generally closed circuit with respect to nutrient-solution supply wherein the solution is re-circulated continuously or intermittently for a period of days or weeks, which is closely monitored through computer attached sensors.

Among the fertilizers applied commonly to the soil, only nitrogen can be recovered totally, but in hydroponics, all the nutrients provided to plants can be retained. It is the most critical aspect to disinfect the drainage solution prior to reuse it.

Micro-irrigation, fertigation and nutrient need diagnosis

Micro-irrigation provides right amount of water, at right time, at right place in right manner. It has proved its efficacy in saving water, improving the yield and also can be responsible for reducing diseases which develop in rather moist conditions but its large-scale adoption is limited to a few states especially for high value crops (Singh *et al.*, 2013). Perhaps high initial investment is one of the limiting factors. The understanding of plant growth behaviour including nutrient requirements and rooting pattern, soil chemistry (solubility and mobility of the nutrients), fertilizers, chemistry (mixing compatibility, precipitation, clogging and corrosion) and water quality factors (pH, salt and sodium hazards) is required for effective fertigation. The existing gap between potential and current status could be bridged through holistic approach with integration of research, development and industry. There is a need to refine technology, improve design and articulate mechanisms for efficient utilization of water through micro-irrigation. Availability of water-soluble fertilizers for fertigation and related policy matters should be addressed on priority. To achieve high nutrient use efficiency, nutrient need diagnosis techniques be intensified to ensure economy in fertilizer use through optimized application. To aid practical application, however, there is a need to correlate nutritional composition of leaf with yield on the basis of extensive nutritional surveys of orchards and comprehensive field experiments.

Nursery management for plant propagation

Production of quality planting material and achieving the high success rate through various propagation techniques are the issues under open conditions. Here, the protected structures appeared as the best means and are being used by many nursery growers for multiplication of healthy planting materials. Keeping in view the high demand for quality planting materials, there exists enormous scope for large-scale expansion of nursery activities under favourable microclimate created in cost-effective low cost structures. Flowers and vegetable seeds especially hybrids are not only costly but also need optimum temperature range of 18-20 °C for their proper germination, further growth and development. Depending upon the situation, the protected structures have proved ideal for good seed germination and seedling production. The nursery raising under protected condition is a paying proposition and is becoming popular. Optimum climatic requirements need to be worked out for all the crops suited to Indian conditions to exploit properly the greenhouse structures. Micropropogated plant hardening, vegetative propagation and grafting/budding, rooting of the cuttings have much higher success in poly house conditions. Efforts also need to be intensified to develop suitable technology for production of healthy and well developed seedlings in plastic perforated trays for Indian conditions. New growing media are available for optimum plant growth under such production systems but techniques are to be refined for hardening of seedlings, including root pruning etc. Greenhouse design, structure and technology for producing high quality planting material are available but require fine-tuning to suit Indian conditions. Commercial grafting techniques with computer control need validation, and adoption for mass production of quality planting materials of horticultural crops, in general and vegetables in particular.

Protected Cultivation of Vegetables

The productivity of vegetables grown under protected conditions using low cost polyhouse at about 3000 m-amsl altitude is 40 to 300 per cent as compared to outdoor cultivations. Record yield of some of the vegetables like cucumber, tomato, summer squash, brinjal, beans, spinach, pea and coriander leaves are much higher than open field conditions. Tomato is the most universally grown greenhouse vegetable in varied agro-climatic regions all over the globe. Tomato varieties grown under green house conditions are of indeterminate type as they continue to grow and set fruits almost for a year-around plant life. Similarly, production of lettuce in greenhouses has become popular and potentially a profitable venture. On the other hand, greenhouse capsicum production is a lucrative industry in itself though it is more labourious than growing tomatoes. It is grown for many of its coloured varieties viz. red, yellow, orange, green and black which are in high demand in fast food restaurants for variety of food preparations, extraction of natural colours and preparing oleoresins and oils.

The differential behavior of varieties are observed in different crops under partially environment controlled greenhouse. There are several instances reported for high productivity under protected conditions for capsicum, tomato, cucumber, gherkins, summer squash, winter squash, watermelon, muskmelon, lettuce, cauliflower, broccoli, chrysanthemum, gerbera, carnation, lilium, rose, strawberry, etc due to favourable air temperature, optimum relative humidity and light intensity in the structure, which had helped in getting good vegetative and reproductive growth. The potential of photoselective light dispersive netting in improving the production of horticultural crops is observed. Among the different coloured nets used for photoselective shading the red net has some productivity advantage over the pearl and yellow nets. The risk of infestation by small pests like aphids, thrips and the incidence of insect borne viral diseases are significantly minimized.

The National Committee on the use of Plastics in Agriculture (NCPA-1982) has recommended location specific trials of greenhouse technology for adoption in various regions of the country. Vegetable crops besides improving biological productivity and nutritional standards also have enormous scope for enhancing profitability and livelihood option. This group of crops has emerged as the core of such agro-economic strategy accelerating the growth. The protected production system has the potential to contribute greatly for achieving high productivity and stability to production in the present scenario of climate change.

Seedling raising

Majority of vegetable and flower crops are grown through transplanting method and traditionally the seedlings are raised in flat bed, raised bed or different kinds of pans in open field condition which makes the seedling prone to the attack of insects, diseases specially the damping off and viruses which hamper the performance of seed production programme and quality of the seed. The healthy and virus free seedlings can successfully be produced under low cost insect proof nursery structures. The same insect proof net house can be utilized during winter by covering with UV stabilized and transparent polythene for growing of cucurbits, tomato, chilli, pansy, petunia, antirrhinum zinnia, cosmos, and gaillardia. To mitigate the effect of high temperature, the structure can be covered with various color shade nets. Large scale production of disease free seedlings of vegetables, ornamentals and papaya could be organized under hi-tech nursery structures with the help of soilless growing media (3 parts cocopeat; 1 part vermiculite; 1 part perlite) in multicelled trays (Sabir *et al*, 2011). Even the nursery of cucurbits can also be raised under such system.

Grafting of vegetable seedlings for healthy seed production

Numerous vegetable crops face the problem of soil borne fungal diseases (*Fusarium wilt, verticillium wilt, damping off and nematodes)* which cause severe damage to the seed crop and reduce the seed production . Grafting is a process involving choice of rootstock and scion species, creation of a graft union by physical manipulation, healing of the union and acclimatization of the grafted plant (Lee and Oda 2003). Therefore, the grafting of seedlings on resistant root stocks can play a vital role to overcome these problems of soil borne fungi and nematodes.

Vegetable grafting is a recent innovation adapted on a commercial scale though grafting of fruit trees is practiced since thousands of years (Sakata *et al.*, 2007). Grafted seedlings are very popular and widely used among farmers in Asian and European countries and is increasing in the US, especially with tomatoes. In Japan, almost 95 per cent of watermelon, oriental melon, egg plant, and greenhouse cucumber and tomato are grafted before being transplanted to the field or green house. In South Korea, the use of grafted seedlings is increasingly popular in the production of many fruit bearing vegetables. The number of vegetables being grafted annually is estimated to be more than 540 million seedlings in Korea and 750 million in Japan. An increase in plant purchases and shortage of skilled workers has made it necessary to automate grafting in Japan. Development of a grafting robot targeted efficient mass production of seedlings especially cucumbers. Grafting robot produced 815 plants per hour (about three times the manual grafting) at a success rate of 97.1 per cent.

Seed production of vegetables under protected structures

Recently there is a technological shift towards production of high quality hybrid seed of important high value vegetables under protected conditions. Selection criteria of protected structures to be used for vegetable seed production certainty depends upon a few important factors viz., climatic conditions of the area selected for seed production, vegetable crop, type of seed crop, season of seed production and targeted quantity of seed production. Generally semi-climate controlled greenhouses are only suitable for hybrid seed production of determinate type varieties/hybrids of standard tomato, cherry tomato, sweet pepper and parthenocarpic cucumber varieties, which are suitable for protected cultivation. Seed yield of such crops can be 3-4 times more compared to their open field cultivation along with high quality of seed. Similarly, zero energy naturally ventilated greenhouses are also equally suitable for hybrid seed production of these high value vegetable crops, where the seed yield is usually 2-3 times more than open fields, but the cost of seed production is only one third (1/3) of the seed produced under semi-climate controlled greenhouse conditions.

Insect proof net houses and low cost protected structures are suitable for quality hybrid or seed production of open pollinated (OP) varieties in large number of vegetables viz., tomato, sweet pepper, chilli, okra, brinjal and cucurbits. The major objectives of seed production under insect proof net houses are to grow virus free seed crops, to protect the seed crops against other major insects/pests like shoot and fruit borer in brinjal, fruit borer in okra and tomato, fruit fly in cucurbits, leaf minor in cucurbits and solanaceous vegetables and red pumpkin beetle in cucurbits. Not only this, insect proof net house also provides protection to the seed crops against mild frost conditions. Some other low cost protected structures like walk in tunnels and plastic low tunnels are used for off season seed production or advancing the season of seed production, but these structures are technically suitable only for northern plains of India during winter season.

Hybrid seed production of vegetables under protected structures

High value low volume vegetable hybrid seeds can be produced successfully under insect proof net houses and naturally ventilated green house conditions. In hybrid seed production genetic purity and seed health of parental line are matters of concern because of the non availability of adequate isolation and attack of various fungi and viruses. The experiments conducted at CPCT at IARI, New Delhi revealed the production of high quality seed of the parental lines of pumpkin Pusa hybrid-1 under insect proof net house condition and compared it with open field condition and observed significant difference for fruit development attributes, seed yield attributes and seed quality attributes in female and male parental lines under insect proof net house conditions. The hybrid seed production of cross pollinated crops viz. pumpkin, bitter gourd and summer squash could be organized successfully under insect proof net house and higher seed yield and the quality can be obtained as compared to open field conditions. An encouraging result had been reported in pumpkin cv. PH-1 and brinjal PH-5 & 6 for hybrid seed production during 2009-10 at CPCT, IARI, and New Delhi.

Thus, the major advantages of vegetables seed production under protected conditions are:-

- Seed yield is generally 2-3 times more compared to open fields in major vegetables. On the other hand quality of seed is always high. The problem of high isolation distance in cross pollinated vegetables can be minimized under protected conditions.
- Problem of synchronization of flowering in parental lines can be minimized.
- Maximum plant population can be maintained along with appropriate ratio of male and female parents for higher hybrid seed yield.
- Seed production under adverse climatic conditions is possible, otherwise it is not possible under open field conditions.
- Duration of seed crops can be extended, which is not possible under open field conditions.
- Field standards can be well enforced under protected conditions and virus free and disease free vegetable seed crops can be grown, which is otherwise very difficult under open field conditions.
- Training, pruning and hand pollination practices are very easily manageable under protected conditions as compared with open field seed crop.
- More number of seed crops can be grown under protected conditions.
- By using soilless media, seed crops can also be grown even under saline and acidic soil conditions.
- Protected conditions provide the best opportunity for organic seed production in vegetables.
- Judicious use of natural resources like soil and water is possible for seed production under protected conditions.
- Virus free and healthy seedling production of the parental lines is possible under protected conditions.
- Grafting technology in seedling can easily be applied to over come the soil borne problems in solanaceous and cucurbitaceous vegetables.
- Quality seed production of vegetables can certainly be adopted as an agri-entrepreneur in various regions of the country and it can generate lot of employment in several areas.

- Seed production cost in vegetables can be reduced by growing seed crop under protected conditions.
- In hybrid seed production, the emasculation practices on the female parents are not required as there are no insect pollinators inside the protected structures.
- Complete protection of the crop from rodents and birds compared to open field seed crops.
- Seed crops can not be damaged by erratic rains at the time of their maturity alike open field seed crops.
- Off type plants, objectionable weeds or plants affected by designated diseases can not pose any problem in the seed crops raised under protected conditions.
- Seed viability and seed vigour can be extended through integrated nutrient management in seed crops under protected conditions.

Quality seed production under protected structures especially hybrid seed production of vegetable crops has tremendous potential being change in climate pattern, and increasing incidence of pest and diseases are creating hindrance in attaining higher quality of seeds. Since, hybrid seed production under protected structures has shown higher yield and better economic returns due to tremendous decrease in pest and disease and increase in their growing period which ultimately help in getting higher seed yield and better quality. Secondly, hybrid seed production of vegetables could be organized under North Indian Plain conditions as a commercial activity for ensuring availability of quality seed to the farmers at reduced cost.

Protected floriculture

The general awareness about greenhouse floriculture is not altogether missing. Some floriculture farmers know of its benefits and a few had also adopted it availing financial assistance. Due to lack of extension contact, floriculture farmers are also not aware of the government programmes for encouraging greenhouse technology. Poor quality of flowers produced by the traditional technology in comparison to the protected cultivation, as also other economic advantages of the later, are some of the important factors which make greenhouse floriculture potentially attractive. However, lack of adequate finance particularly with the small flowers and unavailability of specific extension material seen to be important constraints.

The benefits of greenhouse floriculture of small scale farmers in a specific area can be better utilized, if farmers and farm women are organized into Self Help Group (SHG). For the purpose of greenhouse floriculture the SHG may comprise 10-20 farmers located close to each other, either from the same village or neighboring villages, having common interest of starting greenhouse floriculture on small scale. Members of the SHG may persue the activity individually with the basic objective of optimizing their income levels. The SHG may be encouraged and supported by the extension agency in terms of providing technical knowledge identifying common problems, guiding sources of quality planting material and other inputs and providing package of practices for production and post harvest management.

Growing of flowers in India has been a time-honoured and traditional activity largely for religious purposes, perfumes and landscape. Now it is poised for a transformation with increase in demand, innovative technology, and policy environment and above all, growing consciousness and demand for quality flowers. Introduction of greenhouse technology, new cultivars and support for the post harvest handling have led to a new wave in the growth of floriculture in the country. Innovative technologies, policy support, entrepreneurship have helped in establishing floriculture as a means of enhancing the accessibility to food for small and marginal farmers through high return per unit of area. Many of the new flower crops were introduced and have become popular. The greatest advantage of protected floriculture is the availability of quality flowers for prolonged period which are of the major requirements of International flowers trade. Flowers, which are termed as high value low volume crops like rose, gerbera, carnation, tuberose, lilium, orchids, anthurium, callalily, gypsophila, alstroemenia, iris, lisianthus, etc are preferred flower crops for protected cultivation.

The Oriental Floratech, a TATA enterprise at Pune is perhaps the first professionally managed enterprise to venture into the commercial floriculture in 1991. Encouraged by the success of this unit, many corporate houses have ventured into the blooming business. The Eassar group (Eassar Agrotech), Deccan Florabase and Neha Floritech have established production units in Pune in 1994. Special attention is needed in management of light, humidity and temperature which help to produce good quality flowers for both domestic and export market.

Plant Protection under Protected conditions

With changing horticultural practices new pests like thrips,scales white flies, mites, phytoplasma, viruses and viroids are emerging not only under open conditions but also under protected structures for cultivation of vegetable and flower crops (Table 6). Some of these are exotic pests. Research efforts are required to develop models for application of IPM practices for important crops in different locations under protected conditions. To protect crops from such pests, greater vigilance and stringment quarantine measures are required. With increasing awareness of the ill effects of environmental pollution, biological control of pests will be the most favoured and desirable approach for minimizing losses caused by pests, like insects, fungi and bacteria. Transgenic crops alongwith biological control strategies will be the main armoury in fighting the pest menace. Some initiatives in these directions have been taken in India, but considering the range of crops and agroclimatic zones requiring high-tech approaches, a much greater effort is required for utilizing the full potential of biological control and transgenic crops. Due to enormity of pest problems, need-based application of pesticides will have to continue as part of IPM. Present-day indiscriminate use of pesticides, which results in undesirable levels of residues in the produce and contamination of water and soil, will have to be stopped through appropriate policy measures. Success of such policy will depend on the availability of 'safe' pesticides like botanicals. The country has good information about availability of basic materials like neem, but more emphasis is required for developing economical and environmentally compatible technologies for commercial production of safe pesticides.

Table 6: Pest scenario under protected environment (Sabir *et al.*, 2011)

Pests	Common Name	Scientific name	Host
Aphids	Melon aphid Green peach aphid Potato aphid Fox glove aphid	*Aphis gossypii* *Myzus persicae* *Macrosiphum euphorbiae* *Aulacorthum solani* *Acyrthosiphon lactucae*	Tomato, Capsicum Cucumber, Lettuce Capsicum, Tomato, Lettuce Lettuce, Tomato Capsicum, Tomato Lettuce
Caterpillars	Cotton bollworm Asian cotton leafworm or Tobacco cutworm Cabbage looper	*Helicoverpa armigera* *Spodoptera litura* *Trichopulsia* sp.	Capsicum, Tomato Tomato, Capsicum Cucumber Lettuce
Leaf-miner	American serpentine leaf-miner	*Liriomyza trifolii*	Tomato, cucumber
Mites	Broad or yellow mite Carmine spider mite Vegetable mite Spider mite	*Polyphagotarsonemus latus* *Tetranyshus chnnabarinus* *Tetranychus neocalidonicus* *Tetranychus articae*	Capsicum Tomato Cucumber Tomato, Capsicum, Cucumber
Thrips	Chilli thrips Western flower thrips Banded greenhouse thripsMelon thrips Onion thrips	*Scritothrips dorsalis* *Frankliniella occidentalis* *Hercinothrips femoralis* *Thrips palmi* *Thrips tabaci*	Tomato Cucumber, Tomato Many Vegetables
Whiteflies	Sweetpotato whitefly Silverleaf whitefly Greenhouse whitefly	*Bemisia tabaci* *Bemisia argentifolii* *Trialeurodes vaporariorum*	Capsicum, lettuce Tomato, lettuce Tomato, cucumber, capsicum, beans, lettuce
Powdery mildew		*Erisiphe cichoracearum* *P. xanthii, L. taurica*	Cucumber, lettuce, capsicum
Gray mould		*Botrytis lactucae* *Botrytis cineria*	Lettuce Tomato

(*Contd.*)

Pests	Common Name	Scientific name	Host
Fusarium wilt		*Fusarium oxy. f.sp. lycopersici* *Fusarium oxy. f.sp. cucumerinum*	Tomato Cucumber
Fusarium crown and root rot		*Fusarium oxy. f.sp.radicis lycopersici*	Tomato
Damping off/root rot		*Pythium ultimatum* *Pythium aphanidermatum* *Phytophthora spp.* *Rhizoctonia solani*	Cucumber, lettuce Capsicum
Downy mildew		*Pseudoperenospora cubensis* *Bremia lactucae*	Cucumber Lettuce
Root knot nematode		*Meloidogyne incognita*	Tomato, capsicum, cucumber

Contribution of protected cultivation in arid high altitude vegetable production

In early sixties, arid high altitudes of Ladakh, Lahaul and Spiti grew root vegetables like radish, turnip, carrot, beet root; potato and mongol palak (beet leaf). After Chinese aggression (1962), induction of Indian defence forces in these areas necessitated local production of different vegetables. Defence Research and Development Organization (DRDO) through its laboratory, Field Research Laboratory now called Defence Institute of High Altitude Research, Leh did pioneering research. With the help of protected agriculture technologies it is possible to grow all types of vegetables there during agriculture season (May to September or mid October). Perhaps first glass house in high altitudes of the country was erected in Leh (3825 m amsl) in 1964. Since then DRDO has extensively worked on protected cultivation and developed appropriate production technologies in Himalayan high altitudes of India for the welfare of the soldiers and the society. During long frozen winters of Ladakh one could not have thought of vegetable production but for protected cultivation practices developed by DRDO every farmer and others are producing different types of leafy vegetables almost round the year. (Singh *et al.*,1999).

Some of the major contributions made by DRDO in developing protected vegetable production technologies are as follows:

1. Protected vegetable nursery and crop production

Early production of vegetable nursery under different protected structures during March and April (minimum atmospheric temperature is sub-zero) and transplanting them in May and June with and without plastic mulch extended agriculture period and made possible cultivation of cabbage, cauliflower, knoll-khol, broccoli, brussel's sprouts, tomato, brinjal, chili, capsicum and onion. Use of plastic mulch enabled early, quality and higher yield of these vegetables. In mulched crop, low pressure (gravity/slope) drip irrigation and fertigation are possible as experimented by DRDO. In this way, most of the vegetables are being grown on large scale making the area surplus in cabbage, which is unbelievable. The early and late production of vegetables with the help of protected technology are also standardized which extended the availability period of locally produced vegetables.

2. Cucurbits production in cold desert

In Ladakh, till early 1990s cucurbits cultivation in open was impossible. But growing seedlings in poly pouch under poly houses during April-May and transplanting them in open field with plastic mulch made it possible to grow almost all cucurbits in Leh. This has not only improved vegetable basket in the area but also added variety to food basket of local inhabitants and soldiers. Commercial production of cucurbits in cold desert of India is now possible through protected cultivation. Sarda melon imported in large quantity in the country can be produced in these areas with ease. Production of off season (August and September) muskmelon and watermelon in open fields has also become possible. An early crop of cucurbits like squash and longmelon is also grown in poly houses.

3. Vegetable production in sub-zero atmosphere

During winter these areas remain cut off from main land due to heavy snow fall. Only air communication is in use during winter . Air transportation of bulky and perishable commodities like vegetables is not only expensive but also very difficult. In Ladakh sector, the Army alone spends several millions of rupees only on transportation of vegetables. The cost of transportation is more than the cost of vegetables, hence local production through protected cultivation is being successfully promoted here. This is being encouraged by harnessing thermal and photovoltaic solar energy thus enhancing moderate temperature of green houses. The geothermal energy sources available in the area are potential source of heating greenhouses and remoteness of these sources is causing hindrance in their exploitation

Farmers in Ladakh did not believe in the idea of producing vegetables during long frozen winters but now every farmer is practicing it under one or the other protected structures. Ladakhi poly house is the most favoured structure and supported by the state Government/ Leh Council. Locally produced leafy vegetables under protected structures are seen in Leh market during frozen winters. Vegetables like mongol palak, mustard, lettuce, methi (fenugreek), coriander, leek, Chinese cabbage, mint, amaranth, karam sag, dum, kale, swisschard, celery, parsley, radish, green onion, cabbage, knol-khol and orich/ustahak are produced in sub-zero atmosphere and this research has become unique in the world.

4. Vegetable seed production

Seed production of biennial crops like temperate varieties of cole crops, root crops, and onion took about two years or 18 months in these areas. In first year, the normal crop is grown and stored underground during long winters. In second year, during summer they are planted for seed production. The protected agriculture technology has now made it possible to produce seeds of these varieties in half the time by raising early crop under protected structure and transplanting them in open fields for seed production. Pusa Himani radish, long day onions and Nantes carrot respond well to this technique. Production of seeds of temperate varieties of vegetables in India is a problem due to lack of consorted research and development efforts.

Policy issues and future strategies

A policy framework to generate inclusive growth for a country like India has to be consistent with the agro climatic and social conditions of the area. Sectors like horticulture (Flower and Vegetable cultivation) have comparative advantage in the region due to its agro-climatic conditions. Most importantly however, the chances of success in these specific sectors depend on the development of physical infrastructure. Thus the inclusive growth policies must emphasize the development of following sectors in the country:

1. Development of need based polyhouse designs for different areas.
2. Ensuring easy availability of U.V. stabilized plastic sheet and nets.
3. Development of suitable varieties specifically for polyhouse cultivation.
4. Work on eco-friendly nutrient and pest management.
5. Government incentives for popularization of the technology.

The large capital and information requirement necessary to carry out the marketing activities efficiently implies that only a large organization is able to promote this activity. This requirement can be fulfilled by setting up a Horticulture Marketing Board, either in the public sector or in a public-private partnership mode. Some of the necessary activities of such an organization would include, (i) setting up of input retail outlets (ii) collection of the produce from the farms pre cooling and provision of warehousing so as to address the problem of inadequate storage capacity (iii) value addition by setting up of food processing units and cold chains (iv) strengthening *mandis* and procurement agencies' network for better marketing facilities. Such an organization can help the small and marginal farmers to develop the capacity to regularly supply high quality vegetable products. Once this target is achieved, the organization should focus its attention on the building up of export capabilities for these vegetable products. Major constraint is the increased pressure on cultivable lands near metros where vegetables are generally grown. It is quite necessary to improve the productivity of vegetables adopting protected cultivation in the country in general and high altitudes in particular. Protected cultivation of vegetables in high altitudes of Himalaya is practiced successfully indicating its potential to deal with conditions created by climate change scenario in the country. Protection against adverse climatic conditions for plant growth has become the universal necessity. There is need to develop area specific, most appropriate, efficient and affordable protected structures with cheaper and durable cladding materials. Emphasis would be shifted to development of suitable varieties and hybrids of vegetables for protected cultivation under organic and inorganic production protocols. Horticultural nurseries would be produced under protected structures both at individual farmer and commercial nurseries level. Tools and machinery for protected cultivation need to be developed and made common. Vertical or multitier farming of horticulture crops would be developed to make use of protected space. Plastic mulching coupled with drip irrigation in high value low volume crop production is going to be a common practice because of their proven advantages.

Success Stories in Different Parts of India

The protected cultivation is a time tested technology which is in infancy requiring urgent efforts for its adoption. There are some examples where protected cultivation has empowered the practitioners of it not in terms of money only but pride too which are known to be lacking in agriculture. A number of farmers are adopting protected cultivation practices (net house, low-tunnel technology, bed planting of vegetables to make judicious use of irrigation water and use of sprinkler system for vegetable nursery production) etc. Their presence on the agriculture scene is symbolic of creating a new ‘scientific awakening’ among farmers as a source of inspiration and motivation and has helped in improving their income vis-à-vis socio-economic status. These stories are rays of hope in the event of prevailing conditions of shrinking land and water resources, frequent spell of inclement weather, popularly known as climate change.

DIHAR, Leh waging a war against harsh climatic conditions, developed appropriate greenhouse technologies for ensuring year-round availability of fresh vegetables of various types for army and local people. The farmers’ cooperative society is ensuring supply of vegetables to army and meeting about 55 per cent of their total requirement of fresh vegetables in Ladakh sector. This has developed fraternization between army and civil population with improved socio-economic conditions of poor farmers. A total of 165 greenhouses of various types have been established till now in the region. (Sanwal *et al.*, 2004).

Many farmers in Uttarakhand tried cultivation of cucumber and tomato in their polyhouses and are successful in their endeavour. The farmers reported five-fold higher production in polyhouse cultivation as compared to open with lower incidence of insects, pests and diseases. Polyhouse cultivation gave higher economic return with lower cost of production, giving more benefit per unit area. The produce had better quality, consumer preference, nutrition and keeping quality. The varieties of tomato and cucumber developed at Pantnagar gave better production as compared to those available in market. The varieties available in market are producing satisfactory yield only up to the temperature of 30-35oC .

Amba Dutt Pandy, village Bhagartola, about 50 km on Almora-Pithoragarh road near Jageshwar used to grow all crops under rainfed conditions and was not able to adequately irrigate his modest holding of 0.2 ha. Therefore, not being further much interested in agriculture he started working as a labourer. Scientists of VPKAS examined the terrain and formulated a long term solution of water harvesting from the springs into tanks lined with low-density polythene. He was given training on vegetable cultivation and improved techniques with initial support to implement the work. He constructed six low-cost polyhouses, occupying a total covered floor area of 600 m^2, out of locally available pine wood and UV stabilized sheet and insect-proof net for growing off-season vegetables. He has different LDPE tanks for water storage (capacity of 100 thousand litres) being used for irrigating the crops inside polyhouses. Now, he is earning more than Rs. 1, 00,000 annually. He is following organic farming system and using hybrid seed of vegetables. Pandey received the best Farmer Award from Agriculture Minister during 2010-11 for his work and awareness in the field of Agriculture.

Devinder Singh, a farmer of village Mushkabad in Ludhiana district, has 7 acres of owned land and 3 acres of leased land. To obtain good quality and higher yield of vegetables, he established permanent framework for training vegetables like cucumber, tomato, capsicum, bitter gourd and bottle gourd on bamboo sticks. While attending a lecture on finalizing the new design of net house at PAU, Ludhiana, he became so inspired from performance of bamboo polyhouse that he established polyhouses in an area of five acres with the help of National Horticulture Mission. He grew coloured capsicum and sold to a well established firm. He also used drip irrigation, plastic mulches and low tunnel technologies for water saving and increasing the productivity. He visited Italy, Spain, Germany and England to acquire information on new technologies and explore marketing contacts in vegetable production. His work was recognized throughout Punjab. He was honoured with S. Ujagar Singh Dhaliwal Memorial Award by PAU, Ludhiana for the self-cultivating vegetable grower. He was also conferred with National Award for vegetable Grower by IARI, New Delhi, at Pusa Krishi Vigyan Mela in 2009 and Chief Minister Award for Innovative Farmer in Horticulture (vegetable) at PAU Kisan Mela, in March 2009.

Arvind Beniwal, a farmer who was inspired by scientists of CIPHET, Abohar, on strawberry cultivation dared to take this as a commercial venture in Palla village near Delhi on contract farming basis. His strawberry produce is high in demand in five-star hotels and Embassies in Delhi. He is a source of inspiration for hundreds of progressive farmers in the NCR region. He has already aired his strawberry success stories on DD National and All India Radio several times. Now, he is recognized as STRAWBERRY ICON among farming community.

Balbir Singh Kamboj, from village Nayagaon, after postgraduation in Agriculture joined seed industry in 1991. Getting motivated from the fast coming up polyhouse culture in the state, he joined Flower Growers' Association and established the first polyhouse for Gerbera production with the help of National Horticulture Board and Directorate of Horticulture. With the help from APEDA and a part of as member Flower Growers' Association, he is selling his good quality flowers in niche markets in Delhi, Lucknow, Kanpur and Jaipur ensuring proper profitability.He delivers lectures for district and state level programme to motivate farmers about remunerative polyhouse business.

Horticultural crops besides improving biological productivity and nutritional standards also have enormous scope for enhancing profitability and livelihood security. This group of crops has emerged as the core of such agro-economic strategy accelerating growth. This is the resultant of development of knowledge/ innovations at large scale, and, further adoption of such technologies have played vital role in development of agriculture in general and horticulture in particular. Earlier practices and processes became the foundation for emergence of new technology and development of horticulture sector. Protected cultivation provides an opportunity to produce the crop round the year, enhance crop quality and high productivity. The production of crops using protected structures is certainly a production system that involves high input costs and levels of management and is subject to risk, just like any other production system. With proper planning, attention and dedication to all aspects of the green house production system, the opportunity to increase the factor productivity of different inputs associated with vegetable crop production can be harnessed for maximization of profits. The protected cultivation system has the potential to contribute greatly for achieving high productivity and stability to horticultural production in the present scenario of climate change. Keeping in view the increasing demand of off season and high value horticulture crops in several big cities of the country, there is an urgent need for diversification from the traditional agriculture by production of high value crops under different protected conditions for increasing their productivity and quality not only for getting high returns but in a Horti-Business model. Nursery raising under protected cultivation is highly suitable for such Horti-business model or can be adopted as an agri entrepreneur business in major horticulture crops growing areas of the country by unemployed youths who are graduate in agriculture or post graduate in horticulture. Low cost protected technology like plastic low tunnels or walk in tunnels, shade net houses can be used for year round cultivation for getting high returns from off season produces. Similarly, insect proof net houses can be used on a large scale for safe crop cultivation by way of minimizing the use of pesticides in cultivation and virus free quality seed production in large number of crops in a horti-business model. Growers practicing judicious use of pesticides for safe production can be instrumental in providing quality products without polluting the environment. Some of the technologies like low pressure drip irrigation and low cost nursery raising under protected environment can also be replicated in other parts of the country among the resource poor and tribal farmers.

References

Hickman G W. 2011. A review of current data on international production on vegetables in vegetables in greenhouses. wwwcuestaroblecom, 73 p.

Lee J. M. and Oda, M. 2003.Grafting of herbaceous vegetable and ornamental crops. *Hortic Rev* **28**: 61 -124.

Montero J. I., Munoz P., Anton A. and Igesias N. 2005. Computational fluid dynamic modeling of night time energy fluxes in unheated greenhouses. *Acta Horticulturae* **691**(1): 403-9.

Sabir N, Deka S, Sumita R, Tanwar R. K, Bambawale O , Singh B, Hassan M and Mukul Kumar. 2011. Integrated pest management for greenhouse cucumber: A field validation under north Indian plains. *Indian journal of horticulture* **68**(3): 357-63.

Sabir N, Sumita R, Singh B, Hasan M, Anupama Chilana, Deka S, Tanwar R. K, and Bambawale O M.2011. Superabsorbent hydrogels for efficient biocontrol of root knot nematodes for healthy tomato nursery. *Current Science* **100**(5): 635-7.

Sakata Y, Ohara T. and Sugiyama M. 2007. The history and present status of the grafting of cucurbitaceous vegetables in Japan. *Acta Horticulturae* **731:** 159-70.

Singh, D K., Padiyar Shweta and Chaudhary H. 2010. Biochemical of parthenocarpic gynoecious cucumber lines, hybrid monoecious varieties and wild relatives. *Indian J Hort.***67**(3) : 343-347.

Singh, D. K. 2013. Development of suitable crop varieties/hybrids for protected cultivation in India. (In) National seminar on Advances in Protected Cultivation Technical Session: Crop Production Technologies organized by Indian Society for Protected Cultivation, New Delhi on 21st March, 2013.

Singh, D. K. 2013. Prospects of vegetable crop breeding under protected conditions in India. In National seminar on Advances in Protected Cultivation Technical Session: Crop Production Technologies organized by Indian Society for Protected Cultivation, New Delhi on 21st March, 2013.

Singh, D. K., Sarkar, M., Singh Amar, Singh, D. P. and Fatmi U. 2013. Recent advances in irrigation techniques for polyhouse cultivation of vegetable crops. In National seminar on Advances in Protected Cultivation Technical Session: Crop Production Technologies organized by Indian Society for Protected Cultivation, New Delhi on 21st March, 2013.

Singh, D. K., Singh, Ajay Kumar, Singh Balwant, Sabir, Naved and Mehta, K.S. 2013. Assessment of parthenocarpic cucumber genotypes under naturally-ventilated greenhouse conditions in mid-hills of Kumaun Himalaya. In National seminar on Advances in Protected Cultivation Technical Session: Crop Production Technologies organized by Indian Society for Protected Cultivation, New Delhi on 21st March, 2013.

Singh, D. K. 2013. Search of new genes for ideotype vegetable crop breeding for polyhouse cultivation. In National seminar on Advances in Protected Cultivation Technical Session: Crop Production Technologies organized by Indian Society for Protected Cultivation, New Delhi on 21st March, 2013.

Singh, N.; Diwedi, S. K. and Paljor, Elli, 1999. Ladakh Mein Sabjion Kei Sanrakshit Kheti. Regional Research Laboratory of DRDO, Leh. Pub. By D.R.D.O., Leh. Pub. By D.R.D.O.56 A.P.O. p 21-25.

Sanwal,S K , K.K. Patel and Yadav D. S. 2004. Vegetable Production under Protected Conditions in Neh Region: Problems and Prospects. Envis Bulletin: *Himalayan Ecology* **12**(2):7. r

2

Space Saving Horticulture

Binoo Bonny, Jyothi Bhaskar, M. Prabhakar and K.V. Peter

Abstract

Limits to crop production are shrinking land area under cropping, shortage and higher cost of irrigation water, fertilizers, pesticides, electricity, labour wages and harvest and post harvest losses due to non-availability of cold chains and storage spaces. Diversification of crops to high value products is yet to gain momentum. Restrictions in marketing due to APMC acts are to be reviewed and drafted farmer-producer friendly. Forcing of horticultural crops-vegetables, fruits, flowers, spices, aromatic crops-to market fresh produces during off season is getting attention and many such units are established as a part of urban and peri-urban horticulture. Ready to assemble forcing structures are now available. Hi-tech nurseries, integrated soil and water management through fertigation, pests and disease management by bio-pesticides, predators and parasitoids, pheromones and light traps are successfully demonstrated. Space saving horticulture focuses on maximization of productivity and minimization of inputs costs in an eco-friendly and sustainable way.

Keywords: Food Security Bill, Precision farming, High Density Orcharding, Hydroponics, Aeroponics, Nutritive value, Hunger Index, Pruning and Training, Inputs management, Yield gaps

Introduction

Hunger is not about calorie-deficiency or mere lack of food. Instead it encompasses all about providing a nutritionally balanced wholesome meal to ward of the ill effects of an imbalanced and improper diet. This conceptualization has led to the emergence of food and food based approaches in development paradigms of recent years. Escalations in the cost of food, food related diseases and environmental degradation also contributed in food systems gaining prominence (Lang et al, 2009; Mitchell, 2008). Subsidized and free distribution of food grains through Public Distribution System (PDS) and enactment of Food Security Bill which entitle food as a right, are all means of achieving the avowed goal of hunger-free(zero hunger as in Brazil) India and world at large. These food-based approaches recognize the central role of food as the primary tool for improving the quality of the diet and for addressing and preventing malnutrition and nutritional deficiencies. This brings paramount importance to agriculture and good agricultural practices (GAP) in generating quality produce at affordable rates combined with the long term challenge of producing sufficient quantities in improving nutritional and livelihood security at large. With increasing population, the gap in the per capita availability and projected demand is to widen further as is evident from Table 1. The projected demand for the major food crops in India as given in the table illustrates the growing demand for horticultural crops like fruits and vegetables among both high income groups (HIG) and low income groups (LIG) compared to cereals, pulses and other food grains. This can be attributed to the changes in food consumption pattern that remain skewed away from food grains and growing health consciousness among all sectors of population.

Table 1: Demand for various foods in India (Thousand tones)

Food	Year 2000		Year 2015		Year 2030	
	LIG	HIG	LIG	HIG	LIG	HIG
Rice	84417	84255	101886	101441	114499	113893
Wheat	63375	62545	74607	72411	83045	80087
Maize	10466	10281	12196	11714	13522	12876
Other Coarse Grains	21803	21419	25407	24403	28171	26825
Total cereals	180061	178500	214096	209969	239238	233681
Pulses	16599	17028	21303	22578	24515	26312
Food grains	196659	195528	235399	232547	263752	25993
Potato	19905	20716	26394	28911	30760	34370
Edible oils	8151	8324	10355	10863	11870	12581
Vegetables	83388	91165	123824	151861	150823	193562
Fruits	47688	51774	69678	84099	84336	106126
Milk	76932	82451	109092	127805	130502	158325
Meat	5335	5918	8196	10396	101181	13534
Eggs	1880	2086	2889	3664	3566	4770
Fish	5507	6108	8460	10731	10444	13971

Source: Mittal (2006)

The trend remains almost similar for production and availability of fruits and vegetables in India as presented in Table 2. It is quite evident from the table that the gap in the requirement and availability is getting bridged over the years due to production gains. This increase in production was possible because of the release of over 1588 new cultivars in 120 crops with much higher share of product as a proportion of total biomass (increased harvest index), scientific management of external inputs of water, nutrients and crop protection agents, and massive investment in expansion of irrigation facilities and market linkages. However, these achievements are not reflected in the Global Hunger Index (GHI) estimate for India. Contributions of the three components of GHI viz. proportion of undernourished population (%), prevalence of underweight children under 5 years (%) and under five mortality rates (%) remain quite dismal for India (Table3). This is more pronounced among the vulnerable sections of the society especially women and children which indicate that the agricultural production gains have not been reflected in alleviating hunger and malnutrition. Moreover, it is estimated that the agricultural production needs to be increased by 60% of the present by 2050 to meet the projected demands (FAO, 2012). The exigency arises from the fact that this needs to be attained with only 10 % more of available land, less water and other resources. It has been reported that in many areas the yields are plateauing and room to further increase the harvest index is small (Fischer *et al*, 2010). The prospects for significant breakthroughs in genetic improvement of photosynthesis or abiotic/ biotic stress tolerance also remain elusive (Fisher, 2007). Therefore, the focus needs to be on bridging the yield gaps and meeting the nutritional challenges through intensive cultivation practices of short duration, nutritionally rich crops, wherein horticultural crops holds the key.

Table 2: Vegetable and fruit requirements of (India based on population)

Year	1991	2001	2011	2020
Population(Million)	846.42	1028.74	1210.2	1400.0
Vegetables(Million tones/yr)	86.50	105.14	123.68	220.00
Requirement				
Production	58.53	88.69	146.55	-
Fruits(Million tones/yr)	37.07	45.06	53.01	98.00
Requirement				
Production	28.63	43.00	74.88	-

Table 3: Global Hunger Index (GHI) and its three components of India in comparison with neighboring countries and the world

Country	Proportion of undernourished in the population (%)		Prevalence of underweight in children under five years (%)		Under five mortality rate (%)		GHI with data from 1988-92	GHI with data from 2003-08
	1991-92	2004-06	1990-92	2003-08	1990	2008	1990	2010
India	24.0	22.0	59.5	43.5	11.6	6.9	31.7	24.1
Bangladesh	36.0	26.0	56.5	41.3	14.9	5.4	35.8	24.2
China	15.0	10.0	15.3	6.0	4.6	2.1	11.6	6.0
Pakistan	22.0	23.0	39.0	25.3	13.0	8.9	24.7	19.1
World	-	-	-	-	-	-	19.8	15.1

Source: Global Hunger Index, IFPRI, 2009

Horticulture Crops in Nutritional Security

Horticultural crops are essentially high value crops which ensure maximum returns to the growers with multiple scope for value addition. It has supported agricultural diversification through better land use, creation of employment and overall nutritional security. Moreover, horticultural commodities like fruits, vegetables, nuts, tubers, spices, aromatic crops etc are very high in nutritive value and provide more nutrients even if consumed in small quantity. In other words horticultural products are nutrient rich packets. Therefore, leveraging the benefits of horticultural crops for their contribution to food security and poverty alleviation needs to be discussed in the light of the food and nutrition profile of major horticultural crops. It is more because the horticultural products are rich sources of vitamins, proteins, carbohydrates and minerals (Table 4) and have significant role in solving nutritional crisis of developing countries. They are referred to as protective foods and assume great importance in the nutritional security. According to estimates, the per capita consumption of fruits and vegetables in India is around 46g and 130g against the Indian Council of Medical Research (ICMR) recommendation of 92g and 300g respectively. Seasonal fruits and vegetables are encouraged as they bring a variety into diet. Yellow and orange color vegetables and fruits are rich in Vitamin A, α, β carotenes, zea-xanthins and crypto-xanthins, where as dark-green coloured are good sources of minerals and phenolic, flavonoid as well as anthocyanin anti-oxidants. Moreover, a major shift in consumption pattern of fresh and processed fruits and vegetables holds great promise for huge influx of investment in the horticulture sector with diversification and value addition holding the premium.

Table 4: Nutritional Value of major Horticultural crops

Crop	Energy (Kcal)	Protein (g)	Carbohydrate (g)	Fat (g)	Calcium (mg)	Fe (mg)
Beet root	43	1.7	8.8	0.1	10	1
Carrot	48	0.9	10.6	0.2	80	1.03
Colocasia	97	3	21.1	0.1	40	0.42
Onion	50	1.2	11.1	0.1	47	0.6
Potato	97	1.6	22.6	0.1	10	0.48
Chilli (green)	29	2.9	3.0	0.6	30	4.4
Coriander	288	14.1	21.6	16.1	630	7.1
Black pepper	304	11.5	49.2	6.8	460	12.4
Turmeric	349	6.3	69.4	5.1	150	67.8
Beans	347	24.9	60.1	0.8	60	2.7
Green Peas	432	43.2	20.9	19.5	240	1075
Cabbage	27	1.8	4.6	0.1	39	0.8
Celery	37	6.3	1.6	0.6	230	6.3
Lettuce	21	2.1	2.5	0.3	50	2.4

The tremendous potential in improving the productivity of major vegetable crops as detailed in Table 5 needs to be targeted in achieving this. The estimates indicate that in most of the vegetable crops, there is scope for more than doubling the present yields, even up to five times as in tomato. The basic challenge is how this can be achieved with the limited land, water and other resources. It is in this context the concept of space saving Horticulture assumes significance.

Table 5: Comparison of average Productivity of major Vegetable crops in the World and in India

Vegetables	World's average productivity	Productivity in India (q/ha)	Scope to enhance the productivity (q/ha)	Multiple times
Eggplant	NA	130.80	500.00	3.84
Cabbage	216.20	156.80	450.00	2.87
Cauliflower	136.24	147.00	300.00	2.04
Chilli(green)	82.92`	8.80	70.00	7.95
Cucumber	145.12	67.50	130.00	1.93
Beans	68.83	22.00	80.00	3.63
Okra	NA	85.00	200.00	2.35
Onion	137.96	141.90	400.00	2.81
Pea	60.83	72.90	150.00	2.05
Radish	NA	123.90	250.00	2.01
Tomato	250.96	156.00	1000.00	6.41

Source: Pradeep *et al* 2008

Space saving horticulture

In both developed and developing countries, the availability of cultivable area is limited. Demographic pressures, abiotic stresses and high opportunity costs of land and water limit the option of bringing more area under cultivation. These factors have made farming non-remunerative and unsustainable, shying away the farmer from the vocation. In this context wherein the resources for farming such as land, water, labour are limited the only way out is the cultivation of high value crops under space and resources saving structures. This enables to fetch high price in the market and also retain nutritional qualities. The high nutrient protective products from horticultural crops grown under these conditions can be a constituent in poverty alleviation programmes of the developing countries from where rampant cases of malnutrition and under-nutrition are reported (Table 6).

Table 6: Child malnutrition among select developing countries

Developing country	Underweight (in percentage)
Bangladesh	67
China	16
Dominican Republic	10
Ethiopia	48
Eritrea	41
India	53
Kenya	23
Lao Republic	40
Malawi	27
Nepal	49
Nigeria	36
Pakistan	40
Rwanda	29
Sierra Leone	29
Sri Lanka	38
Tanzania	29
Zambia	25
Zimbabwe	16

Source: FAO 2010

Intensive farming by resorting to the use of hybrid seeds, high density planting, multitier cropping, mixed cropping, intercropping, hydroponics, aeroponics, bubbleponics, fogponics, precision farming, protected cultivation, vertical gardening, training and pruning can be adopted to meet the high quality and productivity targets in horticultural crops.

Table 7: Container size suited for common vegetables

Vegetables	Container Size
Beans, Snap	8" wide, 8-10" deep
Beets	6"-12" deep
Broccoli	20" deep
Brussels Sprouts	12" wide, 12" deep
Cabbage	8"-10" wide, 12" deep
Carrots	10" wide, 10" deep
Chinese Cabbage	20" deep
Collards	12" deep
Cucumber	20" wide, 16" deep
Eggplant	16" deep
Horseradish	5 gallon or larger
Kale	8" wide, 8" deep
Kohlrabi	12" deep
Lettuce	8" wide, 6-8" deep
Onion	10-12" deep
Peas	12" deep
Peppers	16" deep
Potatoes	1-20 gallon containers
Pumpkins	5 gallon tub
Radish	4-6" deep
Spinach	4-6" deep
Squash	24" deep
Tomatoes	24" deep
Turnips	10-12" deep

Use of hybrid seeds

The trend of F1 *hybrid seed use* in *vegetable crops* is increasing globally in term of species, cultivars and volumes of seed *used.* Yield improvement, agronomic performance and consistency of end-*use* quality are the attributes which make the use of hybrid seeds desirable among farmers. By using hybrid seeds the yield can be increased by 4 to 5 times. In developed countries majority of our main vegetables including tomato,chilli, sweet pepper(paprika), eggplant,okra,corn-sweet,baby-,lettuce,kale, cucumber, squash, pumpkin, melons-musk,netted-, watermelon, carrots, brassicas such as cabbage-white, red, savoy-, cauliflower, broccoli, Chinese cabbage and radish, and onion in developed countries are grown using seeds of F1 hybrid cultivars. The popularity of F1 hybrid cultivars is due to their vigor, uniformity, disease resistance, stress tolerance and good horticultural traits including earliness and long shelf-life expressed and therefore giving consistent stable high yield.

However, the distribution of hybrid vegetable seed production in the world is limited to some specific regions where the climate, weather and availability of good growers are the main deciding factors.

Vegetables can be classified into three categories according to their temperature requirements as follows:

1. Low temperature species such as brassicas, radish, carrot and spinach require a low temperature of 8°-15°C of vernalization to bolt, flower and seed set.
2. Moderate temperature species such as tomato, sweet pepper and zucchini require a temperature of around 18°-20° with around 25°C in the day and 15°C at night for optimum seed production. The diurnal temperature difference is desirous to obtain the best result. Too low a temperature causes low seed set and pollen production, and too high a temperature results in flower abscission, low pollen production, low pollen viability and more of pest and disease problems.
3. High temperature species such as okra, cucurbits, sweet corn and tropical vegetables require a temperature of 20°C and above.

Root Stocks and grafting of vegetables to improve production

Grafting of herbaceous seedlings is a unique horticultural technology practiced for many years in East Asia to overcome issues associated with intensive cultivation using limited arable land for vegetable production (Lee and Oda 2003). In vegetable production, most of the damage from continuous cropping is caused by soil-borne diseases such as Fusarium wilt, bacterial wilt and nematodes. As a countermeasure to the damage caused these, grafting of fruit-bearing vegetables is becoming popular against Fusarium wilt in Cucurbitaceae (Cucumber, melon etc.) and bacterial wilt in Solanaceae (tomato, pepper etc.). This technique can also be used to overcome the vagaries of abiotic stress such as salinity tolerance and drought tolerance. Plants are grafted onto various rootstock species and varieties, by a range of grafting methods. Recently, the tube grafting method has been developed for plugs. This is popular in the manual grafting of tomato, eggplant and cucumber plants. Grafting robots and healing chambers have been developed, and are used in nurseries producing grafted plugs. Since grafting gives increased disease tolerance and vigor to crops, it will be useful in the low-input sustainable horticulture of the future.

Depending on the crop concerned, there are mainly two groups of rootstocks available in vegetable crops viz. Solanaceae rootstocks for tomatoes, aubergines and peppers, and Cucurbitaceae rootstocks for cucumbers, melons and watermelons. Grafting onto a vigorous rootstock can increase productivity quite considerably. In addition it can also ensure better fruit set during the off season and so give a higher production and better quality. This is always a matter of just the right combination of cultivar and rootstock. Rootstock efficacies are influenced by compatibility to the selected scion, existing disease pressure, and climate conditions. Therefore it is very important to test the selected candidate rootstocks at a small scale before introducing the rootstock for larger scale.

High density planting (HDP)

High density planting technique is a modern method of raising horticultural crops, especially fruit crops like mango, papaya, pineapple,guava etc. It is one of the effective measures in improving the productivity by increasing the plant population per unit area through efficient use of inputs. It ensures the best utilization of land and resources including radiant energy. It enables better intercultural operations, plant protection and harvesting. Quality of the produce is also high and suitable for export. However, the fruit plants have lesser life span and canopy management is difficult. Sophisticated techniques which demand skill are needed in maintaining high density fruit plantations.

INM, Micro irrigation and fertigation

In order to increase the productivity and improve the quality best management of nutrient and water supply system in vegetable cropping is very important and in this direction integrated nutrient and water management assumes significance. As water available for irrigation is becoming increasingly scarce and demand for irrigation is constantly raising, a rational use of the more and more limited water resources is absolutely essential. The application of high input technologies such as chemical fertilizers, pesticides, herbicides, improved seeds for increased production has adverse effects on soil productivity and environment quality. To sustain the productivity of different vegetable crops and cropping systems, efficient nutrient management is vital. There is a need to develop more efficient, economic and integrated system of nutrient management for realizing high crop productivity without diminishing soil fertility. Use of integrated plant nutrient system (IPNS) which involves cattle, poultry, animal and plant resources in combination with chemical fertilizer, supplemented with bio-fertilizer such as *Rhizobium* culture, *Azotobacter*, *Azospirillium*, mycorrhiza and phosphate solubilizing bacteria would boost the yield of vegetable crops.

A two-fold strategy may be useful in nutrient management to over come any nutrient stress that affects crop growth and yield. First, to continue to encourage the use of fertilizer for a faster growth in agriculture and second, to popularize the use of recycled crop residues, green manures, use of city or urban compost and biofertilizers as a supplementary source of nutrients, in conjunction with chemical form of plant nutrients. There is need to add plant nutrients to the soil on the basis of soil fertility status. The application of nutrients according to soil-test-based-fertility-assessment ensures better efficiency and improved profitability. Principle guide lines are: (a) basic soil fertility and climate. (b) nature of crop in cropping system. (c) at least 30% of the total nutrient levels NPK in organic form.

Multi-tier cropping

It is a system of simultaneous cultivation of two or more crops of different heights at the same time on the same piece of land and thus using land, water, and space the most efficiently and economically. In India, farmers are known to sow as many as 15 types of crops at a time. The tropical multi-tier system where coconut occupies the upper tier, banana the middle tier, and pineapple, ginger or a leguminous fodder, or herbs occupy the lowest tier is a classical case of this system.

Mixed intercropping

It is one of the most basic forms in which the component crops are totally mixed in the available space. It pertains to growing of two or more crops simultaneously on the same piece of land, without any definite row arrangements. It can also include allied sector components like livestock, poultry, fisheries, apiculture etc. depending on the available resources.

Intercropping

Intercropping refers to growing of two or more generally dissimilar but compatible crops simultaneously on the same land. The most common goal of intercropping is to produce a greater yield on a given piece of land by making use of resources which would otherwise not be utilized by a single crop. Careful planning with respect to the soil, climate, crops, and varieties is required. Usually the base crop is grown in a distinct row arrangement. The recommended optimum plant population of the base crop is suitably combined with appropriate additional plant density of the associated crop, and there is crop intensification in both time and space dimensions. It encourages biodiversity, by providing a habitat for a variety of insects and soil organisms that would not be present in a single-crop environment. This in turn can help limit outbreaks of crop pests by increasing predator biodiversity. Additionally, reducing the homogeneity of the crop increases the barriers against biological dispersal of pest organisms through the crop. Suitable selection of crops also ensures agronomic benefits like structural support against lodging for companion crops, suppression of weed growth, shade protection to light sensitive crops and nutrient enrichment.

Numerous types of intercropping, all of which vary in the temporal and spatial mixture to some degree, have been identified and some of the more significant types include the following.

a. **Row cropping**: It involves the component crops arranged in alternate rows. Variations include alley cropping, where crops are grown in between rows of trees, and strip cropping, where multiple rows, or a strip, of one crop are alternated with multiple rows of another crop.

b. **Relay cropping:** Temporal separation of crops is found in relay cropping. In this ,the second crop is sown during the growth phase, often near the onset of reproductive phase or fruiting, of the first crop, so that the first crop is harvested to make room for the full development of the second. Intercropping also uses the practice of sowing a fast growing crop with a slow growing crop, so that the fast growing crop is harvested before the slow growing crop starts to mature. This obviously involves some temporal separation of the two crops.

Hydroponics

Hydroponics is a form of hydro culture where plants are grown in water containing mineral nutrients without soil. Terrestrial plants may be grown with their roots in the mineral nutrient solution only or in an inert medium, such as perlite, gravel, mineral wool, expanded clay pebbles or coconut husk. The system is based on the 18th century discovery that the plants absorb essential mineral nutrients as inorganic ions in water. In natural conditions, soil acts as a mineral nutrient reservoir but the soil itself is not essential to plant growth. When the mineral nutrients in the soil dissolve in water, plant roots are able to absorb them. When the required mineral nutrients are introduced into a plant's water supply artificially, soil is no longer required for the plant to thrive. Almost any terrestrial plant will grow with hydroponics.

Basically there are two types of hydroponics viz. solution culture and medium culture. Hydroponics where nutrient solution is used as the medium is called solution culture. The three main types of solution cultures are static solution culture, continuous-flow solution culture and aeroponics. Medium culture uses a solid medium for the roots and is named after the type of medium like sand culture, gravel culture, or rock wool culture. The containers used should exclude light to prevent algae growth in the nutrient solution.

Advantages of using hydroponics in growing food crops are that, the water stays in the system even after the crop harvest and can be reused. Moreover, pests and diseases will be lower than in soil medium and can be easily controlled. So also the nutrition levels can easily be controlled and avoid nutrition pollution of environment. Stable and high yields are reported from many horticultural crops grown under hydroponics. This is mostly recommended in places where in-ground agriculture or gardening is not possible.

However, without soil as a buffer, any failure to the hydroponic system leads to rapid plant death and only certain species of plants survive for so long in water before they become waterlogged. Moreover, pathogen attacks as a result of high moisture levels associated with hydroponics such as damp-off due to *Verticillium* wilt are more prevalent. Different plants under hydroponic conditions require different fertilizers and containment systems.

Aeroponics

Aeroponic technology was invented and commercialized by Richard in 1983 as an alternative to water intensive hydroponic systems. The word "aeroponic" is derived from the Greek meanings of *aero-* (air) and *ponos* (labour). It is a system wherein roots are continuously or discontinuously kept in an environment saturated with fine drops (a mist or aerosol) of nutrient solution. The method requires no substrate and entails growing plants with their roots suspended in a deep air or growth chamber with the roots periodically wetted with a fine mist of atomized nutrients. It has been proved to be successful for propagation, seed germination, seed potato production, tomato production, leaf crops, and micro-greens. Excellent aeration is the main advantage of aeroponics. Moreover, any species of plants can be grown in a true aeroponic system because the micro environment of an aeroponic can be finely controlled and the nutrient requirement is less. High-pressure aeroponics is defined as delivering nutrients to the roots via 20–50 micrometre mist heads using a high-pressure (about 5 kg per square cm (500 kPa)) diaphragm pump. Plants grown using aeroponics spend 99.98% of their time in air and 0.02% in direct contact with hydro-atomized nutrient solution. The time spent without water allows the roots to capture oxygen more efficiently. Furthermore, the hydro-atomized mist also significantly contributes to the effective oxygenation of the roots.

Bubbleponics

The term bubbleponics describes a top-fed Deep Water Culture (DWC) hydroponic system. In this, highly oxygenated nutrient solution is applied directly to the root zone of plants. Pumping water from the reservoir up to the roots is called top feeding. The water is released over the plant's roots and then runs back into the reservoir below in a constantly recirculating system. In DWC there is an airstone in the reservoir that pumps air into the water via a hose from outside the reservoir. The airstone helps add oxygen to the water. Both the airstone and the water pump run 24 hours a day. In Bubbleponics, the roots get easy access to water from the beginning and will grow faster in the beginning and a few weeks of grow time can be saved.

Fogponics

In fogponics, water is used in a vapourized form to transfer nutrients and oxygen to enclosed suspended plant roots and is an advanced form of aeroponics. It is same as aeroponics except fogponics uses a 5-10 micron mist within the rooting chamber.

Precision farming

The concept of precision agriculture emerged from the belief that the variability of plant-growing conditions is one of the major contributors to field-scale differences in yield, and the idea that it could be beneficial to vary agricultural inputs according to local changes in soil properties (Robert, 1993). In fact, it forms an Information Technology based farm management system to identify, analyse and manage natural resources variability within field to optimize profitability, sustainability and protection of the land resources. It involves defining a Decision Support System (DSS) for whole farm management with the goal of optimizing returns on inputs while preserving resources. A number of approaches that are currently being used to define management zones include hand drawn polygons on yield maps, supervised and unsupervised classification procedures on satellite or aerial imager, identification of yield stability and geomorphology component that typically dictates the hydrology of the farm field (McBratney *et al.* , 2005 and Whelan *et al.*, 2006). It aims at optimizing field-level management by matching farming practices more closely to crop needs, by reducing environmental risks of farming and boosting competitiveness through more efficient practices.

Protected cultivation

Protected cultivation can be defined as a cropping technique where the micro climate surrounding the plant body is controlled partially/ fully as per the requirement of the plant species grown during their period of growth. It forms an invention that attempts to achieve higher water and nutrient efficiencies, increasing photosynthetic efficiency and reducing transpiration thereby contributing for environmental protection through better use of applied growth inputs and bare minimum use of plant protection chemicals. There are many high value and high yielding vegetable crops such as bell pepper, fresh market tomatoes, slicing cucumbers and melons suitable for growing under cover.

Environmental concerns, whether aggravated by climate change or not, appear likely to constrain future expansion of vegetable production unless appropriate measures to prevent further increase in pollution levels and to mitigate pollution caused by other sources. Protective structure is a facility to protect crop from biotic and abiotic constraints. Structures for protected cultivation include green houses, plastic/net houses and tunnels. Artificial materials such as glass, fibre glass, polycarbonate, plastic and nets are used for the cover on the structures. Design of the structures depends on the availability of materials, purpose for crop protection and the economics.

Greenhouse technology is one of the most practical ways of achieving the goal of protected cultivation. Benefit of growing crops in a greenhouse is the ability to extend growing season and higher crop productivity, both of which are important from the point of reducing the land area used for growing crops. Crops could be grown under inclement climatic conditions when it is not possible to grow in open field. It enables efficient utilization of precious inputs like water and nutrients. Early nursery and early cropping are possible under the system. Higher productivity through enhanced crop duration and quality of produce are also ensured under this. Protected cultivation, in general requires 15 to 20 per cent less water as compared to outside grown systems. Even to further optimize water use and enhance water productivity, it is necessary to develop appropriate irrigation schedules aimed at a more efficient water use while obtaining both quality and quantity of yield. This will result in cost decrease and a reduced use of fertilizers and chemicals through direct and indirect savings.

Carbon dioxide (CO_2) enrichment

The benefits of carbon dioxide supplementation on plant growth and production within the greenhouse environment have been well understood for many years. CO_2 increases productivity through improved plant growth, vigor, early flowering, and better yield components. Photosynthesis is a chemical process that uses light energy to convert CO_2 and water into sugars in green plants. These sugars are then used for growth within the plant, through respiration. The difference between the rate of photosynthesis and the rate of respiration is the basis for dry-matter accumulation (growth) in the plant. In greenhouse production the aim of all growers is to increase dry-matter content and economically optimize crop yield. For the majority of greenhouse crops, net photosynthesis increases as CO_2 levels increase from 340–1,000 ppm (parts per million). Most crops show that for any given level of photosynthetically active radiation (PAR), increasing the CO_2 level to 1,000 ppm will increase the photosynthesis by about 50% over ambient CO_2 levels. For some crops the economics may not warrant supplementing to 1,000 ppm CO_2 at low light levels. For most crops the saturation point will be reached at about 1,000–1,300 ppm under ideal circumstances. A lower level (800–1,000 ppm) is recommended for raising seedlings (tomatoes, cucumbers and peppers) as well as for lettuce production. Even lower levels (500–800 ppm) are recommended for African violets and some Gerbera varieties. Increased CO_2 levels will shorten the growing period (5%–10%), improve crop quality and yield, as well as, increase leaf size and leaf thickness. The increase in yield of tomato, cucumber and pepper crops is a result of increased numbers and faster flowering per plant.

Static solution culture

In static solution culture, plants are grown in containers of nutrient solution, such as glass Mason jars (typically, in-home applications), plastic buckets, tubs, or tanks. The solution is usually gently aerated but may be un-aerated. If un-aerated, the solution level is kept low enough that enough roots are above the solution so they get adequate oxygen. A hole is cut in the lid of the reservoir for each plant.

There can be one to many plants per reservoir. Reservoir size can be increased as plant size increases. A home made system can be constructed from plastic food containers or glass canning jars with aeration provided by an aquarium pump, aquarium airline tubing and aquarium valves. Clear containers are covered with aluminium foil, butcher paper, black plastic, or other material to exclude light, thus helping to eliminate the formation of algae. The nutrient solution is changed either on a schedule, such as once per week, or when the concentration drops below a certain level as determined with an electrical conductivity meter. Whenever the solution is depleted below a certain level, either water or fresh nutrient solution is added,

Nutrient film technique (NFT)

NFT is a hydroponic technique wherein a very shallow stream of water containing all the dissolved nutrients required for plant growth is re-circulated past the bare roots of plants in a watertight gully, also known as channels. In an ideal system, the depth of the recirculating stream should be very shallow, little more than a film of water, hence the name 'nutrient film'. This ensures that the thick root mat, which develops in the bottom of the channel, has an upper surface, which, although moist, is in the air. Subsequent to this, an abundant supply of oxygen is provided to the roots of the plants. A properly designed NFT system is based on using the right channel slope, the right flow rate, and the right channel length. The main advantage of the NFT system over other forms of hydroponics is that the plant roots are exposed to adequate supplies of water, oxygen and nutrients. In all other forms of production, there is a conflict between the supply of these requirements, since excessive or deficient amounts of one results in an imbalance of one or both of the others. NFT, because of its design, provides a system wherein all three requirements for healthy plant growth can be met at the same time, provided that the simple concept of NFT is always remembered and practiced. The result of these advantages is that higher yields of high-quality produce are obtained over an extended period of cropping. It is one of the more productive techniques that can be adopted for green house vegetable production.

Vertical gardening

Vertical farming is an idea adopted from the indigenous people in South America. They have long used vertically layered growing techniques, and even the rice terraces of East Asia follow a similar principle. But, now, vertical farming has been adopted as potential tool of urban horticulture to combat the problems related to rapidly growing global population and increasingly limited resources. The term "vertical farming" was coined in 1915 by American geologist Gilbert Ellis Bailey. Architects and scientists have repeatedly looked into the idea since then, especially toward the end of the 20th century. It is calculated that a single 30-story vertical farm could feed some 50,000 people. And, theoretically, 160 of these structures could provide all of New York with food year-round, without being at the mercy of cold snaps and dry spells. Despite these promising calculations, such high-rise farms till recently existed only as small-scale models. However, recent developments in South Korea and Netherlands have brought the concept more near to reality and the first prototypes have been created. The first foods from a vertical farm have already reached the supermarket shelves of Netherlands. The PlantLab, a 10-year-old company based three floors underground in the southern city of Den Bosch, under vertical garden and has cultivated everything from ornamental shrubs and roses to nearly every crop imaginable, including strawberries, beans, cucumbers and corn. This holds great potential for urban farming.

Training and pruning

Left to themselves, the cultivated plants may attain a shape of their own depending upon the variety and crop. They may not also bear abundantly unless trained. Training refers pruning followed primarily to give a definite shape or form to trees and control the growth. The two practices, training and pruning are often confused and considered the same, since they go hand in hand. But, pruning in the broader sense refers to removal of plant parts with an object to get better yield and quality fruits while training is to give the trees a basic frame work specially done when the trees are young. The primary objective of training and pruning is to develop a strong tree framework that will support fruit production. Properly trained and pruned trees will yield high quality fruit much earlier and live significantly longer.

Container growing

Growing vegetables in Containers is one of the simple means of space saving technologies. There are two main types of container materials, non-porous and porous. Examples of non-porous containers are glazed pots and those made of plastic, metal, and glass. Clay is the primary type of porous material. Wood and pressed fiber pots are considered semi-porous. When determining the type of container you want to use, consider the frequency of watering required, weight, and cost. Soilless culture in bags, pots, or troughs with a lightweight medium is the simplest, most economical, and easiest to manage of all soilless systems. The most common types of media used in containerized systems of soilless culture are peat-lite or a mixture of bark and wood chips. Container types range from long wooden troughs in which one or two rows of plants are grown, to polyethylene bags or rigid plastic pots containing one to three plants. Bag or pot systems using bark chips or peat-lite are in common use. The major advantages over other types of soilless culture: (1) the medium materials have excellent retention qualities for nutrients and water; (2) containers of medium are readily moved in or out of the greenhouse whenever necessary or desirable; (3) they are lightweight and easily handled; (4) the medium is useful for several successive crops; (5) the containers are significantly less expensive and less time-consuming to install- and (6) in comparison with recirculated-hydroponic systems, the nutrient-solution system is less complicated and less expensive to manage. From a plant-nutrition standpoint, the latter advantage is of significant importance. In a recirculated system, the solution is continuously changing in its concentration and its nutrient balance because of differential plant uptake. The container size suitable for common vegetables are given as Table-7.

Conclusion

World reeling under the pressure of human population leaves little space for farming. This has necessitated the formulation of novel ideas for space saving in cultivation. These techniques include innovations like planting of two or three different types of fruit trees that bears fruit at different times in a year in a single pit. In this case, training and pruning of the branches of the trees should be done in such a way that all the branches get sufficient sunlight and also the trees get sufficient nutrients and water. Pruning of apple trees which are slow in growth to form columnar shape results in little branching and require less space and pruning requirements in orchards. Layered planting utilizing the planting space in a small flower border helps to effectively double or triple the available space and time. This offers the double advantages of saving space and extending the season of interest. Thus though a multitude of constraints can reduce farm yield, ranging from infrastructural and institutional ones bearing upon farm gate costs and prices and farmer skills and attitudes, to diverse technical constraints prospects for yield gap closing through space saving techniques are tremendous.

References

Adamcunk, Viacheslav I. 2010. Precision Agriculture: Does it Make Sense? *Better Crops* **94**(3)

FAO 2010. The State of Food Insecurity in the World 2010. Economic and Social Development Department, Food and Agriculture Organization of the UN. Rome

FAO 2012, Food Insecurity in the World, Food And Agriculture Organization of The United Nations, Rome, 2012

Fischer, R.A. 2007. Understanding the physiological basis of yield potential in wheat. *Journal of Agricultural Science* **145:** 990-113.ICRIER Working Paper

Fischer, R.A. Derek Byerlee and G.O. Edmeades. 2010. Can technology deliver on the yield challenge to 2050? *Expert Meeting on How to feed the World in 2050.* Food and Agriculture Organization of the United Nations Economic and Social Development Department

IFPRI. 2009. Global Hunger Index. Washington, DC, IFPRI

Lang,T., Barling,D and Caraher,M.2009. Food Policy: Integrating Health, Environment and Society, Oxford: Oxford University Press

Lee, J.M. and Oda , M. 2003. Grafting of herbaceous vegetable and ornamental crops. *Hortic. Rev.* **28**: 61–124.

McBratney, A., B. Whelan, T. Ancev, and J. Bouma. 2005. Precision Agriculture, *Better Crops.* **6:** 7-23.

Mitchell, D. 2008. A note on rising food prices. *Policy Research*. Working Paper No. 468 Washington Dc World Bank

Mittal, Surabhi 2006. Structural Shift in Demand for Food: India's prospects in 2020. *Working Papers* 184. Indian Council for Research on International Economic Relations, New Delhi.

Pradeepkumar, T., Suma, B., Jyothibhaskar and Satheesan, K.M. 2008. Management of Horticultural Crops, : Vol. 11: Horticulture Science Series: In 2 Parts , Peter, K.V. (Series Editor). New India Publishing Co. New Delhi (www.bookfactoryindia.com).

Whelan, Christopher T., Brian Nolan and Bertrand Maître. 2006. Trends in Economic Vulnerability in the Republic of Ireland *The Economic and Social Review,* **37**(1):91-119.

3

Principles of Hi-tech Horticulture

A.A. Patil, S.M. Mantur, M.S. Biradar, and I.M. Mannikeri

Abstract

Indian Horticultural produce to become competitive both in domestic and international markets, there is an urgent need for use of Hi-tech horticultural technologies. The techniques covered under Hi-tech Horticulture are genetically modified horticulture crop varieties derived from biotechnology and genetic engineering, micro-propagation, protected cultivation, micro irrigation and fertigation, high density planting, organic farming, integrated nutrient, pest and disease management and post harvest management.

Micropropagation or tissue culture technology is based on the principle of totipotency. It provides disease free, uniform sized and year round production of mass multiplication of plants in various stages viz. establishment and stablillization, shoot multiplication, pre-transplant and in vitro rooting and transplanting acclimation and ex vitro rooting. Protected cultivation is the technique of providing to the crop to protect plants from the adverse climate conditions by providing optimum growth and development conditions. High value horticulture crops mainly cut flowers and vegetables are more popular for cultivation in green houses. Greenhouse may be of different types depending on the structural designs followed as quonset, curved roof or gable roof and covering materials used as glass, acrylic, polycarbonate, fiberglass reinforced plastic panels or polyethylene etc. applicators. Micro-irrigation system is used for frequent application of required and measured quantity of water directly above or below soil surface, usually as discrete drops, continuous drops, tiny streams or miniature spray through emitters or applicators placed along a water delivery line (laterals). It may be in form of drip irrigation, bubbler irrigation, micro jet irrigation, sprinkler, micro-sprinkler or misting (fogger) irrigation system. Similarly under fertigation system, special fertilizers in form of liquid or water soluble solids are available through a suitable irrigation system for accuracy of water application.

Keywords: Horticulture, Hi-tech, Micro-propagation, cut flowers, fertigation

Introduction

Growth of horticulture is the cornerstone for augmenting employment opportunities in agriculture sector both directly and indirectly. Of late, there is a marked shift in the cropping pattern towards diversification into high value commercial crops in the country. To make horticulture viable and profitable, the thrust should be on improving the resource use efficiency, so that the relative profitability could be enhanced. Success in horticulture is more technology driven compared to other crops. The time bound removal of quantitative curbs on imports and other barriers to access domestic market under WTO, of which India is a signatory, will require Indian Horticultural produce / products to become competitive both in the domestic and international markets. This would call for use of Hi-tech horticulture technologies. The term Hi-Tech Horticulture was coined in recent years. There are no standard definitions for this. The Department of Agriculture and Co-operation, GOI in 1999 defined Hi-tech Horticulture as "any technology which is modern, less environment dependent, capital intensive and has the capacity to improve the productivity and quality of horticultural crops". The technologies encapsulated in Hi-Tech Horticulture are genetically modified horticulture crop varieties derived from bio technology and genetic engineering, micro propagation, protected cultivation, micro irrigation and fertigation, high density planting, organic farming in horticulture, integrated nutrient, pest and disease management and post harvest management. Hi-tech Horticulture technologies are discussed in detail under following heads:

i. Micro propagation

ii. Protected cultivation

iii. Micro irrigation and fertigation

iv. High density planting

I. Micro propagation

The term micro propagation is used specifically to refer to application of "tissue culture" techniques to the propagation of plants starting with very small plant parts growing aseptically in a test tube or other containers. Tissue culture is based first on the principle of totipotency, i.e. the concept that every living cell has the genetic potential to reproduce entire organism. The technique of *in-vitro* cultivation of plant cell, tissue or organs is primarily denoted to solve two basic problems, first to keep plant cells free from microbes and second to ensure the desired tissue development by providing suitable nutrient media and other environmental conditions. The first problem can be eliminated using modern equipments and careful handling necessary for maintaining aseptic environmental conditions. Second problem remains an area of active research mainly on manipulation of culture media, especially growth regulators and to a much lesser extent on environmental conditions.

Merits of Micro Propagation

1. **Mass multiplication of plants:** Only one plant can be produced from a cutting/ seed, whereas one explant can produce more number of new plants. Very a few stock plants are required to provide the explants needed to produce thousands of new plants. This is the most useful for multiplication of new hybrids, mutants, unique seedlings and or disease free stock plants.

2. **Disease free planting materials:** The diseases transmitted from parent to offsprings often can be eliminated through micro propagation. External contamination can be removed by surface sterilization of explants. Many internal viruses can be eliminated by using the apical meristem which will be free from viruses.

3. **Year round production:** Tissue culture is not limited by the season or the weather. Working condition in the laboratory is ideal for year round production.

4. **Uniform size of plants:** Uniform size of plants helps in establishment of orchard uniformly, which helps in carrying out cultural operations and harvesting.

5. **Easy Transportation and storage:** Easy exchange of plant materials, since many plantlets can be multiplied in a container. *In-vitro* plantlets can be stored for several years and hence can be conserved for long duration.

Factors Affecting Micro Propagation

1. **Genotype and choice of explants**: Regeneration capacity varies with species, cultivars and cultural conditions. In general, those species, the easiest to propagate conventionally have also been the easiest to micro propagate. Most herbaceous plants can be easily propagated by conventional or micro propagation, woody perennials are much more difficult to regenerate and new principles of culture have to be developed. Success in the most woody perennial species is related to their juvenility status. Shrubs are generally

more easily handled than trees. In micro propagation, selection of proper explants is a major requirement. Explants selected should be potential and are to be from active growing phase.

2. **Contamination:** Fungal and bacterial contamination with the explants is one of the major hurdles in developing sterile cultures. Overcoming this problem is a big challenge especially with the perennial horticultural species. Disinfecting explants, tools and working area to remove contaminants from the surface is utmost important.

3. **Culture media:** A basal medium must be selected or established through systematic analysis. The most widely used is the MS medium (Murashige and Skoog). Commercial micro propagation requires optimization of not only the inorganic elements but also of other factors of the medium. This can be done by systematically testing a range of concentration of individual elements combined in various ways with other elements.

4. **Hardening:** Plantlets multiplied through micro propagation need to be hardened carefully before taking them to main field. Acclimatization of regenerated plantlets is therefore very essential. In ornamentals and other annuals, the field establishment is not much problematic. In most cases the regenerated plantlets have to be gradually acclimatized by hardening under greenhouse and shade house.

Requirements of Tissue Culture Laboratory

1. **Washing room:** It is essential for washing, drying and storage of vessels. It should be furnished with wash basin.

2. **Media room:** It is required for preparation, sterilization and storage of culture media. This room should be equipped with wash basin, double distillation unit, hot water bath (i.e. steamer), hot plates with magnetic stirrers, electronic weighing balances, pH meter, refrigerator, cupboard for storage of chemicals and media, filter sterilization unit, centrifuges, vacuum pump, etc.

3. **Sterilization room:** Sterilization room should be adjacent to media room but in a separate corner taking all care of safety. An autoclave unit should be placed in this room.

4. **Incubation room:** Laminar air flow hoods are fitted in this room for aseptic manipulation. A small table having stereoscopic microscope or inverted microscope may be adequate for culture observations. It should be equipped with air conditioners and air curtains at doors.

5. **Culture room:** Culture room should have two door facility, controlled temperatures (usually $25+2^0$ C with air conditioners) and humidity (60-80% RH

with humidifiers), culture racks fitted with light (generally 1000 lux light generated by fluorescent tubes with automatic timers to regulate photoperiod) and a horizontal shaker for agitation of liquid cultures. It is desirable to have a generator set to the culture room in the event of electricity failures.

6. **Polyhouse:** Two types of polyhouses are desirable. A fully automatic polyhouse having fan and pad cooling system along with foggers and drip irrigation and mini sprinkler systems depending on sensors are necessary to provide controlled temperatures (25+2^0 C) and humidity (80-100 % RH) required for development of fresh *in- vitro* plants. The plantlets are further hardened in semi-automatic polyhouses where natural ventilation is provided. In these polyhouses, plantlets are repotted in bigger containers supplemented with organic and inorganic manures. After 2-4 months period, tissue cultured plantlets are transferred to shadehouses and finally to the field with initial shade cover or in congenial atmospheric condition. Polyhouses should be connected with a generator set for general safety.

Preparation for Culture

1. **Choice of explants:** For the success of micro propagation, the kind of explants and from where they are collected are very much critical. They may vary with purpose of the culture, the species and cultivars. In the absence of definite information on a given species or cultivar, one can use available information on a closely related species or cultivar or carry out experimentation with the particular plant. While collecting shoot tip, plant should be relatively in active growth phase. The size of the explants varies from 1-5 mm of meristem tip for meristem culture. Smaller the explants, the more effective is the elimination of pathogens. On the other hand, smaller the explants, the more difficult to establish and the lower survival rate. In general explants of 0.25 to 1 mm can be used. Virus status must be confirmed by indexing.

2. **Sterilization:** All the materials like vessels, instruments, medium, plant material, etc. used in incubation and culture rooms must be free from microbes. This is achieved by one of the following approaches.

   Dry heat: Glass wares, plastic wares and instruments may be sterilized by dry heat in an oven.

   Wiping with alcohol: The surfaces that cannot be sterilized by other techniques e.g. platform of laminar flow cabinet, hands of operator, etc. are sterilized by wiping them thoroughly with 70 % alcohol and the alcohol is allowed to dry. The use of mouth mask is also advocated to avoid contamination.

   Flame sterilization: Instruments like, forceps, scalpel, needles, scissors, etc. are ordinarily sterilized by dripping them in 80-90 % alcohol (ethanol) followed by flaming. These instruments are repeatedly sterilized during the operation to avoid contamination. It is customary to flame the mouth of culturing vessels prior to inoculation or subculture.

Surface sterilization: All *in-vitro* plant materials to be used for culture are treated with an appropriate sterilizing agent to inactivate the microbes and called 'surface sterilization'. The protocol depends on the surface and type of explants which determine the contamination load and tolerance to the sterilizing agent. The sterilizing agents used for surface disinfection are calcium hypochlorite (9-10%), sodium hypochlorite (2%), mercuric chloride (0.1-1.0%), silver nitrate (1%), bromine water (1-2%), hydrogen peroxide (H_2O_2,10-12 %) and antibiotics (4-50 mg/l). Among these, calcium or sodium hypochlorite and mercuric chloride are the most commonly used for 10-30 minutes followed by 4-6 rinses with sterilized distilled water.

Autoclaving: Culture vessels (both empty and containing media) are generally sterilized in an autoclave or pressure cooker at 121^{0}C at 15 p.s.i. or 1.06 kg/cm^2 for 15-40 minutes. Care should be taken to properly stopper all the vessels and open the autoclave only when its pressure gauge indicates zero pressure. Use of special thick and long handgloves are recommended while handling the hot autoclaved material.

Filter sterilization: Some growth regulators e.g. GA_3, zeatin, ABA, certain vitamins and enzymes are heat sensitive. Such compounds are filter sterilized by passing their solution through a membrane filter of 0.45 micron or lower size. Then, filter sterilized heat liable compounds are added to cooled autoclaved media (40^{0}C). The laminar air flow cabinet creates aseptic working space by blowing filter sterilized air through filter at velocity of 1.8 km / hr.

3. **Culture indexing:** Culture indexing test for contamination by specific pathogens in containers is done as follows. Cut off a piece of plant tissue, dice it and place the pieces into 15 ml of nutrient broth optimum for bacterial and fungal growth. Positive results are shown in cloudiness of the medium and should appear with in 4-5 days, if made at initial explants stage.

Stages in Growing the Cultures

General patterns of treatments and handling procedures are well established, many of the requirement of individual stages must be developed for different species and even cultivars. The process can be described as optimizing the variables which affect each stage of the process. These include subculture interval, concentration of ingredients in the medium, relative cytokinin / auxin concentrations, size of propagules, light, temperature, etc. Micro propagation may include a preliminary pretreatment culture step in which the explants are initially placed in a petri dish on the surface of an agar medium of basal salts and sugar but without hormones. Using a liquid medium in the initial stages or frequent transfers in agar medium for several days may leach out the toxic material.

Stage I: Establishment and stabilization: The function of this stage is to establish the explants in culture and to stabilize the culture for multiple shoot development. This may involve stimulation of auxiliary shoots, initiation of adventitious shoots on excised shoots and initiation of callus from the cut surfaces. The medium selected may vary with the species, cultivars and kind of explants used. Control of development is the most often achieved by manipulating levels of auxin and cytokinin. Usually 4-6 weeks are required to complete stage I and produce explants ready to be transplanted to stage II. Some woody plant species can take upto a year for this to occur.

Stage II: Shoot multiplication: In the multiplication stage, each ex-plant has expanded into a cluster of micro shoots arising from a basal mass of callus like tissue. This structure is divided into separate micro shoots, which are transplanted into new culture medium. The kind of medium used depends on the species, cultivar and type of culture. Usually it is essentially the same as that used in stage I, but often the cytokinin and mineral supplement level are increased. Adjustments may need to be determined by experimentation. Different species vary in the optimum size of the micro shoot and method of cutting apart. There is generally a minimum mass of tissue necessary to produce uniform rapid multiplication in the next transfer. Division may be made by vertically cutting the tissue mass into sections, keeping some of the base with each piece. Number of micro shoots produced ranges 5-25 depending upon the species and conditions of culture. Multiplication may be repeated several times to increase the supply of material to a predetermined level for subsequent rooting and transplanting. During multiplication, off type propagules sometimes appear, depending on the kind of plant and method of regeneration. In some plants restricting the multiplication phase to only three propagation is recommended to avoid development of off-type micro shoots. Other procedures that can reduce the potential for variability include reducing hormone concentration and avoiding basal callus which may give rise to adventitious shoots in sub culturing.

Stage III: Pre-transplant and *in vitro* rooting: The purpose of the third stage is to prepare the micro plants for transplanting from the artificial heterotrophic environment of the test tube to an autotrophic free living existence in the greenhouse and on to their ultimate location. This preparation may involve rooting, but it also may involve conditioning of the micro plant to increase its potential for acclimation and survival during transplanting. For example, the agar and/or sucrose concentration might be increased. Light intensity is sometimes increased during this period. For *in vitro* rooting, individual micro shoots are sub cultured into new containers in a sterile medium with reduced or omitted cytokinin, an increased auxin concentration and often a reduced inorganic salt content. Selection for uniformity and rouging of obviously abnormal, aberrant or diseased micro shoots should be made prior to stage III.

Stage IV: Transplanting, acclimation and *ex vitro* rooting: Rooted or unrooted micro shoots are removed from the culture vessels, agar is washed away completely to remove a potential source of contamination and the micro plants are transplanted into a standard pasteurized rooting or soil mix in small pots or cells in a more or less conventional manner. Initially, micro plants should be protected from desiccation in a shaded, high humidity tent or under mist or fog. Several days may be required for new functional roots to form. Once the micro plants are established in the rooting medium, they should be gradually exposed to a lower relative humidity and higher light intensity for hardening.

II. Protected Cultivation

India is endowed with diverse agro-climatic conditions, which help to grow all kind of horticulture crops, almost throughout the year, in one part of the country or the other, but the quality of produce under open field conditions particularly in case of high value flowers and vegetables are not suitable for international as well as standard domestic markets. Sometimes there is no guarantee of consistent production from open cultivation as the crop is exposed to a number of natural environment factors which change frequently.

The need of the hour is to increase the productivity and quality of produce to meet the demand of quality conscious consumers. A break through in production technology that integrates market driven quality parameters with the production system, besides ensuring a vertical growth in productivity is required. One such technology is "Protected cultivation", or generally called "greenhouse technology". It is the technique of providing favourble environmental or growth conditions to the crop. It is rather used to protect plants from the adverse climatic conditions by providing optimum conditions of light, temperature, humidity, carbon dioxide etc, for the best growth of plants to achieve maximum yield and the best quality. A greenhouse is a covered structure which protects plants from wind, precipitation, excess solar radiation, temperature extremes and considerable pests and diseases attack.

Cultivation of crops under greenhouse in the world at present is practiced in more than 50 countries. China has adopted plastic greenhouse technology to a greater extent. The smaller land holdings and short growing seasons in the Netherlands and Japan are the reasons for adopting greenhouse technology. Water conservation is one of the reasons for greenhouse crop production in Israel and Middle-East countries. India has greenhouses, mainly glass houses for the last 4-5 decades for research and off season protection of plant materials, but the interest in commercial application is recent. The choice of crop to be raised in a greenhouse is made on the basis of physical size of the structure and economics of crop production. As a result, high value horticulture crops, particularly cut flowers and vegetables are more popular for cultivation in greenhouse.

Structures for protected cultivation: A greenhouse is a quasi-permanent structure, covered with a transparent and translucent material ranging from simple home made designs to sophisticated pre-fabricated structures wherein environment could be modified suitable for propagation or growing plants. Materials used to construct a greenhouse frame may be wood, bamboo, stone pillar and steel whereas covering materials may be glass, various rigid or flexible plastic materials. Depending on covering material, different terminologies are used in the contest of greenhouse structures.

Glasshouse: A greenhouse with glass as the covering material is called glass house.

Polyhouse: It is a greenhouse with polyethylene as the covering material

Type of structures: The greenhouse design and cost range from a simple plastic walk in tunnel costing Rs.200 / m^2 to a climate controlled, saw tooth greenhouse with automatic heating and cooling system costing more than Rs.3000 / m^2.The selection of greenhouse design should be determined by the grower's expectations, need, experience and above all its cost effectiveness and available market price for the produce. Greenhouses are classified depending on shape, which also determine their cost, climate control and use in terms of crop production. Commonly used structural designs are briefly described below:

1. **Quonset greenhouse:** The roof of this type of greenhouse will be semi round. The roof and sides can be covered with poly ethylene, fiber glass or polycarbonate sheet. This type of structure requires relatively lesser cladding material. In this multispan saw-tooth type structure can be installed with top ventilation.

2. **Curved roof greenhouse:** The semi-circular tunnel greenhouse structure appeared with the introduction of poly ethylene film as the covering material. These structures are simple and easy to construct. But these structures have the disadvantage of poor ventilation because the curved roof is not amendable to provide top ventilation.

3. **Gable roof greenhouse:** This is the most basic structure similar to a hut like structure and is perhaps the first version of a greenhouse with glass as the covering material. The structure allows large opening in the side walls and at the ridge for high rates of natural ventilation. Modern gable shaped greenhouses are multi span units with bay width of 6-12 m.

Covering materials for greenhouses: The covering materials are also called as cladding or glazing materials. The purpose of a greenhouse covering is to allow sunlight to pass through it, so that the energy is retained inside. Glass was the main covering material in the early greenhouses. With the introduction of plastic materials, there are now several alternatives available for greenhouse coverings. A brief description of greenhouse covering materials is given below:

Glass: A clean and transparent material provides the maximum light transmittance to the extent of 90 %. Since it is heavier in weight, it requires elaborate structure for adequate support. It is fragile and can break with minimum shock or vibrations resulting in high maintenance cost.

Acrylic: This material has long service life, good light transmittance (80%), moderate impact resistance, but prone to scratches. Being inflammable and costly, it is not a preferred material.

Polycarbonate: It is available in single or double wall sheets of different thickness. A new polycarbonate sheet has good light transmittance of about 78 % , but reduces with age. It has excellent impact resistance and low inflammability. High cost limits its use on large scale.

Fiberglass reinforced plastic panels (FRP): These plastics consist of polyester resins, glass fibers, stabilizers etc. It has an initial light transmittance of about 80 % and has high impact resistance with a service life ranging 6-12 years. Good quality FRP material for greenhouse covering is not quite assured.

Polyethylene: A clear, new polyethylene sheet has about 85 % light transmittance. Its higher strength and low cost have made it the most popular replacement to glass. An ultra- violet (UV) stabilized plastic sheet can have a service life of minimum 3 years. These sheets are generally available in 7 and 9 m. widths with 200 micron (0.2 mm) thickness. Now a days, plastic film with specific characteristics, like, anti-drip, anti-dust, sulphur resistant, high diffusivity etc are available and should be used depending on the requirement.

Concept of Greenhouse Design: The basic concept of designing of greenhouse includes light transmission, crops to be grown, environment control requirement, frame material and cladding material. Under Indian conditions, quonset type, multi span greenhouse is the most suitable because of its low cost and ease of fabrication. An ideal glazing material should transmit the visible portion of sunlight and filter the ultraviolet light. Ultraviolet resistant low density polyethylene (UVLDPE) single film cladding of 200 micron thickness is sufficient for Naturally Ventilated (NV) greenhouse and fan and pad (FP) greenhouses. Double inflated film of UVLDPE is required for heated greenhouses. The UVLDPE film is economical, 90 % transparency for light with 3 years usable life. The size of individual pieces of film should not be more than 100 m^2 and this should be properly fastened to frame. The use of silpaulin film should be avoided due to its low transparency and less life. Naturally ventilated greenhouses are the most suited for mild agro- climatic conditions as in Bangalore , Pune and coastal regions. Greenhouses with FP system of cooling can be used in dry and warm seasons of Northern plains and they are the most suitable for nursery raising and rooting of cuttings due to high relative humidity inside. Greenhouses with heating system are generally required during winter in northern hilly regions and at nights in plains for better growth of plants. The greenhouse can have more than one kind of environmental control system based on the variation in the climate at a location during the year.

Design of Structure: The structure has to carry one or more loads of dead load, live load, wind load and or snow load depending upon the locality and it has to be designed accordingly. Sometimes, the greenhouse should be able to take the dead load plus live load or dead load plus wind load plus half live load. The greenhouse is the best fabricated out of galvanized iron pipes and the foundation should be able to take designed load.

Design of Environmental control: The design of environmental control system depends upon the meteorological data of location (solar insulation, light intensity, wind speed, wind direction, ambient temperature and relative humidity) and the temperature and RH to be maintained in the greenhouse.

1. **Naturally ventilated (NV) greenhouse:** The temperature in these greenhouses is above the ambient temperature. The basic purpose is not to have excessive temperature in mild climates. The NV greenhouse can also be used to increase temperature in cold climates. The ventilation in the greenhouse takes place by (a) thermal buoyancy and (b) wind velocity. More is the ventilation in the greenhouse, lesser the temperature built up. The calculation of ventilation area required is very scientific, however a greenhouse should have approximately 25 % of floor area as opening area inside walls and an equal area of opening at ridge. The wall and ridge opening have to be through out length and provided with 40 mesh nylon insect proof net. The temperature build up inside the greenhouse can be controlled by controlling the size of wall opening or roof opening by using adjustable opening by rolling or unrolling LDPE film over wall opening or ridge opening using crank mechanism. The irrigation also reduces the temperature inside the NV greenhouse by 2 to 3°C. The size of NV greenhouse has to be less than 50m x 50m to avoid excessive temperature build up. The shade nets when used inside the greenhouse reduce the heat load on plants.

2. **Fan and pad greenhouse:** These greenhouses consist of the wet, cross fluted cellulose pad on one end and the exhaust fans on the other end. These greenhouses are not suitable for high humidity areas. The pad should be provided through out width, the height of pad can be 5 % of the length of the greenhouse. The 10 % of the circulating water on pad should be wasted to avoid build up of salts on pad. The water flow rate should be 9 litre / minute / m length of pad. The length of greenhouse should be restricted to 60 m.

3. **Heated greenhouse:** The heat loss through film is reduced to 60-65 % when double inflated film is used. The total heat loss of the greenhouse is determined by multiplying the exposed area of the glazing, walls, wind factor, tightness of the greenhouse and the temperature to be maintained inside the greenhouse in comparison to ambient temperature. In centrally heated GH, the boiler of required capacity can be installed and steam at 102°C or

water at 85^{0}C can be circulated through GI pipes installed above plants and around the side of the greenhouse.

The unit heaters in which fuel burns inside the combustion chamber, hot fumes pass through the heat exchanger and the fan is used to blow air over heat exchangers are also used in small greenhouses. Infra-red heaters consisting of exhaust fume pipes and reflectors are also used. The solar heater can also be used for heating the water and circulating hot water inside the greenhouse through pipes at night.

4. **Insect proof net house:** Insect –proof nets are available in different intensities of perforations, ranging from 25 to 60 mesh (perforations/cm^2). Nets of 40 or higher mesh are effective means to control entry of most flying insects and save crop from diseases. These structures reduce the risk of vectors causing viral diseases. Higher mesh size, however, reduces the air exchange of the structure, thereby, enhancing temperature and humidity inside the net house. For most flower crops, 30 and 40 mesh nets are used. Now a days, nets with characteristics like photo-selectivity, better UV stabilization and higher mechanical strength are available.

5. **Shade net house** Shade nets are perforated plastic materials used to cut down the solar radiation and prevent scorching or wilting of leaves caused by marked temperature increases within the leaf tissue from strong sunlight. These nets are available in different shading intensities ranging from 25 to 90 %. Leaf vegetables and ornamental greens are recommended to be grown under shade nets whose growth rates are significantly enhanced compared to unshaded plants when sunlight is strong. These nets are also very useful in managing the temperature inside the greenhouses. There are specialized nets having reflective surfaces which are generally called as thermal screens. These nets are effective in saving of energy required for temperature control inside the greenhouses.

Temperature control in greenhouse: Thermostats can be used to control exhaust fans in fan and pad greenhouse or controlling fuel combustion in centrally heated greenhouse. In naturally ventilated greenhouse, it can be controlled by side and top ventilation.

Relative humidity control: Humidistat can be used to control the water circulating pump for pad in FP greenhouse. The foggers or misters can be controlled through humidistat to maintain RH in NV greenhouse and centrally heated greenhouse. High humidity can be maintained in GH used for nursery raising or for rooting of cuttings by misting or fogging. This type of GH is also known as mist chamber.

Light intensity control: The shade nets of required cut off capacity can be used to control light intensity. The HDPE net or the LDPE film with reflective type stripes when used in NV greenhouse, increases inside temperature due to restriction in ventilation. These shade nets decrease the temperature inside the FP greenhouse.

Protected Cultivation of Flowers and Vegetables

As a result of globalization of trade and liberalization of Indian economy, there is immense scope for export of high value flowers and vegetables from India, besides, meeting the increased demand in domestic market. To get higher productivity of superior quality produce, the crop should be grown under protected environment adopting modern techniques like micro irrigation, fertigation etc.

India with a diversified climate grows most of the high value flowers and vegetables under partially environment controlled protected structures or naturally ventilated polyhouses across the county. High value and low volume flower crops like rose, gerbera, carnation, chrysanthemums, anthuriums, orchids, liliums etc., and vegetables like capsicum, tomato, cucumber and exotic vegetables like broccoli, Chinese cabbage, red cabbage, lettuce, cherry tomato, zucchini, pockchoy, celery, brussels sprout, asparagus and leeks can be grown under protected environments, besides, healthy seedling production under cover. Production technologies for protected cultivation of important crops are given below.

Rose

Rose is a leading cut flower grown commercially all over the world. It ranks first in global cut flower trade. This flower has a worldwide consumption of more than 40 billion stems annually. Its cultivation demands special care and attention, so that plants bloom to their maximum potential. To meet the perpetual demand, cultivation of roses in greenhouse has become very commercial.

Soil: For successful rose cultivation, soil should be well drained and a depth of 50 cm with high organic matter is preferred. It requires slightly acidic soil with pH 5.5 to 6.0.

Climate: The temperature between 15 to 27^{0}C is optimum for good growth and temperature below 15^{0}C hinders plant growth. If temperature go more than 28^{0}C, it should be brought down by providing ground level misting. Humidity of 60 to 65 % is optimum for its growth and this can be maintained by operating foggers/ misters for about 1-2 minutes/hr during winter and 1-2 minutes for every half an hour during summer.

Varieties: While selecting the varieties, one should know about their yield, shelf life and demand in the market. Rose varieties are classified into three groups based on the stem length. Long stemmed varieties are Skyline (yellow), Polo (white), First Red (red), Grand Gala (red) etc., medium stem ones are Passion (red), Lambada (salmon), Nobles (pink), Naranga (orange) and Miracle (orange) etc., and short stemed varieties are Golden Gate (yellow), Print Stone (peach), Spinks (yellow) and Sweetness (pink).

Propagation: Good quality budded plants are commercially propagated through budding Plants budded over *Rosa indica or Rosa mannetii* root stocks are better. The budded plants of 4 to 8 months old are used for planting.

Land preparation and planting: Land should be ploughed thoroughly and brought to a fine tilth after removing unwanted materials. Add well decomposed farm yard manure (10 kg/m^2). Addition of coir pith (2 kg/m^2), vermicompost (1 kg/m^2) and neem cake (200 g/ m^2) improves soil structure, besides nutritional status of growing media. Before preparation of the beds ,media should be sterilized with formaldehyde (5%). Raised beds of 25-30 cm height and width of 90-100 cm all along the greenhouse length are prepared. Path of 50-60 cm in between two beds is taken. Planting is done on the bed with paired row system in zig zag fashion by giving the spacing of 40-45 cm between two rows and 20-25 cm between the plants. With this 5-6 plants/m^2 of green house area could be accommodated. While planting, care should be taken that budded portion should be above the ground and soil around plant is pressed gently to remove the air around root system. Irrigation is given immediately after planting with rose head can or hose pipe with rose head and also for initial 15 days. Higher humidity is maintained initially for better establishment.

Canopy Management Practices

i. **Pruning and bending:** Crop canopy management is very important in commercial production of roses. Young plants, if allowed to flower immediately, the entire structural frame work of plant is disturbed. Hence, developing appropriate plant architecture is very essential by adopting pruning and bedding. To develop basel shoots, pruning is carried out after two months of planting and it is done just above the budded portion leaving two to three leaves. Recently bending is followed instead of pruning. Here blind and weak shoots are bent towards the path leaving two buds at base. While bending flower buds, sprouts are removed. Shoots coming after pruning or bending play very important role in flower yield. Hence remove all the shoots by pruning or bending retaining 3 to 4 productive shoots.

ii. **Pinching side shoots:** Remove the shoots developed on root stocks at a regular interval. The side shoots from flowering stem are also removed regularly to get quality flower.

Irrigation: Water requirement of rose plants depends on the type of soil, temperature, humidity and stage of the crop. Generally 7-8 litres/m2/day is required. Irrigation is done by drip irrigation system, having emitters at 30 cm apart on the lateral line. Ground level misters are also adopted to increase humidity.

Nutrition: Total nutrient requirement of rose is N-125g, P and K-75g each, Ca-40g and Mg-8g/m^2/year This dose of nutrient is spilt equally and provided on daily basis thorough fertigation system. Nutritional dose is altered based on nutrient status of the soil which is analysed at regular intervals. Soil EC and pH are also analysed regularly and corrected, if necessary.

Harvesting and yield: Flowers can be harvested 30-40 days after pruning. Flower stems are harvested at tight bud stage when one or two petals start opening. Harvesting stage varies with the variety, market distance, temperature and buyers/ consumers need. For local and domestic market, flowers are harvested at little open stage, while for distant market or export at tight bud stage. Harvesting should be done either in the morning or evening hours. Depending on the varieties, 200-300 flowers can be harvested / m^2 / year.

Post harvest handling: Immediately after harvest, flowers should be stored at 3-5^{0}C temperature in cold storage to remove field heat, reduce rate of respiration and water loss. Based on the stem length, flowers are graded and made into bundles of 20 flowers which are covered with soft corrugated paper by putting rubber bands. These bundles are packed in corrugated boxes and sent to the market. Flowers can be stored in cold house at 2-3^{0}C temperature for about 5 to 7 days.

Grading: After cooling, the stems are shifted to air-conditioned grading room. In the grading room, defective stems with blemishes or symptoms of diseases and pest infestation are removed. The stems are sorted out in different grades of length by colour coded board or automatic grading machines. The quality parameters for grading are :

1. Long stemmed varieties are graded from 40 m onwards with the difference of 10 cm whereas short-stemmed varieties are graded from 40-65cm with the difference of 5 cm.
2. Leaves should be dark green and healthy. There should not be any yellowing due to micro-nutrient deficiency. Leaves should also be free from dust, chemical residue, chemical injury, spider mites or powdery mildew .
3. The flowers should not be bullhead, too open, too tight, with bent neck, damaged by thrips or botrytis etc.
4. Bud size should be representative to the variety.
5. The length of the neck should not be too long.

Plant Protection

Insect pests: Thrips, aphids, red spider mite, bud borer and stem borer. Sucking pests like thrips and aphids can be controlled by spraying oxydemeton methyl, monocrotophos or dimethoate (2ml/l). Bud and stem borers can be controlled by spaying of quinolphos (2 ml/l). Red spider mite is a serious pest which can be controlled by spraying dicofol (2.5 ml/l) or wettable sulphur (3g/l). In case of severity, it can be effectively controlled by spray of abamectin (0.5 ml/l).

Diseases: Powdery mildew, black and *alternaria* leaf spots.

Powdery mildew can be controlled by spraying wettable sulphur (3g/l), black and *alternaria* leaf spots can be controlled by spraying carbendazim (1g/l) or mancozeb (2g/l).

Gerbera

Gerbera (*Gerbera jamesoni*) is one of the important cut flowers commercially grown in several countries of the world. It is commonly known as transvalls daisy or African daisy. There is great demand for gerbera in recent years as it is used in different flower arrangements, bouquets and flower vases. Flowers are in good demand locally during marriage and festival seasons.

Soil: Gerbera requires well drained soil with high organic matter having acidic pH of 5.5-6.0. Deep soil is suitable as gerbera roots grow up to 50-60 cm. Before selecting the land for cultivation, it is better to test the physical and chemical properties of the soil. Soil with EC of less the 1 dS / m is better.

Climate: Though gerbera crop can be cultivated in moderately warmer areas in open sunny situations or in shaded conditions, but to get higher productivity of superior quality flowers, cultivation under greenhouse is feasible. Optimum temperature should be maintained, as it influences the leaf number, stem length and flower yield. Temperature of 23° C is required for flower initiation and about 25 to 27° C is optimum for leaf growth. Temperature below 12° C and above 35° C affect the flower production.

Varieties: Different varieties of attractive colours are available and the demand varies with the consumer preference in day to day market. Based on the consumer demand, breeders release several varieties every year in the market. Hence, it is difficult to name specific varieties. The varieties in demand at present are Savannah, Pink Elegance, Danaellin, Topic Blend, Goliath, Sunway, Rosalin, Dalma, Winter Queen, Amulet, etc,. Generally varieties with light coloured petals having dark center are in good demand .

Land and bed preparation: Land should be ploughed thoroughly and brought to fine tilth. After removing unwanted materials, add well decomposed farm yard manure (10 kg/m^2). Addition of coir pith (2 kg/m^2)/ vermicompost (1 kg/m^2)/ neem cake (200 g/ m^2) will improve soil structure, besides nutritional status of growing media.

Before preparation of the beds, media should be sterilized with formaldehyde (5%). Raised bed of 45 cm height and 70 cm width of greenhouse length are prepared. Path of 30 cm in between two beds is given. Planting is done on the bed with paired row system in zig zag way by giving the spacing of 35 cm between the two rows and 25-30 cm between the plants. With this 6-8 plants/ m^2 of greenhouse area could be accommodated. While planting, care should taken that the crown portion should be just above (1 to 2 cm) the ground level and root ball should not be disturbed while planting. Soil above the bed should not be disturbed while planting. Top soil of the bed should be racked every fortnight to a depth of 3-5 cm and it helps in better root aeration. Humidity of about 80-90 % has to be maintained immediately after planting and also for about 4-6 days. There should not be moisture stress at this stage.

Irrigation and fertigation: The pH of irrigation water should be between 6.5 to 7.0 and EC 1 dS/m. It is better to analyse water before use. Irrigation is given immediately after planting with rose head can or hose pipe and also for about 10-12 days till establishment of plants. High humidity is to be maintained thorough misting or fogging for initial four weeks, irrigation through drip system is given after 15 days of planting. Water requirement is about 5-6 litres/ day / m^2. Moderate moisture should be maintained and based on the season, irrigation can be altered. Excess moisture invites root or crown rot and leads to death of plant.

Nutrients are supplied through fertigation with very low concentration on daily or alternate basis. The quality of nutrients available in the growing media or supply of nutritional dose has direct influence on growth and productivity. Unlike in open conditions, scheduling of fertilizer application inside the greenhouse is tricky and has to be monitored carefully. Fertilizer application based on growth stages in different ratio of NPK such as 2:3:1 at early vegetative stage (15-20 days), 3:2:1 at rapid vegetative stage (up to 45-50 days), 1:2:2 at bud formation stage (50-60 days) and 1:1:3 at the beginning of the flowering (60-90 days) and also after flowering regularly. The nutrients applied through irrigation should not be more than 1g / litre of water. Micro nutrients are applied through foliar sprays regularly and also when deficiency of particular element is noticed.

Harvesting and yield: Total crop duration is about 24-30 months. Based on the climatic conditions, flowering starts 8 to 12 weeks after planting. Peak flowering is observed 7 months after planting. Flowers are harvested when outer 2-3 rows of the disc-florets are fully developed. Flowers are harvested usually in the morning or late in the evening by side ward pulling of the stalk. The heel of the stock should be cut about 2-3 cm above the base and flower stalk should be kept in clean water. Flower yield is about 150-200 / m^2 /year in standard varieties.

Post harvest handling: Flowers should be kept in fresh water immediately after harvest for about 4 hours in cold house at 7 to 8^{0}C. Harvesting stage, stem length, firmness of stalk, etc. decide the flower quality. Individual flower head is put in thin plastic sleeve and a bundle containing 10 flowers are made, about 30-40 such bundles are packed in corrugated boxes.

Grading

Grade	Stem length (cm)	Flower diameter (cm)
I	>50	>10
II	<50	<10

Plant protection: Among insect pets, red spider mite and thrips are important. To control mites, spray 2.5 ml dicofol or abamectin 0.5 ml and for thrips oxydemeton methyl 2.0 ml / litre of water is effective. Root rot and crown rot caused by *Pythium, Rhizoctonia, Phytophthora, Sclerotium* are important and to control these, drenching with metaloxyl 3.0 g / litre can be used. Powdery mildew is another disease which occurs when humidity increases. Spraying of wettable sulphur (3.0g/litre) is advised.

Carnation

Carnation stands next to rose in world flower trade. It is available in different colours, remains fresh for longer period in vase and the best suited for export as the flowers are light in weight and have very good rehydration capacity. It is grown in greenhouse condition in recent years due to its higher demand.

Soil: Sandy loam soil with a pH 6-7 is suitable for its cultivation. Add soil, sand and farm yard manure in 2:1:1 ratio and addition of coir pith (2 kg/m^2) will improve the drainage. Media should be sterilized by using 5 % formaldehyde (10 l/ m^2). After putting formaldehyde to media, completely cover with plastic sheet for about 4-5 days. Then media are flushed with water to drain out the left over chemical.

Climate: Carnation requires day temperature of 20-25^0 C and night temperature of 10-15^{0}C . It requires higher sunshine but during summer, light should be cut down by providing 30-50 % shading net inside the greenhouse. Humidity of about 50-60 % is optimum for plant growth and flower development. Fluctuation in day and night temperature results in calyx splitting and malformation of flower.

Bed preparation: Raised bed of 75-80 cm width and 30 cm height are prepared along with the greenhouse length. Leave 50-60 cm path in between two beds. Add 100 g calcium nitrate, 600g super phosphate, 50g potassium chloride, 50 g magnesium sulphate / m^2 area and mix it thoroughly in the soil.

Varieties: There are two types in carnation viz, standard (sim) and spray. Standard varieties bear a single bold size flower on the stem while in spray type small flowers are produced on the branches of flower stem. A few important varieties of both the types with different coloures are given below.

Flower colour	Standard type	Spray type
Red	Domingo, Master, Grana, Imphala, Aicardi, Goudina	Alister, Darling, Red arrow, Durango
Pink	Cherry solar, Bolagro, Petra, Pink dover, Pinkshiva, Buemonde	Happyness, Resca, Rosini, Cherry bag
White	Dovar, Sansura, Nordica, Mistral, Viking, Alibaba	Vergo, Closeup, Crystal, Amaliya
Yellow	Yellow solar, Tahiti, Solaroro, Kiro Diana, Yellow Firato	Sonia, Ureka, Castillo, Koreno
Orange	Solar, Orange, Pinto,Dali, Firato	Indira, Nikhi, Macarena
Purple	Lavender, Jargia, Biasis	Purple chopin, Rosetti
Variegated (mixed)	Pikaso, Rodrick	Jakobin, Sapora, Challenger.

Planting: Planting can be done thoughtout the year under greenhouse conditions. However, mid winter is suitable for planting. Rooted tip cuttings are used for planting. Planting net of 7.5x7.5 cm size is stretched on the bed at 12 cm height, then planting is done on a bed in 4 rows at a spacing of 15x15cm for standard type and 15x20cm or 20x20cm for spray type. In one square meter greenhouse area, 15 to 20 plants can be accommodated. Planting should not be done too deep to avoid stem rotting as it is highly susceptible to *Fusarium* wilt and *Rhizoctonia* root rot.

Irrigation and fertigation: For a few weeks after planting, it is better to provide irrigation through misting for better establishment. Drip irrigation can be started 3 weeks after planting. Water requirement is about 4 to 5 litres/ day/ m^2. Nutrients should be given at all stages of plant growth regularly. The following doses of nutrients are given through fertigation at different stages of growth.

Stages	Fertilizers (g/m^2/week)				
	N	P	K	Ca	Mg
Planting to end of pinching	4.64	1.62	3.62	2.55	0.55
End of Pinching to until harvesting	4.69	1.50	5.58	2.72	0.51
During harvesting	4.88	1.50	7.03	3.09	0.50

Netting for support: Netting is an important practice in carnation cultivation. To avoid bending and breaking of long stem, netting made of wire or nylon is essential. Nets of different sizes are tied to iron poles provided at 3 m distance on the edge of either side of the bed in 4-5 layers. First layer of size 7.5 x 7.5 cm is provided before planting, then second one is of 12.5x12.5 cm, third layer is of 15x15 cm and fourth is of 20x20 cm size , provided at a distance of 10-15 cm between layers.

Pinching: Pinching is an important cultural operation in the successful production of top quality carnation flowers. After planting, the plant continues to grow a main stem. If left unpinched, this main stem produces a crown flower. Pinching means to remove the tip of the main stem. This allows the side shoots to develop and these side shoots produce the first flush. Pinch the plants leaving 6 pairs of leaves (internodes). This is called single pinching. The plants are usually pinched for the first time about 3 weeks after planting. Second pinching can also be done 3 weeks after first pinching. All the side shoots produced after first pinching can be pinched leaving 4 pairs of leaves. By this, flowering will be delayed. This is called double pinching. For staggered flowering, 50 % of side shoots produced after first pinching will be pinched leaving 4 pairs of leaves. This is called single and half pinching. If temperature is more, second pinching is not preferred.

Disbudding: From standard carnations, the side buds may be removed for main flower to develop. In case of spray carnations, only the central terminal bud is removed to encourage lateral flower buds to develop. Care must be taken to avoid any injury to the main stem.

Harvesting: Flower harvesting starts from 15 to 20 weeks after planting. Harvesting in standard type is done at paint brush stage (just opening of petals in the flower bud). In case of spray type, flowers are harvested when two buds are opened and remaining buds at paint brush stage. Stems are harvested with sharp knife leaving 15 cm stalk at the base and should be kept in fresh water immediately after harvest. While harvesting, care should be taken to see that, side shoots and supporting net should not be damaged. Carnation yield depends on the varieties, growing region, climate, planting time and type of pinching. Every year yield / m2. of green house area will be about 200-250 stems.

Post harvest handling: After harvesting standard carnations are classified based on the stem length and made into a bundle containing 20-25 stems. Then it is wrapped with soft corrugated sheet by putting a rubber band and such 20 bundled are kept in corrugated box of size 122x50x30cm and sent to market. Spray carnations are packed in a bundle containing 5 stems. The flowers after packing are pre-cooled at 0^0C temperature and can be stored for 3 weeks at 3-4^oC storage temperature.

Grading: Before packing, flowers should be graded. The grading is based on the degree of bud opening, minimum flower diameter and stem length. In the United States, the Society of American Florists recommended following grades for the standard carnations in U.S.A.

Grades		Minimum flower diameter (mm)	Minimum stem length (cm)
Blue (fancy)	Tight	50	55
	Fairly tight	62	
	Open	75	
Red (standard)	Tight	44	43
	Fairly tight	56	
	Open	69	
Green (short)	Tight	None	30
	Fairly tight	None	

Plant protection: Red spider mite is an important pest of carnation. To control this pest maintain temperature at optimum level and spray 2.5 ml dicofol or abamectin 0.5 ml or wetteble sulphur 3.0 g / litre of water. Wilt, stem rot and foot rot are important diseases in carnation. To avoid spread of these diseases, provide proper drainage. Drench beds with copper oxichloride 4.0 g or carbendizim 2.0 g / litre of water.

Anthurium

Among tropical cut flowers, anthurium occupies the place of pride. This is a popular flower among flower arrangers. The cut flowers of anthurium are valued for myriad of colours, showy spathe and long vase life. It is mainly used for different flower arrangements and grown as potted plant to keep as indoor plant. There is great demand for quality anthuriums in international market. Recently commercial cultivation is being taken up in Kerala, Karnataka and North Eastern states adopting suitable structures based on the existing agro climatic conditions.

Media and Bed preparation: Great attention should be given while selecting the media as the crop once planted remains for about 6-7 years. Several materials can be used as media like charcoal, coconut fiber, coconut husk, tree bark, paddy husk, coco peat, brick pieces, rock wool, coarse sand, etc. In India coir pith and coconut husk are used as growing media. Media should be well drained, should hold sufficient amount of moisture and nutrients. It should not be easily decayed and cause diseases to the plants. Media containing well sterilized soil, sand and well decomposed farm yard manure in 3:1:1 ratio can be used. Recently farmers cultivating anthurium in polyhouses are using coconut husk which is giving good results. The pH of media 5 to 6.5 is optimum for growth.

Beds are prepared including irrigation systems. The beds of 90 to 120 cm width and 45 cm height with convenient length depending upon greenhouse are prepared. The pathway between two beds is 45-60 cm. It is very important to have good drainage system in the beds. A perforated PVC pipe is passed thorough the center of bed through which excess irrigation water is collected and drained out.

Climate: Anthurium grows naturally in a shaded environment, well protected from adverse weather conditions. When anthurium is cultivated, it must be protected from several weather conditions like wind, rain, cold, heat, sunlight, CO_2, temperature and relative humidity. Plant growth and flower production depend on the temperature. Day temperature of 26 to 30⁰C and night about 18 to 20⁰C are ideal for the crop. Night temperature should not go below 15⁰C, it delays the flowering. High temperature of 35⁰C and relative humidity up to 80-85 % are not problems for the crop. Temperature more than 35⁰C and a very low relative humidity of 20-30 % are not suitable for the crop. Anthurium is a shade loving plant, higher temperature affects the flower and also leads to leaf scorching. Shade requirement varies with varieties and the existing climate. Generally 75 % shading net is used.

Varieties: Anthurium varieties are classified into 3 groups based on the spathe character. Varieties with heart shaped spathe are called 'standard' (commonly grown); 'obake'- type has bi-coloured spathe and 'tulip' type has cup shaped spathe. However, Hawaiian, Holland and other types of anthurium varieties are listed below based on colour of spathe.

Hawaiian: *Red:* Osaki, Kosohara, Koumana, Micke mouse, Hawaiian red, Crinkle red

Orange: Nitta, Sunburst, Sunset orange, Diamond jubilee.

White: Manamist, Morocco, Suchiro, Chameleon, Hidden treasure,

Pink: Blush, Candy strip, Candy queen

Holland: *Red:* Tropical, Avo rosette, Fla red, Arizona, Vio letta, Avo netta

Orange: Fla orange, Avogino, Casino.

White: Acropolis, Uranus, Fla exotica, Bianco

Pink: Anneke, Surprise, Cheers, Magic pink, Snata

Others: *Red:*Mauritius red, Tanaka, Calipso, Modonna, Butterfly, Sweet heart

Orange: Mauritius orange, Farao

White: Mauritius white, Lime, Meringo, Ivory, Grace, Lambada

Pink: Agnihotri, Lady gem, Paradise pink, Passion, Sultan

Potted varieties: Mia Diablada, Condor, Bonina, Titicaca, Florida, Falicita,

Propagation: It is through suckers . Commercially, tissue culture plants are used for planting.

Planting: Planting can be done throughout the year but care should be taken to avoid planting in extremely hot or cold periods. After bed preparation, planting is done at a distance of 30x30cm, accommodating 6-7 plants / square meter. Close planting is to be avoided for better aeration and uniform spreading of plants.

Irrigation and fertigation: Irrigation is done daily twice once in the early morning and again in the noon and one more extra irrigation during sunny days are required. Irrigation should not be done late in the evening. Ideal pH of water is 5.7 to 6.2 and EC 1.2 to 1.3 dS/m

Fertilizers are given daily at very low concentration through irrigation system under greenhouse condition. Anthuriums are fertigated with a solution containing nitrogen 14 ppm, potassium 176 ppm, calcium 60 ppm, magnesium 24 ppm, nitrate 91 ppm, sulphate 48 ppm and phosphate 31 ppm, besides trace elements.

Other cultural operations

i. **Thinning of leaves:** The old leaves of anthuriums grow at the expense of production of flowers. Too many leaves may even harm the flowers. Anthurium plant has to be thus pruned to retain adequate number of leaves, without any adverse effect on flower production and quality. This would also improve health of plants in view of better air circulation. The leaves are to be cut as frequently as possible in order not to lose the equilibrium of plants. Pruned leaves that are healthy can also be sold as cut foliage. The number of leaves retained and the frequency of cutting also depend upon the size and the orientation of the leaves. On an average, 3-5 leaves are sufficient on a plant.

ii. **Removing suckers:** Most anthurium plants naturally produce small suckers at the base of the plant. However, this is highly undesirable for production of flowers. Suckering results in crooked stems and decreased flower size. Moreover, young suckers withdraw energy from parent plant, which does not benefit flower production. It is better to remove the suckers at an early stage. When the suckers become too large, they will cause too much damage to the base of plant when removed. Remove suckers by pulling them out by hand, never use a knife or scissors.

Harvest and packing: The flowers are cut with a sharp knife, leaving about 3 cm of stem on the plant to prevent rotting of stem. After harvest, the basal portion of stalk of flower should be kept in water as soon as possible to prevent drying out. If the flowers are taken to distant markets, a piece of cotton soaked in water is kept at the end of the stem to prevent desiccation and for export purpose each flower stalk cut end is inserted in small plastic test tube containing water. Soft protective materials can be put in between the spathe and spadix to prevent brushing. The spathe along with the spadix is then inserted into a thin polythene cover of appropriate size. This prevents bruises on the spathe due to pressing of the spadices of adjacent spikes. The open ends of the polythene cover are stapled just outside the spathe to prevent movement of the spathe inside the cover.

Corrugated fiber board boxes of 60x30x23cm size are used for packing 90-100 flowers. Harvesting starts 6-8 months after planting. Five-7 flowers/plant/ year for first two years and 10-12 flowers after 2 years of planting can be produced.

Grading: Anthurium flowers are graded based on the spathe breadth which is as follows.

Grades	Spathe width
Extra large	> 15 cm
Large	13-15 cm
Medium	10-13 cm
Small	8-10 cm
Mini	< 8 cm

Plant protection: Mealy bug, scale and spider mites cause damage to anthurium. To control mealy bug and scale, spray malathion 2.0 ml or oxydemeton methyl 1.0 ml or methyl parathion 1.0 ml and for spider mite dicophal 2.5 ml / litre of water. Bacterial blight, anthracnose and foot rot disease are very important. To control bacterial blight streptocyclin 0.5 g, for anthracnose captan 2.5 g or carbandizen 1.0 g and to control foot rot drench beds with mancozeb 2.5 g and thiram 2.5 g / litre of water.

Orchid

Orchid belongs to family Orchidaceae and their flower spikes are valued for varied shape and colour of florets and long keeping quality. Demand for orchid flowers in the international market is increasing every year. India is home to nearly 1250 orchid species. Orchids can be divided into two broad groups; the terrestrials and the epiphytes. Terrestrial orchids grow on the ground (e.g. *Cymbidium, Paphiopedilum*). The epiphytes have aerial roots and grow on trees (*Vanda, Dendrobium*). Orchids exhibit two types of growth, monopodial and sympodial. Monopodials have single non-branching stem, which grows upwards. Typical examples are *Phalaenopsis, Vanda,* and *Arachnis*. The sympodial plants have a rhizome, which grows horizontally, producing new growths. Thus a well developed sympodial plant contains a clump of shoots of varying age and size (eg. *Dendrobium, Cymbidium, Oncidium*).

Propagation: Orchids can be propagated by seeds or vegetative means. The vegetative means can be by division, cutting, off shoots (keikis), air layering and by tissue culture for commercial cultivation.

Growing Conditions

Light: Sun light is essential for growth and development, but shade is important for flowering in orchids. Choose a site, which obtains direct sunlight. It is relatively easy to cut down the light if so required by the use of shading materials. The tolerable light intensities vary with genera. *Arachni and Vanda* can tolerate very high light intensities. It is lower in the case of *Dendrobium* and very low in *Phalaenopsis*. A level of 50 % shade is ideal for most of the orchids.

Temperature: Temperature requirement varies with species and it ranges from 20-30^{0}C during day and 10-25^{0}C during night. Based on the temperature requirement, orchids can be classified as warm (tropical), intermediate (sub-tropical) and cool (temperate) species. Orchids should never be subjected to sudden changes in temperature.

Humidity: Orchids require about 70-80 % humidity. A good thumb rule to follow is that when the temperature is high, humidity also should be high. Providing misting units or foggers or even humidifier will ensure adequate humidity.

Air circulation: Orchids thrive well when the atmosphere is fresh. Air circulation not only keeps the atmosphere clean and fresh but also helps to keep both temperature and humidity at the desired levels. Good ventilation is also necessary to achieve this target.

Media: Varieties of media are used for orchids. Decayed leaves added with humus are ideal media for terrestrial orchids. A potting medium containing equal parts of leaf mould and soil would be desirable. For epiphytic orchids, media used should provide a surface over which the plants can cling to. The medium should be moist but never be soggy. Tree fern fiber, bark, coconut husk, tiles, bricks and charcoal are ideal media for epiphytic orchids.

Planting: Planting in containers and on ground are practiced. All the sympodial and also monopodials, which have limited vertical growth like *Phalaenopsis* can be grown in containers. Miniature orchids also exhibit their best if grown in appropriate type and size of containers. The most widely used containers are clay pots with sufficient holes for drainage and aeration. They are porous and absorb excess water. Plastic containers are light in weight, easy to transport and do not accumulate algae or salts on their sides but they are not porous. Wooden baskets are generally used for *Vanda*. Half or complete portion of coconut husk and wire baskets can also be used as containers.

Irrigation: Mist and foggers are ideal for small units. Micro-sprinkler is ideal for large scale commercial units. Spraying water on the floor and walls helps to maintain high humidity. Water with neutral pH of 6.5 to 7.0 is ideal. Water high in salts is detrimental to the plants. Flushing the pots thoroughly may help to bring down the building up of salts.

Nutrient management: Nutrients, their quantity and frequency of application depend largely on the type of orchid, potting medium, season of the year, growing conditions and stage of development of the plant . Nitrogen is essential for vegetative growth. Phosphorus is needed to initiate flowering process. Potash is required in building up strong roots, which also makes the plants resistant to biotic stresses. Minor nutrients and trace elements are required in very small quantities by orchids. Since most of the cultivated orchids are epiphytic in nature, the media used are highly porous. Hence, foliar feeding is ideal. Here the total quantity required as well as the loss will be low.

Since inorganic fertilizers are applied as foliar spray, the quantity of nutrients used / plant should be relatively low. A fertilizer mixture containing major nutrients and minor nutrients is ideal for general application. The concentration may be adjusted between 0.1 and 0.3 % depending upon the situation already explained. A higher dose of nitrogen leads to better vegetative growth, which may be required during early stage. An increase in the proportion of phosphorus will be required to induce flowering when the plants are at maturity. The ratio of NPK suggested is 3:1:1 during pre-flowering stage and 1:2:2 during flowering stages. Since small quantities of fertilizers are recommended for orchids, they are to be applied more frequently. The frequency of application can be twice or thrice a week under tropical conditions. Organic manure is applied once a week.

Harvesting: Depending on varieties, tissue culture plants reach harvesting stage in 1-2 years. Orchid spikes are harvested when more than 50 % florets are open. From third year, 5-6 spikes can be harvested / plant and from sixth year 10-12 spikes can be produced. After harvesting, all flowers should be washed with clean water before subjected to grading, pre cooling and storage. Grading is made based on spike length, number of florets / spike and floret size.

Plant Protection

Mealy bug and scale	:	Malathion 2.0 ml or diamethote 1.5 ml
Snail	:	Metaldihyde 5.0 % bate or beer (fermented alcohol) or putting salt around beds.
Thrips	:	Oxidemeton methyl 2.0 ml or diamethoate 1.5 ml
Black root	:	Captan 2.0 g
Leaf spot	:	Captan 2.0 g or mancozeb 2.0 g

Seedling Production Under Protection

Production and productivity of crops particularly annuals depend on use of healthy seedlings. Growing plants under a cover protects the crop from damaging rains and scorching radiation, temperature and pests and diseases. Use of protected structures for raising seedlings as a commercial venture is a recent phenomenon especially in the tropics. The most commonly used method to reduce radiation and temperature is by shade net covering and may be with fogging or misting facilities. These are ideal for the production of vegetable seedlings as they are comparatively more economical and feasible. Use of low cost structures involving stone pillar or wooden poles; shade net and 40 mesh nylon net to prevent some vectors which transmit viral disease are in practice.

In India, many young entrepreneurs and progressive farmers are coming forward to take up quality seedling production and supply to the vegetable growers based on the requirement. Seedling production facilities using plug or container seedling production under protected structures viz. shadehouse, nethouse, polytunnel, etc. provide healthy and uniform growth of the seedlings. Since the hybrid seeds are expensive, this method reduces cost by minimizing seed waste. It is a commercial 'venture to produce the seedlings of tomato, capsicum, cauliflower, cabbage, chillies and brinjal. Mostly hybrids are used for raising seedlings in the containers. In the commercial nursery, the most commonly used pro-trays are 98 celled and for bigger size seeds like melon and cucurbits 50 celled, the most common growing media used is sterilized coco peat which helps to prevent diseases especially the seedlings damping off. Coco-peat has high water holding capacity and good texture. Since it is low in mineral nutrients, the media need to be supplemented with nutrient solution.

Procedure for raising seedlings: Use pro-trays having 98 cells for raising seedlings of capsicum, tomato, chilli, cabbage, cauliflower, brinjal and exotic vegetables and 50 celled trays for melons and cucurbits. Plastic pro-trays are disinfected by dipping in thiram or copper fungicide solution prepared by mixing 3 g chemical/ litre of water. Bleaching powder solution can also be used for this purpose. Fill the seedling trays with a sowing medium like coco peat, potting mixture (soil+compost+sand) or vermcompost can also be used with soil in equal proportion. For optimum germination, use well drained sterile medium. Make depressions/holes in the center of each cell in pro tray and sow one seed / cell and cover with the same media. During winter or whenever temperature is low, keep about 10-15 trays one over the other, cover with a plastic sheet and leave for about 3-4 days. This helps in increasing temperature of media and germination will be early. In the protected structures like net house, shade house or greenhouse prepare raised beds of 1 m wide with slope on both the side and leave 0.5 m space for path. Mulch the beds with plastic sheet and arrange the seed sown trays in double row over mulched beds and water daily. Mulching of beds helps in avoiding penetration of seedling roots into the soil, which will damage while lifting trays. In case seedlings are raised under net house or shade house, rain protection should be made over the beds used for keeping trays. Over the beds use low tunnel structures made by PVC pipe of 3cm diameter and cover with poly sheet. Water the trays thoroughly every day or as needed using a fine sprinkler. Drench trays with 0.3 % 19:19:19 NPK water soluble fertilizer twice at 10 and 20 days after germination for healthy growth of seedlings. In case, if there are any insect pests or diseases follow plant protection as per the requirements. Seedlings will be ready for transplanting in 25-35 days based on the crops.

Capsicum

Capsicum commonly called as sweet pepper or bell pepper is the most popular salad vegetable grown in greenhouse. It is rich in vitamin A, C, and minerals. Capsicum cultivation is gaining popularity in peri-urban production systems because of easy access to urban markets. Capsicum yields in open field cultivation ranges 20-30 t/ha, whereas in greenhouse the yield range is 100 to 120 t/ha.

Climate: Capsicum is a cool season crop and day temperatures less than 30^{0}C is favourable for growth and yield. Higher temperature results in rapid plant growth and affects fruit set. Lower night temperature (20^{0}C) favours flowering and fruit set. Shading and misting are required during summer to avoid temperature build up in greenhouses. Moderately high RH (50-60%) is preferred, which can be maintained by managing the ventilation.

Varieties: Capsicum hybrids with indeterminate growth habit are suited for greenhouse cultivation, as these hybrids grow to a height of 2 m and above to utilize the vertical space of greenhouse. Commercial green hybrids like Indra, Green gold and coloured hybrids like Orobelle and Bomby are suitable for greenhouse cultivation with a yield potential of up to 120 t / ha from a crop of six months duration.

Nursery: Seedlings can be raised as described earlier and about 400 g seeds required for one ha area.

Growing medium and beds: Sandy loam soil and well-decomposed farmyard manure mixed in equal proportions are the best suited for capsicum cultivation. In heavier soils, mixing sand up to 25 % is required to provide proper aeration in the root zone. Loosen and work the soil to fine tilth. Prepare the beds of 100cm width and 15 cm height leaving 50 cm path between the beds. Add 20 kg well decomposed farmyard manure/ m2 of bed for the first crop. Mix thoroughly and level the beds. For subsequent crops, apply 5 kg FYM/ m2 of bed. Apply 50:50:50 N: P_2O_5: K_2O kg/ha to the growing beds before formaldehyde fumigation. Apply neem cake and trichoderma formulation (200 kg+2 kg/ha) just before planting but soon after formaldehyde fumes are exhausted completely.

Sterilization: For disinfecting growing beds, drench the beds with 5% formaldehyde (4 l/m^2) and cover with black polyethylene (400 gauge) sheet. Close all ventilation spaces. Three to four days after formaldehyde treatment, remove the polyethylene cover. Two days after removing the polyethylene cover, rake the bed repeatedly to remove trapped formaldehyde fumes completely before transplanting. Disinfecting is done once in a year.

Laying of drip line: Place one inline drip lateral at the center of the bed. Inline drip lateral should have an emitting point for every 30-cm interval with a discharge of 2 litres/ hour. Before covering with the polyethylene mulch, check the emitting points for uniform discharge of water.

Mulching: Use 100 micron thick black polyethylene film of 1.2m width to cover the planting bed. Make holes of 5 cm radius on the mulch film as per the recommended spacing (45 cm x 30 cm) then cover the planting beds with mulch by securing the edges of the sheet by burying in the soil.

Transplanting: Water the bed to field capacity before transplanting. Seedlings of 35-40 days old, vigorous and uniform in size are selected for planting. Transplant the seedlings at recommended spacing of 45 x 30 cm at a shallow depth of 2-2.5 cm. Water the beds with rose can or hose pipe with rose head immediately after transplanting and every day until the plants get established. Mist the greenhouse in the seasons of low humidity. Drip irrigation is given daily to replenish 50 % of open pan evaporation.

Fertigation: Fertigate 150:150:150 N:P_20_5:K_2O kg/ha using water-soluble fertilizers. Start fertigation from third week after transplanting. Fertigate twice or thrice a week for 18 weeks, by splitting the dose of fertilizer equally. Micro nutrients can be used as per requirement.

Pruning and training: Prune capsicum plants to four stems. Prune at weekly intervals starting from 15 to 20 days after transplanting. Capsicum plants show dichotomous branching. The tip of the plant splits into two at 5th or 6th node at first and these two branches again split into two giving rise to four branches. Maintain only these four main branches. At every node, the tip splits into two giving rise to one strong branch and one weak branch. Remove the weak branch and retain strong branch. Train the plants along a plastic twine. Each branch should have separate plastic twine to train along.

Harvesting: Harvesting of capsicum fruits starts 60-70 days after transplanting and continues up to 170 to 180 days at 10 days interval. Mature green or 10-15 % colour turned capsicum fruits weighing 150 to 250 g are harvested. Keep the harvested fruits in cool place and avoid direct exposure to sunlight. Yield of 100 to 120 t/ha can be expected from one crop of six months duration.

Plant protection: Red spider mite and thrips are the major insect pests, spider mite can be controlled by spraying dicophal 2.5 ml or 3.0 g wetteble sulphur/litre of water and thrips by spray of dimethoate 2.0 ml or oxidemeton methyl 2.0 ml. Powdery mildew is a common disease which can be controlled by spraying wettable sulphur 3.0 g or carbondizim 2.0 g/litre of water.

Tomato

Growing tomatoes in greenhouses can increase the yield and greenhouse grown tomatoes will have superior quality. Fluctuation in the price of tomato is a regular feature. Through greenhouses, production of tomato can be achieved round the year and off season production is more profitable. In addition to the quantum jump in yield and superior quality, substantial reduction in use of pesticide makes it an eco-friendly proposition to grow tomato in naturally ventilated greenhouses/ nethouse/ shadehouse.

Climate: Day temperature of 28⁰C and night temperature of 18⁰C are ideal for its growth. Fruit set is affected at temperature higher than 35⁰C and a relative humidity of more than 90 %.

Varieties and hybrids: Tomato hybrids with indeterminate growth habit are suited for greenhouse cultivation. Indeterminate tomato plants grow continually, producing flowers and fruits over a long period of time. Commercial hybrids like SH 7711 are suitable for greenhouse cultivation, with a yield potential up to 180 t / ha from a crop of six months duration.

Nursery: Seedlings can be raised in pro trays as described earlier. About 100 g seeds are required for one ha area.

Growing medium and beds: Sandy loam soil and well decomposed farmyard manure mixed in equal proportions are the best suited for tomato cultivation. In heavier soils, mixing with sand up to 25 % is required to provide proper aeration in the root zone. Soil has to be brought to fine tilth. Beds of 100 cm width and 15 cm height leaving 50 cm path between the two beds are laid out. FYM 20 kg/ m2 of bed area is added for the first crop. For subsequent crops, 5 kg FYM/m2 has to be applied. Apply N, P_2O_5 and K_2O at the rate of 50:50:50 kg / ha to the growing beds before formaldehyde fumigation. Apply neem cake and trichoderma formulation (200 kg+2 kg/ha) just before planting but soon after formaldehyde fumes are exhausted completely.

Sterilization: For disinfecting the growing beds, 5% formaldehyde (4 litre/ m^2) will be applied and covered with black polyethylene sheet. All ventilation spaces need to be closed after the application of formaldehyde. Three to four days after formaldehyde treatment, polyethylene cover has to be removed. Two days after removing the polyethylene cover, the beds are raked repeatedly to remove trapped formaldehyde fumes completely before transplanting. Sterilization can be done once a year.

Laying of drip line: At the center of the bed, one inline dripper lateral has to be placed. Inline dripper lateral should have an emitting point for every 30 cm interval with a discharge of 2 litre/ h. Before covering with the polyethylene mulch, emitting point has to be checked for uniform discharge of water.

Mulching: Black polyethylene film of 100 micron thickness and 1.2 m width can be used to cover the planting bed. Holes of 5 cm radius are made on the mulch film as per the recommended spacing (60 cm × 45 cm) then cover the planting beds with mulch by securing edges of the sheet by burying in the soil.

Transplanting: Beds have to be watered to field capacity before transplanting. Seedlings of 25 days old, vigorous and of uniform size are selected for planting.

Transplant the seedlings on each bed in two rows following a spacing of 60 x 45 cm. at a shallow depth of 2-2.5 cm. Plants are watered with rose can immediately after transplanting and every day until the plants get established. Misting is done in the seasons of low humidity. Drip irrigation is given daily to replenish 50 % of open pan evaporation.

Fertigation: Fertigation is carried out using water soluble fertilizers with a $N:P_2O_5:K_2O$ dose of 250:250:250 kg / ha for six months duration crop from third week after transplanting Fertigation is carried out twice or thrice a week for 18 weeks. Based on the requirements, micro nutrients can be used.

Pruning and training: The tomato plants are pruned to two stems / plant. Pruning usually starts 20 to 30 days after transplanting. Plants are pruned at weekly intervals. The main stem of tomato plant branches into two after the first flower cluster which are the only two branches (stem) retained and all other branches are removed. Plants are twined along the plastic twine. Separate plastic twine has to be provided to each branch. Remove older leaves periodically.

Harvesting: Harvesting of tomato fruits starts 70 to 80 days after transplanting and continues 170 to 180 days. Harvesting of fruits is done at weekly interval. Fruits should be harvested at colour break stage. A marketable fruit yield of 170 to 180 t / ha can be realized from a crop of 6 months duration.

Plant protection: Red spider mite and thrips are the major insect pests. Spider mite can be controlled by spraying dicophal 2.5 ml or 3.0 g wetteble sulphur / litre of water and thrips by spray of dimethoate 2.0 ml or oxidemeton methyl 2.0 ml. Powdery mildew and leaf spot are common diseases of tomato. Powdery mildew can be controlled by spraying wettable sulphur 3.0 g or carbondizim 2.0 g / litre of water and leaf spot by using mancozeb 2.5 g / litre of water.

III. Micro irrigation and Fertigation

Micro Irrigation

The water requirements of plants are influenced by many factors. The main factors are climate, including temperature, relative humidity, radiation, wind; and vegetative growth, influenced by age and type of plant and leaf shape. A combination of these factors or each one separately, may change the evapotranspiration rate thus changing the water requirements. Irrigation water management is the key input for successful crop production. Efficient irrigation requires uniform application of water at proper time and proper amount. The irrigation system should therefore be designed to ensure satisfactory adoption to the soil and crop conditions so that the desired amount of water may be applied as efficiently as possible. Different irrigation methods are adopted because of variation in soil type, topography, water supply, crop and other management practices. Due to improper methods of irrigation 27 to 42 % of irrigation water is lost by deep percolation beyond the root zone of the crop. This is not only a loss of valuable water but also loss of plant nutrients due to leaching. Therefore selection of proper irrigation method is the most important aspect. The micro irrigation has to be efficient in saving water and also in improving the yield. Micro irrigation means frequent application of required and measured quantity of water directly above or below soil surface, usually as discrete drops, continuous drops, tiny streams or miniature spray through emitters or applicators placed along a water delivery line (laterals).

Types of Micro Irrigation Systems

1. **Drip irrigation system:** It is a standard set of components with the emitting devices which are pre-fitted or inserted during installation along the lateral lines to get continuous flow of water to form the wetting strip on the soil surface or subsurface around the root zone of crops called as drip / line source tubing. Line source tubing may be of built in drippers, inline drippers, integral drippers, porous tube line, pressure compensating drip tube etc.

2. **Bubbler irrigation system:** It is a standard set of system in which the application of water to the soil surface as a small stream or fountain, where the discharge rates from the point source bubbler emitters are greater than for drip or line source emitters because emitter discharge rate normally exceeds the infiltration rate of the soil.

3. **Micro jet irrigation system:** It is a standard set of system in which the application of water in the form of discrete or fan or full circle or half circle or quarter circle or part circle spray on the surface of soil from low height or low angle through air around the crop. Micro jet includes moving part and having greater discharge rate and greater coverage than dippers and bubblers.

4. **Micro sprinkler system:** It is a standard set of system in which the application of water in the form of discrete or fan or full circle or half circle or quarter circle or part circle spray on the surface of the soil from stake height less than 1 m or low angle through air around the crop. Micro sprinkler incorporates moving part and having greater discharge rate and greater coverage than drippers, micro jet and bubblers.

5. **Sprinkler system:** It is a standard set of system in which the application of water in the form of discrete or fan or full circle or half circle or quarter circle or part circle spray on the surface of soil from riser height 1 m or more and high angle through air around the crop. This system includes moving part and having greater discharge rate and greater coverage than drippers, micro jet, bubbler and micro sprinklers.

6. **Misting or fogger irrigation system:** It is a standard set of system in which the application of water in the form of the full spray to create hu-

midity is done in the air at low level or overhead. Misters or foggers do not incorporate moving part and having relatively low discharge rate and smaller coverage as compared to micro jet and micro sprinkler.

Components of Micro Irrigation System

Distribution Lines: Distribution lines consist of a network of graduated pipe sizes starting with a single, large main line followed by smaller sub main and lateral lines. The buried main and sub main lines are normally permanent and are made up of PVC and HDPE. Main and sub main lines range from 40 mm to 110 mm size depending upon required water flow to the laterals. Both main and sub main lines should have outlets with valves for periodic flushing. Lateral lines are usually made of LDPE and LLDPE and range from 12 to 16 mm diameter.

Control System: Main control for the micro-irrigation system is meant to organize, to measure, filter and to regulate pressure. This includes pump, primary filter, pressure regulator, pressure gauge, and usually chemical/fertilizer injection equipment.

Filter system: Proper water filtration is essential and is achieved by using either screen, sand or centrifugal filters either individually or in combination.

1. **Screen filter:** This is used where there is problem of clay particles in the water. Screen filters are the most frequently used equipment for removing particles. The method of water entry circulation and exit varies among the manufactures. These filters have back wash arrangements. Corrosion resistant stainless steel and plastic materials are the more commonly used substances for the screens. Development of an abrupt pressure across the filters means that the screen is beginning to clog. In contrast, if no pressure change occurs for long periods, the screens or seals are broken or mesh size is too large.
2. **Sand filter:** Pressure type, high flow sand filters are more common ones used in the systems. Almost the full depth of the sand is used in the pressurized filters.
3. **Centrifugal filters:** It is used when the source of water is canal or tank and it removes suspended materials with specific gravity greater than water.

Advantage and Disadvantage of Micro Irrigation

It saves water, improves productivity and quality of the produce, reduces labour cost, salt concentrations in the root zone, weed and disease incidence, allows use of water soluble chemicals and fertilizers and saves on fertilizers. Water is applied at frequent intervals in micro-irrigation and hence growth of the crop is uniform. As there is no water/moisture-stress, the crop growth is not affected at all. Water is also controlled and only the required quantity of water is given to each plant based on evapotranspiration requirements.

High investment at the initial stage, frequent clogging of emitters, pilferage of components are a few of its disadvantages.

Fertigation

After realizing the importance of micro irrigation system, the next most essential input to boost productivity and quality is the fertilizer. Fertigation is a simple word coined with union of two words, fertilizer and irrigation, indicating application of fertilizer through an irrigation system. The fertilizers are applied daily or alternate days as per the requirement of crops to the root zone.

Suitable forms of fertilizers for fertigation: The availability of 100 % Water Soluble Solid (WSS) or liquid fertilizer is essential if it is to be applied through fertigation. The fertilizers , traditionally available in the market, are not fully soluble in water and it also contains some non-soluble solids which make them unsuitable for application through micro irrigation systems. For fertigation, the special fertilizers are available in the following forms.

1. **Liquid fertilizers:** These are the solutions which contain one or more plant nutrients in liquids form, but these are not popular in India.
2. **Water soluble solid fertilizers:** These are 100 % water soluble fertilizers, generally containing two or more major nutrients and micro nutrients. These fertilizers completely dissolve in water leaving no precipitate. Therefore, there is no chance of clogging of emitters due to fertigation.

Suitable conditions for fertigation: There are a number of modern techniques suitable for the application of fertilizers, like sprinkler and drip irrigation. The most important criterion for the suitability of an irrigation system for fertigation is accuracy of water application, which largely depends upon proper designing and installation of irrigation system and availability of correct equipment for injecting fertilizers. The success of fertigation not only depends on good irrigation or proper use of equipments, but it also requires adequate information on crop water requirements, soil moisture characteristics, and crop physiology to enable accurate irrigation scheduling and therefore, the maximum water use efficiency. Efficient fertigation requires access to information on soil fertility and an under standing of nutrient balance in soil.

Advantages of Fertigation

1. The accurate placement of fertilizers in the active root zone with less loss through leaching.
2. The application of nutrients in an available form can maintain the optimum level of nutrients required for crop growth.

These two points offer potential for much higher fertilizer application efficiencies than with conventional application methods, representing a potential saving of 20-30 % in fertilizer use.

3. High frequency fertigation avoids high peaks in salt concentration in the soil solution, thus maintaining a high osmotic potential and aiding nutrient uptake.
4. Use of acidic fertilizers for fertigation on high lime soils can reduce the pH within the zone thus reducing the problems of nutrient fixation.
5. Once information is gathered regarding the peak requirement for nutrient uptake, fertigation can be scheduled to apply the correct proportion of nutrients for optimum yield. Timely application of fertilizer can help to optimize crop yield.
6. The automation of fertilizer application through fertigation can save labour cost and reduce mechanical damage to crop and soil structure by labourers entering the field.

Fertigation methods: There are a number of techniques used to introduce fertilizer into the irrigation system. This may be drip irrigation. micro sprinklers, micro jets or sprinkler irrigation. In the beginning as per the decided schedule, system has to run for 15-30 minutes, then fertigation is done, again followed by irrigation for 15-30 minutes. While fertigating, it should be ascertained that the optimum concentration is maintained taking into consideration the crop reaction. For sensitive crops like tomato and strawberry, it should be round about 500 ppm and for other crops, maximum is 1000 ppm.

Fertilizer Application Equipments

1. **Ventury:** This system of injection involves very low cost and works by creation of a partial vacuum in the system and suction of the fertilizer solution into the irrigation stream. The vacuum is created by diverting a percentage of water flow from main line and passing it through a constriction which increases velocity of flow thus creating a drop in pressure. When the pressure drops, the fertilizer solution is sucked into the ventury through a suction pipe.

   This is a simple and inexpensive method of fertilizer application, but it has some disadvantages. The pressure loss across a ventury is high, about 1/3 of the pressure existing near inlet of filtration system. Secondly, precise regulation of flow is difficult because the rate of injection is very sensitive to the pressure and rate of flow in the system and it is used for smaller area.

2. **Fertilizer tank:** In this systems, a part of the irrigation water is diverted from the main line to flow through a tank containing the fertilizer in a fluid

or soluble solid form, before returning to the main line. The pressure in the tank and the main line is the same, but a slight drop in pressure is created between off take and return pipes for the tank by means of a pressure reducing valve. This causes water from the main line to flow through the tank causing dilution and flow of diluted fertilizer into the irrigation stream.

Concentration of fertilizer decreases gradually, until it reaches the level of the irrigation water. It is observed that with liquid fertilizers, it takes four tank volume displacements to empty the tank. If solid fertilizer is used, at least ten volumes displacements are needed to dissolve all the materials. Size of the tank may range from 50 to 100 litres depending on area to be irrigated.

Fertilizer tank is simple in construction and operation and also low in cost. There is no need of external power supply. However, the varying concentration of nutrients causes bulk of the fertilizer to be applied at the beginning of irrigation cycle. The tank has to be refilled with solution for each irrigation cycle.

3. **Fertilizer injector:** With this method, a pump is used to inject fertilizer solution into main pipe. The solution is normally pumped through the container, and the type of pump used depends on the power source. The pump may be driven by an internal combustion engine or electric motor. Out of them, electric pump is more in use. With these injection pumps, fertilizer may be supplied to the irrigation water at more or less constant rate. Pumping rate and concentration of the solution can be adjusted to attain the desired level of fertilizer application. Changes in water flow, power failure or mechanical failure may cause deviations from the planned concentration.

   Nowadays, hydraulic pump, being operated on line pressure is in common use. Such type of injector needs no external power supply. Its linear hydraulic motor is powered by hydraulic pressure of irrigation system. Injector unit should be resistant to chemicals. Since the maximum injection rate of the pump is proportional to the pressure in the main line, the required injection rate is easily adjusted by trotting the injection line using the valve. Injector consists piston and to and fro motions of piston create suction from container. Piston movement is due to hydraulic pressure of main line. Injection rate may vary from 1-320 litre / hr with corresponding increase in main line pressure.

   Liquid and 100 % water soluble fertilizers are to be mixed in bucket full of water. Fertilizers and water can be mixed in fertilizer tank. Then this mixture to be mixed for 5 minutes and then this can be used through one of the above methods and equipments.

Fertigation scheduling: While deciding the fertigation schedule, following points need to be taken into consideration.

1. Total nutrient requirements of crops to produce target quality yield.
2. Part of the nutrients supplied in the form of organic manure.
3. Possible savings in the nutrient doses due to higher efficiency of added fertilizer through fertigation.
4. Relative requirement of nutrients at different growth stages of crop.

The requirement of nutrients varies with stage of crop growth. Therefore it is essential to know the nutrient requirements of the crop based on the growth stages. High level of nitrogen is required during vegetative growth stages to facilitate accumulation of dry matter in different plant parts. Its requirement is comparatively low during reproductive stage. The requirement of phosphorus is more during reproductive phase as it induces flowering, fruit set and seed development. Potash requirement is high during crop maturity stage.

Fertilizers Used in Fertigation

1. Straight fertilizers

Urea phosphate (17:44:0), mono ammonium phosphate (12:61:0), mono potassium phosphate (0:52:34), potassium nitrate (13:0:46), sulphate of potash (0:0:50+18%S), calcium nitrate, magnesium nitrate and calcium magnesium nitrate.

2. NPK formulations

20:10:10+Trace Elements(TE)+Mg, 19:19:19+TE, 16:08:24+TE+Mg, 13:40:13+TE, 20:20:20, 6:12:36, 13 :05:26, etc.

3. Chelated micronutrients

Fe EDTA (13%), Zn EDTA (15%), Fe EDDH-A (6%), chelated combination of micronutrients, etc.

IV. High Density Planting

Increasing the yield / unit area rather than / plant is the concept of high density planting in perennial horticulture crops. High density planting has a potential to increase yield and reduce cost of production of several horticultural crops. It is already exploited successfully in banana, pineapple, kinnow, orange and to some extent in apple and mango. The productivity for a given land area is related directly to fruit bearing foliage and its photosynthetic capacity. These concepts have converged themselves in the form of high density planting of fruit crops. The continuing decline in the availability of suitable orchard lands, the raising input and labour costs and the mounting demand for fruits have given further impetus to develop technologies to increase productivity / unit area of land, for which the theme of high density planting comes to fore front. High density orchard system is a very intensive form of fruit production with great relevance to food and nutritional security. This differs from traditional way of cultivating fruit crops where very a few plants are accommodated / unit area with a large canopy. These conventional orchards require low establishment cost, involves less labour, but the economic yield can be expected only after 8-10 years and in some cases, it is still more. Low and irregular production of fruits is also a problem. The quality is also affected due to meager sunlight penetration into the dense canopies and large sized trees. This is also a problem for orchard management practices. High density orchard on the other hand, accommodates more plants / unit area. It facilitates better utilization of solar radiation and increases bearing surface / unit area.

There are several ways of achieving this objective like selection of dwarf variety, use of dwarfing root stocks, pruning and growth regulators for canopy management. The use of dwarf trees and checking excessive vegetative growth enhance the harvest index and enable early economic yield , thereby increasing productivity and profitability. It is more amenable to modern input use efficient technologies like drip irrigation, fertigation, mulching, mechanical harvesting and horticultural operations like pruning, plant protection that enhances labour use efficiency. Better fruit quality due to optimum exposure to sunlight and easier harvesting of fruits with less injury resulting in better post harvest life are added advantages of high density orchards.

The high density orchard is very much practiced in temperate fruits crops particularly in apple because of use of dwarfing root stocks. It was introduced in apple in Europe during 60's with the development of Malling and Malling-Merton root stocks. The principles of high density orchard, developed over the years for temperate fruit crops, can be attempted for tropical woody perennials, but needs modifications and systematic long term experiments. These modifications are required since it has to be remembered while attempting to restrict size of tropical fruit trees that the evolutionary development of these trees in tropical forests where they originated normally give rise to large trees whose survival depend upon vegetative vigour rather than prolific production of fruits and seeds. Other characteristic of these trees are that they are mostly evergreen, profuse branching and with multiple apical meristems which helped their survival in the original habitat.

High density planting could be maintained throughout by using dwarf varieties and dwarfing root stocks. The normal varieties can also be used in high density planting by following pruning. Pruning can be practiced only in those crops which respond to pruning like guava. The other way of adopting high density planting is to plant at a closer spacing according to the crops. In case of mango, normal spacing adopted is 10m x 10m. In high density planting, 5m x 5m or still less spacing could be followed. Once the plants canopy is covered, alternate plants can be removed in a row. After certain years when plants grow completely, all the plants in the alternate rows can be removed. This helps in getting early yield. This system can be practiced for crops where yield is affected by pruning.

References

Choudhary, M.L. 2005, Souvenir of International Conference on Plasticulture and Precision Farming. National Committee on Plasticulture Application in Horticulture, New Delhi, November 17-21, 2005.

Iyer, C.P.A. and Kurian, R.M.,2006, High density planting in tropical fruits. Principles and practices. International Book Distributing Co., Lucknow

Jadav, A.S., 2002, Compendium of summer school on Protected cultivation. Organized by Hi-Tech Horticulture and Vegetable Project, College of Agriculture, Pune

More, T.A., 2001, Hi-Tech Horticulture. Mahatma Krishi Viswa Vidhyalaya, Rahuri, MS

Omr Zeidon, 2005, Tomato production under protected cultivation. Mashav, Cinadco, Israel

Prabhakar, B.S., 2002. Compendium of short course on Greenhouse production of flowers and vegetables. Organized by IIHR, Bangalore.

Rajeevan, P.K., Valsalakumari, P.K., Geetha, C.K. and Sudhadevi, P.K., 2004. Reflections on Indian Floribusiness. Directorate of Extension. KAU, Thrissur.

Singh, D.K. 2004, Hi-tech Horticulture. Agro-tech Publishing Academy, Udaipar,

Singh K.P. 2009, Souvenir of National Conference on Floriculture for Livelihood and Profitability. Organized by ISOA, at IARI, New Delhi, March 16-19, 2009.

4

High Tech Interventions in Horticulture Crops

P. Soman

Abstract

High Tech interventions begin with acute desire and demand for high productivity and profit from Agricultural activity. High Tech cultivation implies judicious use of space, time and inputs; nothing more and nothing less. It is interesting to note that the different components that were listed in each case above are known to scientists and extension workers and to a lesser extent to farmers also. Piecing them together to formulate a practise is the novel thing. The major components of high tech cultivation assume the highest priority by virtue of the fact that the country has shortage for water and fertilizer and both are essential for any type of crops. Different components of high tech intervention can be used for growing crops, also for the growers to enhance production and quality .High tech intervention includes a number of components like- uniform virus free planting material, drip fertigation technology, assistance to post harvest management, and training and guidance to farmers, value addition opportunity, technical support to the farmers etc. These high tech interventions are introduced to other crops- a host of vegetables, fruits like Pomegranate, Cashew and Guava, and agricultural crops like, Cotton, Pulses, Sugarcane, and even cereals like, maize, wheat and rice. Another important aspect of such interventions is that only if concomitant interventions in marketing of produce are made, the whole exercise would bring profit to the farmers.

Keywords: High Tech, horticultural crops, water management, fertigation

Technologies for cultivation of crops have been changing over the years. Adherence to the principles of conservation, high input- use efficiency and higher productivity with optimum inputs and the very idea of crop production as a business venture; all these are leading to the increased adoption of high technological (high-tech) crop production practices.

In India the tempo of adoption of high-tech methods has been receiving a shot in the arm by the promotion of such technologies through governmental interventions. The trend of governmental support to the small farmers to acquire technologies like drip irrigation, fertigation and protected structures etc. has been spreading from state to state.

Among the different inputs to crop production water and fertilizer, the two major components of High-tech cultivation assume the highest priority by virtue of the fact that the country has shortage for both and both are essential for any type of crops. For example, our country with a landmass of 2.4% has only 4% freshwater of the globe to cater to 17% of world population. For the sake of comparison, USA with a population of 4.5% has 7% land mass and 6% freshwater (IWMI, 2009).

Indian Agriculture at Cross Roads

Agriculture in India is climate restricted; 48% of the geographical area of the country is receiving less than 1000 mm rain and the rest 1000-2500 mm. The difficulty is that the rainfall is occurring in 3-4 months duration, making it imperative for rainwater harvesting, storage and irrigation. But the available water for irrigation is insufficient to cover the net cultivated area. Only 48% of cultivated area is presently irrigated. Irrigation cover can not be increased as the total available water of 1143 BCM/year would not extend further. By 2050 our water need (both irrigation and total need) would cross the availability level. This is a grim situation. It is made more so by the need for increasing food production. To achieve the increased food production of 494 million t by 2050, our irrigated area should increase from 79 million ha to 146 million ha. This is impossible to happen as the water is limited. Production can not be increased by increasing in area alone; area will increase only by 2 million ha during 2010-2050. So we are into a very difficult situation.

The only way out is to identify water conserving irrigation methods and achieve high productivity through input efficient technologies. Incidentally, the technology of drip irrigation and fertigation serve better in this scenario. It offers a way of irrigating more land with less water (water saving); more yield with less water (food security) and more food production with less energy use (energy security). In all high-tech agriculture is pushed ahead by these *twin technologies*. The use of protected structures for growing crops offers yet further advantages for the grower to enhance production and quality.

What are Hightech Interventions?

There are several steps and factors where high –tech methods of cultivation can be introduced. As long as the objective is to produce *more from less,* the different measures to enforce precision in operation and provision of inputs would be justified.

By definition, high-tech farming or precision farming is

- A management strategy which identifies, analyses and manages within field variability for increased profit and reduced environmental impact
- Farming by the meter, farming by satellite, micro- site specific management and prescription farming.

Precision agriculture is the application of technologies and principles to manage *spatial* and *temporal* variability associated with all aspects of agricultural production for increasing productivity and environmental quality. The success in precision agriculture depends on the accurate assessment of the variability, its management and evaluation in space-time continuum in crop production; i e. *precision in quantities and timeliness for each operation.* It is nothing but exact agriculture to extract the highest returns from farming activities.

The agronomic feasibility of precision agriculture has been intuitive, depending largely on the application of traditional recommendations at finer scales. The agronomic success of precision agriculture has been quite convincing in crops like vegetables, banana, sugar beet, sugarcane, tea and coffee. The potential for economic, environmental and social benefits of precision agriculture is largely unrealized because the space-time continuum of crop production has not been adequately addressed.

Variability in Everything Connected to Crop Cultivation

An analysis of a farm land and the situation of Indian farmer will show very clearly variations within land for every factor, one can think of. As shown below (By grid sampling and analysis of a large number of samples taken from the grids) the available N in the soil is highly variable from grid to grid in one field.

This field should be fertilized with N fertilizer NOT uniformly but variably using VRT (variable rate technology).

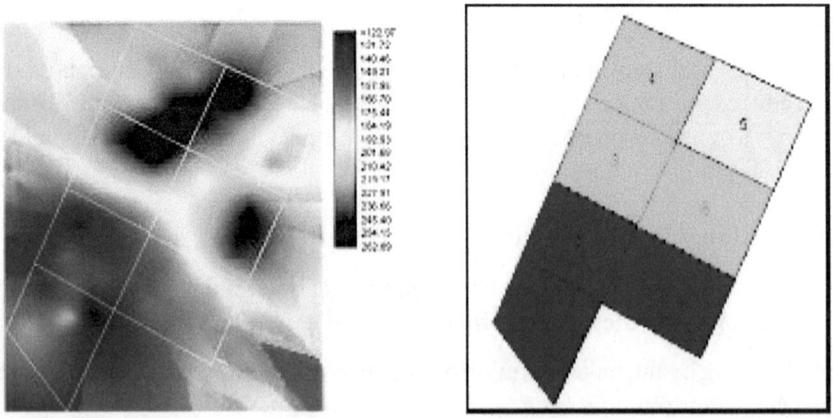

Fig 1. The Colour-gram for intra- field variability in available N in a farmers' field (data from Precision farming Project of Dharmapuri Dist. in Tamil Nadu)

(Courtesy), Tamil Nadu Agriculture University, personal communication)

Spatial Variability in Available N

Thiru. K. Vadivel, Periyamittahalli, Beneficiary farmer

Likewise variability can occur in other soil nutrients, soil physical state, texture, weed incidence, germination of seed, water holding capacity, disease incidence, insect pest incidence etc. All these variability would contribute to optimal to sub optimal crop performance in the same field in different grids. How can we overcome this variability and get the entire surface area of the field to perform uniformly high. Precision farming is an approach to make this happen.

The assessment of existing variability is only one aspect of PF. In our farming situation, even this type of assessment cannot be done easily. In developed countries such assessments are done using modern technological handle. For example use of GPS (global positioning system);this involves dividing a field into several small and equal divisions using the sub-inch accuracy of GPS. To do this, the tractor is fitted with a dish antenna to receive signals from satellites, which are recorded on a tractor-mounted computer. Soil samples are mechanically taken from each sub-division,a process technically known as "Grid Sampling" and analysed for each nutrient. The data are then superimposed on to the grids and colour-grams will clearly provide a message to the grower on deciding how much fertilizer should be applied to each grid based on what is actually available.

Second most important factor after the assessment is application of the input (here N fertilizer, an example) according to its availability as seen from the grid data. This procedure is termed as Variable Rate Technology (VRT). The following are the technologies behind ensuring *Precision* in farming.

- Global Positioning System (GPS)
- Yield Monitoring in every square meter of the land
- Variable Rate Technology (VRT) for application of inputs
- Remote sensing for assessment and quantification
- Geographic Information System (GIS) for rapid study and correction measures

VRT for example, is achieved through tractor-mounted computer guided spreader equipment capable of reading the variability of fertility from colour-grams. Fertilizers are then automatically applied at variable rates only to where they are needed as indicated by the colour-grams.

However, in Indian situation so far we have not been applying these technologies as such. Though GPS and Remote Sensing are utilized in large scale resource assessment, they have not found a place in small farms yet. Similarly computer guided tractors and variable fertilizer spreaders are still a long way far for us.

Let us now look at the way PF is getting practised in our small farm situations. The affordable level of precision in input management-fertilizer, irrigation water, pesticides and management- timing of operations, and ensuring of quality for all inputs –seed, nursery seedlings, fertilizers, pesticides etc. are brought in, in many of the precision farming projects that are implemented.

An important component of this package is training and capacity building among the growers. Agencies implementing this type of project and public institutions sponsoring/funding them should ensure that growers are adequately trained and supported.

In a situation like ours where the application of these high level technologies in small Indian farms is less than practical, the following interventions at the farm level are found to yield appreciable results.

• Soil Preparation	:	Humus level, microbial load, aeration & drainage, fertility restoration
• Seed	:	Variety and Quality
• Nursery/seedlings	:	Physically, physiologically uniform seedlings with intact roots
• Crop geometry	:	Single ,double ,triple and tetra rows
• Irrigation	:	Strictly as per crop requirement at each stage
• Fertigation	:	Varying nutrients at critical stages of growth
• Growth management	:	Regulation of flowering, training/pruning of the canopy
• Plant protection	:	Monitoring System, IPM
• Field Level PHM	:	Harvest, handling, sorting and grading and labeling crates
• Crop plan	:	Rotation/Relay/Intercrop systems to fully use time and space.

Interventions with tested and proven technologies and inputs in the openfield or protected (green house) cultivation in any one or more steps as listed above would result in higher yields and higher financial returns to the grower. The whole concept centres round the idea that space, time and resources are limited and hence endeavour to hike productivity to maximum in relation to these limitations.

Protected cultivation can also form a part of PF but its need has to be assessed specifically; the climate, soil or market if impose restrictions in achieving higher yields and profits, then introduction of exotic crops, growing in off -season and producing several fold yield per unit area can be implemented through protected cultivation.

Availability of several different technologies for input management and high yielding input responsive crop varieties offer a golden era for high tech cultivation of several crops. It is up to the grower or the supporting agency to collate these technologies and give an appropriate package to the grower. In the process, growers can be encouraged to form co-operatives/groups and can be trained to manage their produces and plan the farming with an entrepreneurial spirit.

Why do we need High-Tech crop production?

It is estimated that world population will grow from 6.0 billion in 2000 to 7.5 billion by 2020. While there is an ever increasing demand for food, total arable land will not grow from current level of 3080 million ha. In order to ensure global food security and nutrition demand, productivity of land has to be increased by a minimum at 40-50% in next 20 years by using appropriate technologies.Population in India has topped 1.21 billion and it is growing by 17.64% per decade (Min. Agri, GOI website). Ensuring food security to the billion plus population of our country requires high-tech crop production programs.

In short, we are living in an era when;

- Cultivable land is limited
- Irrigation water is insufficient
- Energy availability is limited
- Human Labour for Agriculture is scarce

Agriculture production at the existing levels of productivity per unit land; water and energy is woefully inadequate to meet the demand for food, feed and energy (bio fuel).

Table 1. Projected Food grain Demand and Irrigated Crop Area

	unit	2010	2025	2050
Food grain demand	million t	247	320	494
Net cultivated area	million ha	143	144	145
Total cropped area	m ha	193	204	232
Total irrigated crop area	m ha	79	98	146

Ministry of Agriculture, GOI

Another factor is the acute attrition of people from agriculture activities as occurring presently. Cropping is no longer an interesting vocation to keep the young people in the farm. The introduction of high-tech technologies associated with possible mechanisation of cultural operations is a solution to this problem. These by itself may not stop anti-agriculture migration of the population. One needs to look at farming as an enterprise, and the farmer an entrepreneur running a profitable business. Promotion of drudgery saving and profit generating technologies will become essential in this scenario.

Individual Cases of High-Tech Interventions

The following are actual cases where high-tech cultivation practises were introduced after considerable R&D and field level demonstrations by Jain Irrigation Company. These technologies are taken up by farmers from different parts of the country whole heartedly. Some among them have become entrepreneurs themselves.

Banana

The objective is enhancing the productivity of Banana through Tissue Culture technology.

Banana (*Musa paradisiaca*) is one of the most important staple food crop for millions of people in developing countries. It is a parthenocarpicfruit and therefore renders them sterile and unable to produce viable seeds. Owing to this condition, there is no other option except to resort to vegetative method of propagation using suckers, which is commonly known as "corm". The major disadvantages of using suckers for vegetative propagation, are the lower yields, longer crop duration (14-15 months for one crop), carry- over of diseases and pests etc. To overcome these constraints, *in-vitro* micro-propagation of banana under stringently controlled aseptic conditions in the laboratory is the most ideal way to produce high quality planting material.

As far as tissue culture technique is concerned, the key issues relate to adoption of a very precise protocol for the micro-propagation of banana, followed by primary and secondary hardening that it renders the tissue-raised plants to perform well in the field adapting to the local soil and climatic conditions. This is perhaps the most important criteria for developing disease-free and true-to-type plantlets with higher farm yields and excellent quality that can be accepted world-wide (Vasane and Kothari, 2009)

Of all the varieties tested, Grand Nain (belonging to the dwarf Cavendish group) proved to be the most suitable one. This introduced variety, owing to its quality attributes, was found to be ideal both for processing as well as for fresh fruit export.

An R&D project was set in motion for developing and standardizing the technology for production of high quality Tissue Culture planting material and testing and developing an appropriate cultivation technology.

In Jalgaon district of Maharashtra, banana is grown in an area of approximately 48,000 hectares and is therefore aptly termed as the 'Banana Bowl of India'. This district produces approximately 76% and 18% of Maharashtra's and India's banana production, respectively. The area suffered low yields of banana due to poor quality planting material, uneven and off- type suckers and high incidence of virus diseases. Therefore, the main objective of the project was to transform Jalgaon district's productivity to a higher level.

Outcome of the field trials showed 1.8 m x 1.5 m spacing to be the best, employing drip irrigation and fertigation practices, besides choosing Grand Nain as the best variety.

Acceptance by Banana Growers

The Tissue cultured banana plants and the package for cultivation have received an overwhelming response from the farms not only in Jalgaon district, but from different states of India as well. It is reflected in the ever increasing demand which is currently estimated to be 60 million.

The various components of the high-tech package were

- Elite uniform virus free planting material
- Intervention in Land preparation, crop geometry
- Introduction of drip and fertigation technologies
- Interventions in Crop management till harvest and packing
- A precise sucker management technology and opportunity for ratoon
- Assistance in Post Harvest Management (PHM) and technical support for packing house and ripening facilities
- Training and guidance to farmers

- Introduction of banana puree technology in commercial scale for value addition.

Table 2. Key differences between traditional sucker planting vs elite TC banana plants

Traditional Sucker planting	Elite Tissue Culture Banana plants
Haphazard Planting	Defined Crop Geometry
Flood Irrigation	Drip Irrigation
Broadcasting Solid Fertilizers	Fertigation
Yield per plant is 13-15 kg	Yield per plant is minimum of 30 kg
High disease incidence	Low disease incidence
Long gestation crop period	Short gestation crop period

Performance

Hi-tech farming also generates employment opportunities in the villages as it needs large number of farm hands for harvesting, loading, transport, unloading etc. Employment is generated for agriculture technicians who assist the farmers in maintaining their equipments/gadgets/systems of irrigation & fertigation stations etc.

Economic impact of TC Banana Cultivation Among Rural Farmers

The Tissue Culture plants that we give have consistently provided better yields to the farmers, from a banana bunch of 15 kg (with conventional planting material) to almost 25 kg on an average (with tissue culture planting material). This has improved not just the overall yields but has reduced the growing period giving farmers higher returns in a shorter period of time. With the high quality tissue cultured plants, farmers are able to harvest 3 crops in a period of 30 months as compared to only two crops with conventional sucker planting in same period.

Drip irrigation and fertigation augment the yield increment. Some control can be exercised on maturity period affording advancing /postponing harvesting to fetch good prices in the market. Banana processing adds value to the Banana fruit produced by farmer.

Tissue culture plants increased productivity to two and half times with that of the sucker planting material. Total revenue generation and net profit from one planting cycle (i.e. 30 month) have gone to Rs 3.06 and 1.80 lakh per acre, respectively as compared to Rs 1.25 and 0.52 lacs per acre, respectively in conventional cultivation.

Mango

India produces world's highest tonnage of mango (13.73million tones from an area of 2.15 million ha) with an average productivity of 6377 kg/ha. Mango, so far has been grown either this as a crop with least management efforts and without inputs like irrigation, fertilizer etc. Therefore the productivity is low rendering it un-remunerative. Mango cultivation can be revolutionized by the adoption of Ultra High Density Plantation (UHDP) and associated practices. Jain Irrigation Systems Ltd (JISL) heralds a new era where mango becomes a quick, high, stable yielding and sustainable cash crop. With increased demand of mango, shrinking land availability, increasing the productivity per unit area is imperative. Besides, effective utilization of natural resources like water, nutrients, fuel, etc. is also important. Hi-tech cultivation techniques like UHDP combined with micro irrigation, fertigation, crop regulation, proper training and pruning play important role in achieving these objectives. JISL has established commercial UHDP orchards in its R&D farms and their performance have been very promising. There is immense scope to increase the productivity to 5-12 trac depending on varieties by using UHDP and other improved agro-techniques.

Mango tree is believed to have evolved as a canopy layer species in the tropical rainforest of south and south-east Asia. Mature specimens can attain a height of 30m and can survive for more than 100 years. The root system consists of a long, vigorous taproot and abundant surface feeder roots. Both male and perfect flowers are found within a single inflorescence; the pistil aborts in male flowers. The ratio of male to perfect flowers is strongly influenced by genetic, environmental and cultural factors. Flowers are cross-pollinated by flies and honey bees.

Mango is well adapted to tropical and subtropical climates. Though Mango can grow up to 1400 m altitude, commercial mango orchards are raised at places upto 600m altitude. Its growth and fruiting are influenced by temperature, rainfall, wind velocity and altitude. Annual mean temperature ranging from 21°C to 27°C is ideal for successful crop, though it can grow in a temperature range of 5 to 44° C. Higher temperature at fruit development enhances fruit size and quality. Mango can perform well in areas with average annual rainfall of 250 to 2500mm. However, frequent rains and high humidity will result in incidence of pests and diseases.

Cultivation Practices for UHDP

Mango can be grown in a wide range of soils except clayey or extremely sandy or rocky calcareous, alkaline or water logged soils. It prefers a soil pH of 6.5 to 7.5. For successful establishment of orchard, soil depth of 2.0 to 2.5 m is required. Soil amendments may be used to correct the soil to get required pH. Soil test is prerequisite for deciding quantity of soil amendment.

Mango Varieties

Field experimentations in Jain R&D farm were conducted with Alphonso, Ratna and Sindhu and several other varieties of South and Central India. Based on our experiences and various scientific reports, we believe that the most of varieties in India can be grown successfully under UHDP. As the Cultivation Practices are very intensive, we prefer to recommend high value varieties.

Table 3. Mango varieties suitable for UHDP

State	Varieties
Andhra Pradesh	Alphonso, Alampur Baneshan, Banganapalli, Totapuri (Bangalora), Mallika
Bihar	Bombai, Himsagar, Langra, Chausa
Goa	Fernandin, Mankurad
Gujarat	Alphonso, Kesar
Karnataka	Alphonso, Bangalora, Neelum, Mallika
Madhya Pradesh	Alphonso, Kesar, Ratna
Tamil Nadu	Alphonso, Banganapalli, Imampasand, Totapuri (Bangalora), Neelum
Uttar Pradesh	Bombay Green, Dashehari, Langra, Lucknow Safeda, Mallika, Chausa

Under UHDP, Mango is planted at 3m x 2m which accommodates 674 plants per acre. Pits should be marked at 3m x 2m before pit digging and pits of l x l x l m are to be dug at marked places.Alternatively one meter deep and one meter wide trench can be prepared at every three meter. The pits should be allowed to wither for some weeks before filling with mixture of planting media. The media consist of 40-50 kg native soil, 0.5 -1.0 kg Single Super Phosphate (SSP), 0.25 kg neem cake, 20 kg compost or 10 kg vermi-compost and 10-15 g of Thimet or 20g Furadon. Plantations are to be raised by using grafted saplings. Plants must be procured well before planting. While planting, soil around each graft should be pressed well.

Epicotyl grafts are recommended for UHDP as the training of the tree starts at a very early stage itself.The ball of earth around the base of the graft should be retained intact and the graft joint should remain just above the ground level.

Drip irrigation should commence after planting. The newly planted grafts are to be supported by a bamboo stick to avoid damage by wind. Similarly these are protected from desiccating winds by shade crops like sunhemp raised around the sapling and which can be cut and used as a mulch later.

The critical components of UHDP technology are the management of inputs: irrigation water and fertilizer and besides the canopy management. These two inputs are provided through drip irrigation system. In several states in India, mango is not considered as an irrigated crop. In the event that UHDP gets acceptance from Indian mango growers to enhance productivity and income, it is imperative that they should adopt drip irrigation. By adopting the recommended practices, the farmer would be changing from subsistence level to a cash crop situation. But to achieve this he should adopt the technology wholly and not in parts.

Table 4. Crop water requirement of Mango planted at 3 x 2 m crop geometry

Month	Evaporation (mm)	Water Requirement lt/plant/day			
		1st yr	2nd yr	3rd yr	4th yr onwards
Jan	2.87 - 40.60	0.5 - 0.75	2.0 - 2.6	4.0 - 5.8	7.0 - 10.3
Feb	3.38 - 5.90	0.75 - 1.0	2.5 - 4.1	6.0 - 9.3	10.0 - 16.5
March	3.79 - 7.29	0.75 - 1.3	3.0 - 5.3	6.5 - 12.0	11.5 - 16.4
April	6.69 - 8.38	1.3 - 1.5	5.0 - 6.5	11.3 - 14.0	20.0 - 25.0
May	7.54 - 9.32	1.5 - 2.0	5.8 - 7.0	13.0 - 16.0	23.1 - 28.0
June	5.97 - 7.45	1.0 - 1.2	3.5 - 4.5	7.5 - 10.0	13.0 - 23.1
July	4.24 - 7.47	0.5 - 1.2	2.0 - 4.8	4.5 - 10.7	8.0 - 19.0
Aug	3.22 - 7.84	0.5 - 1.2	1.5 - 4.8	3.5 - 10.8	6.0 - 19.2
Sept	3.57 - 7.78	0.5 - 0.6	1.0 - 2.2	2.5 - 4.9	4.0 - 8.7
Oct	4.42 - 7.47	0.4 - 0.5	1.5 - 2.0	2.2 - 3.0	3.9 - 5.0
Nov	3.48 - 3.84	0.4 - 0.5	1.6 - 2.0	3.5 - 4.0	6.2 - 6.5
Dec	3.15 - 3.90	0.4 - 0.5	1.5 - 1.6	3.5 - 3.6	6.0 - 6.5
Average	4.65 - 6.02	0.8 - 1.0	2.8 - 3.3	6.3 - 7.5	10.8 - 13.3

Crop water requirement will vary according to the location. The table serves only as indicative.

For the bearing trees (3rd year onwards) irrigation to be given at survival level during September 3rd week to October to induce flowering (stress period). The rainfall events are very erratic and therefore not adjusted on a daily basis. The general recommendation is that if rain fall exceed 10mm in any one day, suspend drip irrigation for the next 2 days.

The online drip system is more suitable for Mango. The drip laterals are spaced at the relevant row spacing. Each tree is provided with one dripper of 4 lph during initial two years and 2 drippers of 4 lph from 3rd year. When 2 drippers are installed they should be placed 45cm away from the trunk.

Application of fertilizer

The recommended fertilizer dose for UHDP Mango in soils having medium nutrient content is given below. Estimation of fertilizer requirement should be based on soil analysis.

Table 5. Fertigation schedule for UHDP Mango (to be applied at weekly interval)

Fertigation Schedule and Quantity (kg/dose/acre)						
Age	Month	Number of Doses	Urea	H_3PO_4	MOP	$MgSO_4$
1yr	July-Sept	12	1.4	0.5	0.8	0.000
	Jan-May	20	1.7	0.6	0.9	0.000
2 yr	July-Sept	12	2.7	1.2	2.3	0.278
	Jan-May	20	1.6	0.7	1.4	0.167
3 yr	15 June-Aug	12	4.5	2.3	3.5	0.555
	Sept	4	1.4	1.2	3.1	0.000
	Jan-May	20	3.2	1.2	1.5	0.333
4 yr onwards	15 June-Aug	12	7.2	3.5	4.6	0.833
	Sept	4	2.2	1.7	4.2	0.000
	Jan-March	12	5.1	1.7	3.2	0.833

- If phosphorus has to be given as solid fertilizer like SSP, it can be given in two equal split doses along with organic manure.
- If chloride injury is found or if the water contain high chloride then MOP should be replaced by Potassium nitrate or sulphate.

Training & Pruning

When the plant height reaches 45-60cm, the terminal bud should be pinched at 5-6 cm below the apex to encourage growth of auxiliary buds. After the growth from auxiliary buds 2-3 vigorous shoots are to be retained in different directions at 15-20cm interval which will develop as primary branches.

After 4-6 months growth of primary branches, they should be headed back at 45-60cm length to allow further growth of 4-5 secondary shoots on each branch, which in turn will form tertiary shoots in due course. Ultimately the tree will present a dome shaped hedge.

Pruning is very essential and critical operation of UHDP to maintain fruiting shoots and to contain the canopy. Pruning must be completed as soon as possible after harvest preferably before 15th June in Central and Southern India. Tertiary branches have to be headed back in such a way that the plant height can be maintained at 1.5 m and having 10-15 tertiary shoots. Excess tertiary shoots have to be thinned out to avoid overcrowding. Cut ends needs to be applied with Bordeaux paste or 2% Copper Oxychloride (COC) suspension. About one month after pruning, thinning of newly emerged shoots is essential to avoid excess shoots and overcrowding. On each tertiary shoots, 3-4 new shoots are to be maintained. Dried panicle along with 2-3 leaves and dried shoots/branches must be removed at the time of pruning.

Crop Regulation

Some varieties of mango are prone to alternate bearing. However bearing can be regulated by UHDP management combined with pruning and chemical treatments.In recent years, application of Paclobutrazol (PBZ) has shown positive response. PBZ should be applied from the third year onwards, in the month of September (in Central & South India) or October (in North India).

Rejuvenation Pruning and Top Working

By rejuvenation we can improve productivity of old orchards. Combined with top working, improved or new variety can be grafted in existing trees. During this operation, an existing conventional orchard can be converted to UHDP by planting additional saplings. The old and non- productive branches are cut at the base or the origin. Leave one branch to allow for survival growth and cut all other branches at the point of origin. This is the best done after the fruiting season (in central and southern part of India) or at the end of winter (in northern India) which will allow for sprouting after 30-45 days. Apply COC paste on the cut surface to prevent fungal entry into the trunk. Drip irrigate as per schedule. Rejuvenation and top working will make the old tree juvenile again and help in canopy management of the orchard. Once top working is practiced, inter cropping can once again be practiced in the orchard.

Benefits of UHDP and Associated Technologies

- Increases productivity upto 2-3 times.
- Reduces orchard gestation period to three years as compared to conventional orchards (7-9 years)
- Makes orchard crops as profitable as other cash crops
- Reduces water used for irrigation upto 50%
- Increases fertilizer uptake by plants when fertigation is practiced.
- Suitable for small, medium and semi-large farmers as its management intensive and yields good quality fruits which can get better market price.
- It makes sensible bankable project to offer financial support due to assured high early returns.

Table 6. Advantages of Ultra high density planting (UHDP) of mango

Particulars	Planting Type		
	Traditional (40 trees /ac)	Medium Density (200 trees/ac)	Ultra High Density (674 trees/ac)
Gestation Period (years)	10-12	5	3
Duration to reach full potential (Years)	15	8	6
Yield potential	Medium	High	Very High
Orchard management activities :			
Pruning	Very difficult	Manageable	Easy
Spray operation	Difficult	Manageable	Easy
Spray efficiency	Very low	Fairly high	High
Harvest	Very difficult	Possible	Very easy
Control on fruit quality	Impossible	Possible	Easy
Expected yield at maturity (t/ac)			
1.High volume varieties	4-5	7-8	10-12
2.Low volume varieties	1.5-2.0	2-3	4-5
Expected income/yr (lakh Rs/ac)			
1.High volume varieties (@Rs 8/kg)	0.32-0.40	0.56-0.64	0.80-0.96
2.Low volume varieties (@Rs15/kg)	0.23-0.30	0.0.3-0.45	0.6-0.75
Commercial orchard life (years)	Up to 50	30-35	20-25

The cost of setting up an UHDP plantation is Rs 80,000.00/ac as against 50,000.00/ac for traditional (40 trees/ac) and 60,000.00/ac (200 trees/ac)and the yearly recurring expense comes to Rs 30,000.00/ac in the UHDP (Soman, 2009).The actual yields harvested in UHDP in years 4 and 5 are highly indicative of the projections above; a high volume variety like Ratna yielded 8.4 and 9.0 t/ac and a low volume cultivar, Alphonso yielded 1.7 and 2.4 t/ac.

Conversion of conventional garden into UHD plantation

1. Top work the existing old trees. Allow regrowth in a controlled canopy fashion.
2. Plant new saplings in the inter row and intra-row space to get a final spacing of 3 m x 2 m.
3. Train and prune the trees, old and new, as per a schedule.
4. Provide drip fertigation to all the trees.

High-tech components in Mango cultivation

- Quality planting material –epicotyl grafts
- Changing Crop geometry and population
- Training of tree at very early stage to enforce a tree architecture

- Intensive pruning after harvest to contain the canopy to 1.5 m height every year
- Drip irrigation and fertigation
- Enforcing a stress period
- Hormone application to convert alternate year bearing to annual bearing.
- Fruit pruning to maintain optimum fruits per tree
- Hand harvest for quality assurance
- PHM guidelines
- Value addition opportunity by pulping/diceing

Fig. 2. Ultra High density Mango plantation in flowering , age 2 years 7 months

Fig. 3. Ultra high Density Mango plantation at fruiting, -age 2 years 10 months

Onion

Onion is one of the important vegetable crops grown in India. In terms of area, India ranks first in the world with over 480,000 ha. Globally, the country occupies the second position after China in onion production with a production share of around 14 per cent.Productivity, however, at around 13 t/ha, is much lower than the world average of 17.3 t/ha.

Onion is produced in several states in India, with Maharashtra being the leading producer accounting for about one-fourth of the country's onion production. India largely produces red onions used as fresh produce in the domestic market. White onions (used for dehydration) are grown on commercial scale only in the states of Maharashtra and Gujarat and that too only in a few districts. Red onions are not suitable for dehydration mostly due to low solids, low pungency levels and high reducing sugars. The local white varieties available in the country are low in solids <12.5% TSS, prone to premature bolting during Rabi (30-40% bolting) and hence not suitable for dehydration

Onions for processing focus on high yields and other desirable processing characteristics such as high total soluble solids content, snow-white colour, narrow stem-base, bolt resistance during bulb crop etc.. The woody stem of the stalk of the umbel reduces the quality of the dehydrated product. Thus the company has to take up focused R&D to breed for High TSS and bolting resistant white onion.

During 1996, about 45 cultivars of short-day onion from different parts of the World were introduced and evaluated for their yield potential, soluble solids content and other processing attributes.

Seed Production

Cultivar Cv. JV12, was selected for further detailed investigation. Extensive trials were carried out during the subsequent four year period on water, fertilizer, disease and pest management of JV 12 variety. Based on the results, a package of agronomic practices has been standardized. The R&D team then went ahead to produce the seeds of this variety within the country under our own conditions and management. Quality seeds of the elite selection JV12 are being produced on commercial scale since 2000 in Himachal Pradesh to get cool short day periods.

High-tech Precision Farming Practices

The introduction of hi-tech precision farming technology in onion cultivation has been a great success. The company was successful in revolutionizing the farming practices with an objective of ensuring supply of quality produce. The high tech practices were introduced in 2001-02 with the principal objective to enhance productivity and uniformity with better agronomical practices and also to improve the quality of onion bulbs (Balasubramanyam, Dhake and Moitra, 1999).

Under contract farming, the high-yielding onion seeds are distributed to the registered growers and the seeds are sown and transplanted in the farmers' field by the farmers using their own resources. Company's R&D and extension wing help the cultivators from time to time through frequent field-visits, guidance and input supplies.

An important feature of the planting material provided by the company is the high germination rate of about 80-85%.

Availability of Production Technology/Equipments and their Adoption

A prescribed set of practices was adopted by the contract farmers, specifying the fertilizer requirements and spraying time. Also, they have educated farmers about cropping patterns. Earlier farmers in the region were largely cotton growers. After working with the company they have shifted to rotation with legumes. Two key advantages from growing legumes are nitrogen fixation which helps to enhance soil fertility and legume cycle is shorter than cotton and more appropriate for planting onions. The farmers feel that overall, they are benefited from the legume, onion rotation. The overall benefits of soil fertility and income are higher in legume – onion crop rotation. The productivity for the farmers working with the package was between 10 to 14 tone/acre (average productivity 12 tone/acre), whereas the average productivity in the case of traditional production is around 5-7 tons/acre.

An important aspect of technology is that the farmers working with Jain Irrigation System Ltd (JISL) are using drip irrigation. Use of micro-irrigation technology enables higher productivity by ensuring adequate availability of water. In the long run, the farmers are using much less water than the traditional flood irrigation approach, which is better for the environment, as well as ensures lower cost and higher productivity for the farmer. If the farmers cannot pay for irrigation upfront, the company either arranges a loan for them, or provides drip irrigation on credit, for which the cost is deducted from the sales of onions.

Success Stories and Impact of Extension Activities in Onion

The focus was to achieve higher farm yields by providing optimum inputs; i.e "more from less" concept. The technology package helped in transforming the lives of many onion farmers in and around Jalgaon in particular by way of the unique contract farming activity which has now become a revolution. The contract farming has now spread like wild fire among the farming community because of the company's Minimum Guaranteed Price (MGP) concept which insulates the farmer's earnings from fluctuating market prices.

Farmer Earnings

The gross margin for the farmers working with the technology package and selling for processing is much higher than the margins for farmers selling to the traditional market.The cost of production (variable cost) for farmers working with the high tech practises is Rs. 17,800 per acre, whereas the same is Rs. 11,800 per acre for the farmers following traditional practices and selling to the traditional market. The main difference is due to the higher costs for irrigation system and associated water costs.

There is however, a significant difference in the productivity levels of both the groups. While the average productivity of the farmers selling to the traditional market is app. 7 tone/acre, it is almost twice at 12 tone/acre for the contract farmers with drip irrigation.

Onion is a staple commodity in India and the price varies significantly given the market conditions. Often times, because of market glut, the prices crash to very low levels, for instance, as low as Rs. 1.50/kg. The pricing policy is such that on the lower side, the company has fixed a Minimum Guaranteed Price (MGP) of Rs. 3.00/ kg for its contract growers, while on the higher side, the prices are linked to the market price. The MGP of Rs. 3.00/kg is the break even cost calculated based on all costs including fixed costs. If the market price fall below the minimum level, the contract growers get the MGP. Based on the yield and prices, the value of production in case of contract farmers is Rs. 36,000 to Rs. 84,000/acre whereas the same is Rs. 14,000 to Rs. 49,000/acre for farmers selling to the traditional market. Owing to the higher productivity and higher the minimum price support, the margin for high tech farmers is Rs. 18,200 to Rs. 66,200 per acre. In the case of farmers selling to the traditional market, the margin is Rs. 2,200 to Rs. 37,200. However, in case of farmers selling their produce to the open market, other overheads such as gunny bags, commission for agents, labor charges, and market fee and handling charges, all amount to 10 % and eventually the margin is only between Rs. 1,980 to Rs. 33,480. (Punjabi 2008)

Evidently, JISL contract farmers have higher profits per acre. Further, since the minimum price is announced ahead of the planting season, they know the value of their output before production and hence can invest in inputs required for higher productivity. If the farmer do not know what price he will get at the end of the season, there is hesitation to investment in inputs, which further reduces earnings by low productivity.

The High Tech Components in Onion Production

- High quality seed with high germination percentage
- Soil surface management –bed and furrow
- Change in Crop geometry and plant population
- Drip irrigation and fertigation
- Pest and disease monitoring by qualified extension workers

- Timely pest management
- Association of farmers with a Processor
- Processor assuring MGP and supporting farmers
- Harvest and Post Harvest management under technical supervision.
- Technical support for rotation crop –selection and cultivation

Fig. 4. Traditional onion cultivation-planting

Fig. 5. High Tech adopted onion crop

References

Balasubramanyam V.R., Dhake, A.V. and Moitra, P. 1999. Improvement of onion cultivation. Agro India, February:24-27

IWMI, International Water Management Institute, 2009. Revitalising Asia's irrigation: To sustainably meet tomorrow's food needs. IWMI, Sri Lanka.

Punjabi, M. 2008. Supply chain for exports of dehydrated onion: Case Study of Jain Irrigation Systems in India. FAO, New Delhi.

Soman, P. 2009. Ultra high density plantation in mango-a rewarding experiment. Paper presented at a National seminar "Recent improvements in cultivation of Mango", Tamil NadAgriculture University, Periyakualm Campus, Madurai.

Soman, P. and Narayanan, S. 2010. Role of micro irrigation in enhancing crop productivity-an industry perspective. (In) Micro-irrigation, Economics and Outreach; International Water management Institute, Hyderabad and Agricultural Economics Research Association, New Delhi 110002; pp 285-294.

Vasane, S.R. and Kothari, R.M. 2009. An integrated biotech approach to sustainable cultivation of banana var. Grand nain. (In) A Textbook of Molecular Biotechnology (Eds. A. K. Chauhan & Ajit Varma) I.K. International Publishing House Pvt. Ltd, New Delhi, pp. 507-541.

5

Technological Advances in Irrigation Management for Protected Cultivation

P.K. Singh, Ganga Joshi and D.K. Singh

Abstract

Protected cultivation of horticultural crops especially vegetables provides the best way to increase the productivity and quality of vegetables especially during off-season, which also fetches better market price. Yield of some of the vegetables like tomato, capsicum and cucurbits can be increased manifold compared to their open field cultivation. Irrigation management is an important aspect of protected cultivation technology for improving the input use efficiency such as water and nutrients for the cultivation of crops under protected environment such as polyhouse, net house, poly tunnels and nursery production. Various other techniques such as use of mulch can be used to further enhance the water and nutrient use efficiency along with improving the quality of produce and precise resource savings. Automation further reduces the manual intervention and increases the input use efficiency.

Keywords: Protected Cultivation, Micro irrigation, Irrigation scheduling

Introduction

Plants depend on water and within the protected structure with no rainfall to supply it naturally and the added warmth makes the soil dry out more quickly.

However, near zero wind velocity under many protected environments such as polyhouses, poly tunnels, low tunnels, etc., the water requirement of many crops decreases significantly (Rana *et al*, 2010). Traditional watering-cans are, of course, the simplest solution to the problem. They are particularly useful in providing very specific, targeted watering for individual plants, particularly if they are being grown in pots or containers. However, as a means of achieving mass watering, they are far too labour-intensive to be practical for anything other than the smallest of greenhouses. Many gardeners are inclined to try to overcome the limitations of watering can by using a hosepipe instead,

However, as a means of achieving mass watering, they are far too labour-intensive to be practical for anything other than the smallest of greenhouses. Many gardeners are inclined to try to overcome the limitations of the watering can by using a hosepipe instead, but this approach itself has its problems – chiefly in terms of achieving the delivery of water where it is really needed, rather than on the leaves and soil surface. The traditional method of irrigation as in case of open field condition creates lot of diseases under protected environment because of the increased level of relative humidity along with increased temperature. Therefore, the appropriate irrigation techniques having the characteristics of frequent application of water in small quantity as per crop need at right location in right manner is essential for obtaining the maximum benefit of protected cultivation. Micro irrigation has the ability to provide the right amount of water, at right time, at right place and in right manner. Similarly, nutrients may also be applied precisely along with irrigation water using the fertigation device along micro irrigation. In this way significant amount of two important agricultural inputs (water and nutrients) could be saved significantly besides, controlling the environmental (non-point source / agricultural) pollution. Further, it has the potential to increases crop yield. There is an additional positive environmental impact from precision irrigation in that farm runoff, a major source of water pollution, can be reduced. The major limitations associated with precision irrigation are high initial cost; operation and maintenance need skilled work force.

Protected Cultivation

Protected cultivation practices can be defined as a cropping technique wherein the micro environment surrounding the plant body is controlled partially/ fully as per plant need during their period of growth to maximize the yield and resource saving. With advancement in agriculture, various types of protected cultivation practices suitable for a specific type of agro-climatic zone have emerged. Among these protective cultivation practices, green house/poly house cum rain shelter is useful for the hill zones. The greenhouse is generally covered by transparent or translucent material such as glass or plastic. The greenhouse covered with simple plastic sheet is termed as polyhouse. The greenhouse generally reflects back about 43% of the net solar radiation incident upon it allowing the transmittance of the "photosynthetically active solar radiation" in the range of 400-700 Nm wave length. The sunlight admitted to the protected environment is absorbed by the crops, floor, and other objects. These objects in turn emit long wave thermal radiation in the infra red region for which the glazing material has lower transparency. As a result, the solar energy remains trapped in the protected environment, thus, raising its temperature. This phenomenon is called the "Greenhouse Effect". This condition of natural rise in greenhouse air temperature is utilized in the cold regions to grow crops successfully. However, during summer season due to the above stated phenomenon, ventilation and cooling are required to maintain the temperature inside the structure well below 35°C. The ventilation system can be natural or a forced one. In the forced system, fans are used which draw out 7- 9m^3 of air/ sec/ unit of power consumed and are able to provide 2 air changes / minute.

Why Protected Cultivation?

- The vegetables/any other produce harvested from the protected cultivation have better quality in terms of the fruit size, per fruit weight, TSS, colour, texture and other quality parameters.
- The productivity of the crop under protected cultivation increases manifold (2-10 times) as compared to conventional system.
- Off-season cultivation of the vegetables/other crops is the most important aspect of protected cultivation. The winter season vegetable crops such as cauliflower, coriander, spinach, etc can be grown during rainy season. Similarly, summer season vegetable crops are successfully grown during winter and the winter season leafy vegetable can be grown during summer.
- Under protected environment, incidence of insect and disease are minimized due to its isolation from open field. Use of micro irrigation and off-season cultivation also minimize the disease and insect incidence. Less infestation of insects and diseases reduce use of pesticides.
- The protected cultivation offers efficient use of two most precious resources land and water. Other inputs such as fertilizers, chemicals and labour are also efficiently utilized under such environment.
- The year round cultivation of single/multiple crops with high quality produce and high productivity. The off-season cultivation of vegetable offers better market price and increased employment generation round the year for small and marginal farmers in hills and peri-urban areas.

Irrigation Techniques for Protected (Greenhouse/Shade Net House and Walking /Low Tunnel) Cultivation

It is very difficult to obtain full benefit from protected cultivation without micro irrigation and fertigation. An efficient irrigation system, preferably micro irrigation combined with fertigation system must be an essential component of protected cultivation. Therefore, an irrigation system is essential for growing plants in a greenhouse. Plants rely on water to live and grow and because a greenhouse will not allow natural rainfall in, artificial means for irrigation become necessary. A variety of irrigation methods exist, and each method has benefits and drawbacks. Choosing the best irrigation method depends largely on the size of protected structure and the types of plants growing inside. Often the most effective irrigation comes from a combination of methods.

Types of irrigation systems for nurseries

A plant nursery requires an irrigation system to water plants effectively and simultaneously (cacti nursery image by Doug Stacey from Fotolia). comCreating a nursery to house your plants can involve a significant effort and commitment. Necessary decisions include selecting plant species, soil types and building materials. Also consider the amount of time and money available to invest in such a project. Nurserymen should think about which watering or irrigation system best suits their nursery type and size.

Overhead sprinkler irrigation systems

Nurserymen using overhead sprinklers typically have two options. The first option, rotary sprinkler heads, contains a rotating nozzle that sends a torrent of water over plants. The second option, stationary sprinkler heads, sends a rapid flow of water against a plate. The impact disrupts the steady stream of water, turning it into a continuous spray that waters plants.

Although overhead sprinkler systems are the most common option in nurseries, they are not very efficient. They require high pressure pumps that consume large quantities of energy. Overhead sprinklers also waste about 80 percent of the water emitted. In nurseries containing plants with large or broad leaves, these plants encourage water waste when leaves redirect water away from plant containers rather than into the soil. Some gardeners compensate for water loss by installing slanted plant beds that channel water into ponds where it can accumulate and be recycled back into the nursery, although this may also recycle bacteria, sodium, fertilizer or pathogens as well.

Micro irrigation systems

Unlike overhead sprinklers, micro irrigation systems are highly efficient and can function using low pressure. However, soil, algae and chemical fertilizers can clog emitters for which various types of filters are provided. Three types of micro irrigation systems are used in nurseries. One type of micro irrigation, known as the capillary mat system, uses tubes that carry water into a mat. The mat becomes saturated with water, providing containers sitting on top of the mat with a supply of water to soak up through plant root systems. Although capillary mat irrigation uses 60 percent less water than conventional overhead sprinkler systems, they can cause salt accumulation in the soil over long periods of time.

The second type of micro irrigation system is known as a micro sprayer, micro sprinkler or spray stake system. Considered one of the most efficient nursery irrigation systems, micro sprayers use a tube to carry water directly into the soil from a water source. Not only does this eliminate water waste that is deflected off broad plant leaves, micro sprayers carry water directly to the plant's root system. Although micro sprayers cost more than overhead sprinklers when installed in small plants, they operate efficiently in larger plants with more foliage and heavier canopies.

The third type of micro irrigation is known as the spaghetti tube system. This nursery irrigation method uses narrow tubes to bring water into the plant container. A miniature weight at one end of the tube ensures that it stays in the container. Water travels from one pore to another, through a capillary system. Consequently, gardeners must use a high quality, uniform soil for maximum efficiency. When using the spaghetti tube system, gardeners should keep soil moist at all times; dry soil will lead to poor water distribution.

Capillary sandbeds

Unlike sprinkler and micro irrigation systems, capillary sandbeds do not involve any electricity. Containing wood panels, a plastic liner, sand, a small water reservoir, drainage pipe and valve, capillary sandbeds are built to slant slightly started, allowing water released into the raised end to slowly travel to the lower end. Providing an even and continuous water supply, capillary sandbeds involve less maintenance. Plants grow evenly, relying less on fertilizer and pesticide. However, capillary sandbeds do attract weeds. Gardeners can purchase products to reduce the occurrence of weeds or they can remove them manually. Capillary sandbeds also have high installation costs.

Types of mist irrigation systems recommended in poly house

Overhead irrigation can produce constant humidity in poly-tunnels. Flag this photo A misting sprinkler system produces a very fine spray or mist over plants. Greenhouses, poly-houses and poly-tunnels can use either overhead or bench misting. The best type for your needs depends on what you are growing, the size of your poly house — essentially plastic over a framework — and on the growing conditions in your area.

Related Searches: Overhead systems

In overhead misting systems, lines or sprinklers are installed under the roof framework of your poly-house, and this "rain" water goes down on your plants.

This type of irrigation system is easy to automate and produces high humidity. This high humidity allows you to protect crops against frost damage. For the best coverage, space overhead sprinklers to around 50 to 60 percent of the wetting diameter of the sprinkler.

Bench misting

Bench misting uses a central line of sprinklers or hoops placed at or just above the level of your plants. Bench misting requires plants to be placed on raised benches, and these must be made of materials that are impervious to water, such as a metal. You can also use a self-contained misting bench, in which the bench is partially covered with a "roof." This allows you to have just a single misting bench in the poly-house, and a different watering system in the rest of the poly-house.

Benefits

Both types of misting system are well suited to plants that need to be kept moist, such as seedlings, and to reducing temperatures in a poly-house. Misting is commonly used for propagation and for growing tropical plants that require constant humidity. Some misting systems can also be used to spray fertilizers evenly and finely. Fertilizer applied this way is more easily absorbed into the plants than fertilizer applied on the soil. By allowing you to vary the humidity within the poly-house, mist systems also allow you to vary the temperature and control the growing conditions.

Drawbacks

It can be easy to over-water with either type of mist irrigation system. To prevent this, you can use a timer to turn the water on and off. Misting nozzles have very fine holes that can clog up if hard water is used. Misting also works best if poly-house is completely enclosed, as a breeze can disrupt the fine spray and cause areas to remain dry. Misting may not be suitable for all types of plants, so if poly-house contains many different types of plants, with different water tolerances, one may need to water each type of plant individually, or use individual benches, rather than use an overhead mist irrigation system.

Drip irrigation system

Drip irrigation is one of the most efficient methods of watering, typically operating at 90 percent efficiency rate. Runoff and evaporation are at a reduced rate when compared to other irrigation systems such as sprinklers. In drip irrigation, tubes that have emitters run alongside the plants receiving irrigation. The water leaves the tubing through the emitters by slowly dripping into the soil at the root zone. This method of irrigation minimizes leaf, fruit and stem contact with water resulting in reduced plant disease. It reduces weed growth by keeping the area between plants dry. Irrigation through drip method can be set to run automatically or controlled manually.

Drip watering is an excellent way to water in greenhouses and tunnels as it keeps the humidity low leading to less pest and disease problems. Water is directed to exactly where it is needed either with an individual dripper, especially for pot grown plants, or inserted into a pipe for beds. Ideal for raised beds

Ebb and flood tables

Ebb and flood tables are frequently used in greenhouse where growers raise plants hydroponically. But an ebb and flood table may also be used for plants that the grower is growing traditionally. In an ebb and flood table, the plants are kept in a tray that has a hose nozzle at one end to admit water into the tray and a drain at the other end where the gardener can let water out of the tray. Plants that are kept in individual receptacles in the tray wick up water from the base of their containers. Gardeners who wish to recycle water and eliminate waste make use of ebb-and-flow systems for their greenhouses.

Hand watering

Hand watering is the basic method of irrigation for a greenhouse. Watering cans or hoses with nozzles allow you to tailor watering to the needs of individual plants. This is also the most labor-intensive method for plant irrigation and may not be effective for large greenhouses.

Irrigation Scheduling

Greenhouse drip irrigation scheduling

Availability and decreasing quantity of water for agriculture highlights the objective of optimizing productivity, with adequate and efficient irrigation, that replenish the root zone soil water deficit and maximize the applied water that is stored in the rooted soil profile and used afterwards by the crop, in order to reach best yields (Castilla, 1990). As crop responds more to soil water level and irrigation regime than to method of irrigation, information developed for other irrigation methods is applicable to drip systems, in general.

Four basic questions must be answered in pursuing precise irrigation scheduling

1. When to irrigate? (Frequency)
2. How much water to apply? The amount of water to be applied must replenish the evapotranspirated water, once corrected by the application efficiency (as far as the soil-water content variations are unimportant, due to the high

frequency of drip irrigation). When saline water is used, the applied water must cover the leaching requirements (Ayers *et al*, 1976; Doorembos *et al*, 1976; Stegman *et al*, 1980; Vermeiren *et al*, 1980). Other components of water balance are normally unimportant in drip irrigated greenhouses (unless rainfall penetrates inside, as it is the case in flat-roof perforated plastic greenhouses).

3. Where to irrigate (point/line/ strip/disc source)
4. How to irrigate – drip (surface or subsurface) bubbler, micro sprinkler, mist etc.

Evapotranspiration (ET) in protected environment

Evaporation of water requires energy. Availability of energy depends on microclimate of the protected environment, solar radiation being the primary source of energy in the ET process. In an unheated greenhouse, the energy used in ET process can reach upto 70% of incoming solar radiation (Hanan, 1990). The amount of ground area covered by the crop is this most relevant factor affecting ET. Evaporation (E) from the soil surface is high following irrigation, but decreases rapidly as the soil surface dries. Transpiration (T) will increase with the rise of intercepted radiation (and subsequent increase of ground covered by the crop), while soil E will decrease (as the crop progressively shades the soil surface). Other energy sources (greenhouse heating hot air flow) can increase ET.

Crop evapotranspiration (ETc) or crop water requirements can be related with a reference value ''reference evapotranspiration" (ET_0) which is defined as "the rate of evapotranspiration from an extended surface of 8 to 15 cm tall green grass cover of uniform height, actively growing, completely shading the ground and not short of water". (Doorembos *et al*, 1976)

$$ET_c = K_c \times ET_0 \quad (1)$$

The crop coefficient (Kc) is "the ratio between ET_c and ET_0" and depends basically on the crop characteristics, the sowing or planting dates, development rate of the crop, length of the cycle, climatic conditions and the irrigation frequency, especially at the beginning of the cycle (Doorembos *et al*, 1976).

In greenhouses, class A pan evaporation method, as well as the radiation (FAO) and Priestley-Taylor methods, have been proposed as more reliable for ET estimation for periods of several days (Castilla *et al*, 1990-B). The difficulty of accurate measurement of wind inside the greenhouse (Castilla *et al*, 1990-B) limits the use of Penman method. The ease of management of the evaporation pan, without sophisticated equipment, is remarkable, but a proper pan placement is necessary.

The crop coefficient figures for different vegetable crops under greenhouses have been estimated (Castilla, 1989; Castilla *et al*, 1990-B; Martinez *et al*, 1990; Veschambre *et al*, 1980).

When using the class A pan method:

$$ET_o = K_p \times E_o \quad (2)$$

K_p = pan coefficient

E_o = pan evaporation

$$ET_c = K_c \times ET_o = K_c \times K_p \times E_o = K \times E_o \quad (3)$$

Where $K = K_p \times K_c$ (4)

Recent studies show that K_p inside the greenhouse is approximately 1.0 (Sirjacobs, 1986; Castilla, 1986; Castilla *et al*, 1990-B), higher than open-air values (Doorembos *et al*, 1979). The crop coefficient evaluation and values for different vegetable crops are presented in Table 1. Recent research in the Almeria area confirms the K values detailed in Table 1, pointing that K_p is around 0.8-0.9, but the quantified values of K_c are higher than those described in the literature (Doorembos *et al*, 1976), being the products of both coefficients ($K_p \times K_c$) similar to those indicated in Table 1.

Table 1. Crop coefficient values for different vegetable crops in a plastic greenhouse in Almeria, using drip irrigation.

Days after sowing or transplanting	Tomato	Capsicum	Cucumber	Melon	Watermelon	Beans	Eggplant
1-15	0.25	0.20	0.25	0.20	0.20	0.25	0.20
16-30	0.50	0.30	0.60	0.30	0.30	0.50	0.35
31-45	0.65	0.40	0.80	0.40	0.40	0.70	0.55
46-60	0.90	0.55	1.00	0.55	0.50	0.90	0.70
61-75	1.10	0.70	1.10	0.70	0.65	1.00	0.90
76-90	1.20	0.90	1.10	0.90	0.80	1.10	1.10
91-105	1.20	1.10	0.90	1.00	1.00	1.00	1.05
106-120	1.10	1.10	0.85	1.10	1.00	0.90	0.95
121-135	1.00	1.00	-	1.10	0.90	-	0.85
136-150	0.95	0.90	-	1.00	-	-	0.80
151-165	0.85	0.70	-	-	-	-	0.80
166-180	0.80	0.60	-	-	-	-	0.80
181-195	0.80	0.50	-	-	-	-	0.80
196-210	0.80	0.50	-	-	-	-	0.80
211-225	-	0.60	-	-	-	-	0.80
226-240	-	0.70	-	-	-	-	0.60
241-255	-	0.80	-	-	-	-	0.60
Total	318	322	156	349	290	146	393

The net irrigation requirements (IR_g) must replenish the crop evapotranspirated water (ET_o), as rainfall and other components of water balance are normally unimportant in greenhouses in the Mediterranean area.

The gross irrigation requirements (IR_g) must increase the IR_n, in order to compensate the irrigation efficiency and to leach salts.

$$IRg = \frac{IRn}{Ea(1-LR)} \tag{5}$$

Where, Ea = irrigation efficiency coefficient (smaller crop root zone to be used by the crop/applied water).

$$Ea = Ks \times Eu \tag{6}$$

Ks is a coefficient (smaller than 1) which expresses water storage efficiency of the soil (0.9 in sandy soils, 1.0 in clay or loam soils).

Eu is a coefficient (smaller than 1) which reflects the uniformity of water application (a properly designed and well managed drip system should reach Eu values of 0.85-0.95). This coefficient should be measured for each system regularly (Vermeiren *et al*., 1980).

LR: minimum amount of leaching needed to control salts with drip irrigation

$$LR = \frac{ECw}{2(maxECe)} \tag{7}$$

Where, EC_w: = electrical conductivity of the irrigation water (dS/m)

EC_e: = maximum electrical conductivity (dS/m) of the soil saturation extract due to crop withdrawal of soil water to meet its evapotranspiration demand.

Typical max EC_e values are 12.5 in tomato, 10.0 in cucumber, 8.5 in pepper and 6.5 in bean. Recent research shows that the leaching requirements could be lower than the indicated values (Stegman *et al.*, 1980).

A study conducted at Department of Irrigation and Drainage Engineering, GBPUA&T, Pantnagar showed that (Fig 2 to 5) the water requirement for capsicum varies from 0.90lpd/plant to 0.8519lpd/plant for open conditions, 0.067lpd/plant to 0.474lpd/plant for 35% shading conditions, 0.0675lpd/plant to 0.4605lpd/plant for 50% shading conditions and 0.052lpd/plant to 0.423lpd/plant for 75% shading conditions (Fig 1). The water requirement for tomato varies from 0.090pd to 93lpd/plant for open conditions, 0.067lpd/plant to 0.63lpd/plant for 35% shading conditions, 0.067lpd/plant to 0.62lpd/plant for 50% shading conditions and 0.059lpd/plant to 0.46lpd/plant for 75% shading conditions (Fig 2). The water requirement for cucumber varies from 0.051lpd/plant to 0.81lpd/plant for open conditions, 0.0385lpd/plant to 0.545lpd/plant for 35% shading conditions, 0.038lpd/plant to 0.529lpd/plant for 50% shading conditions and 0.030pd to 0.403lpd/plant for 75% shading conditions (Fig 3).The water requirement for summer squash varies from 0.0945lpd/plant to 0.608lpd/plant for open conditions, 0.038lpd/plant to 0.409lpd/plant for 35% shading conditions, 0.0385lpd/plant to 0.397lpd/plant for 50% shading conditions and 0.030pd to 0.302lpd/plant for 75% shading conditions (Fig 4) (Rana and Sah, 2010).

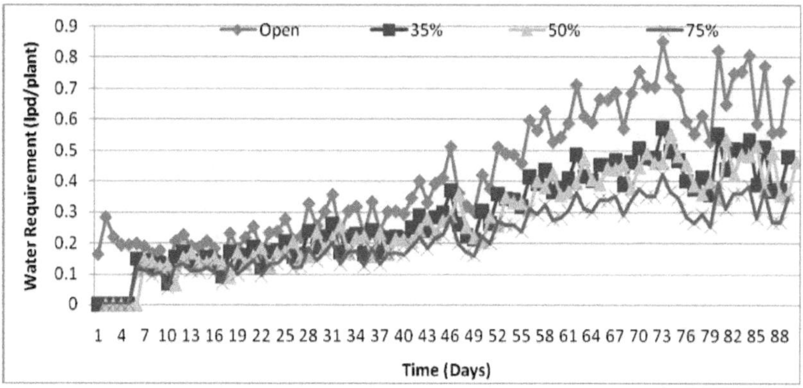

Fig 1. Water requirement of Capsicum under open field and shadenet

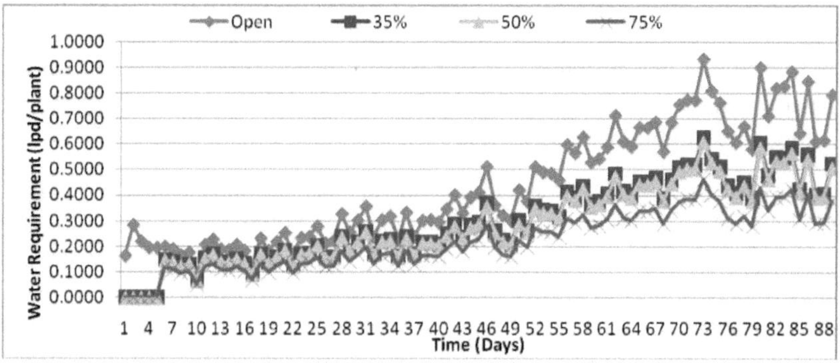

Fig. 2. Water requirement of Tomato under open field and shadenet

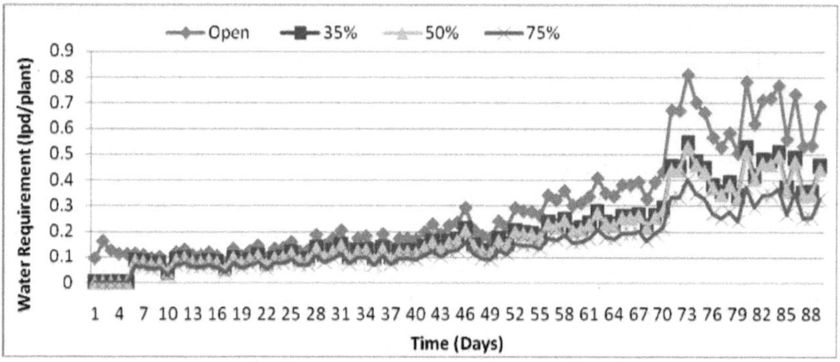

Fig. 3: Water requirement of Cucumber under open field and shadenet

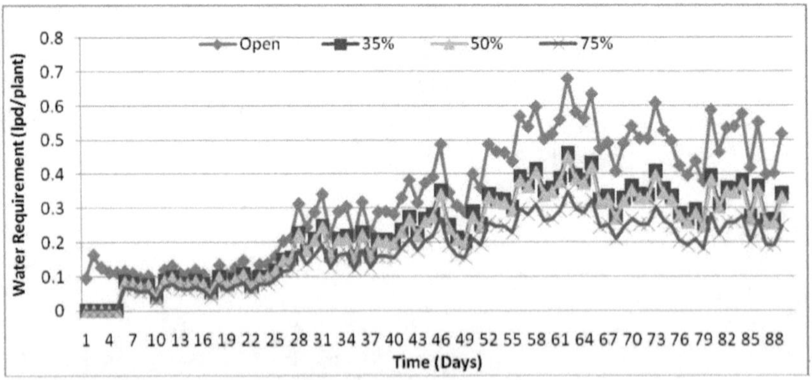

Fig. 4: Water requirement of Summer Squash under open field and shadenet

Drip Irrigation Frequency

In conventional irrigation, soil water depletion must be maintained below certain thresholds (available soil water depletion) in order to avoid crop transpiration reductions, that can induce yield decrease. A proper irrigation frequency will avoid excessive depletion. In drip systems, good management will always be based on very high irrigation frequency, even several times each day (Villele, 1984), specially when using saline water, being the water storage role of the soil unimportant relative to conventional irrigation methods.

Different plant and soil parameters have been suggested to schedule the irrigation frequency. A wide range of plant based measurements more or less sophisticated (sap flow, stem diameter, water potential, plant temperature...) have been suggested to detect stress, using the plant as a biosensor. The leaf water potential method, reliable when used in conventional irrigation systems, is not practical in drip irrigated vegetable crops. Plant temperature based methods of water stress detection are more accurate in greenhouse than in open field (Stanghellini, 1993) but they must be developed and locally adapted. The crop water stress index, based on the higher temperature of the crop when suffering from water stress, has been suggested as a more reliable method, but it is not easy to use. The spatial variability of soil-water contents in drip-irrigated soils limits the interest of methods based on soil water content measurements ("available soil water depletion"). The soil water matrix potential measurement, in drip irrigated soils, using tensiometers is a reliable way for monitoring soil water conditions in the wetted zone, in order to fix the irrigation frequency and to confirm the adequacy of the applied water amount. Two tensiometers, at least, should be placed in each observation point, installed at two depths, a few cm away from the emitter, depending on the soil water distribution and rooting patterns. Other methods for monitoring soil water potential, as gypsum blocks, need a good calibration depending on the composition of soil solution.

Wetting the soil profile with preplanting conventional irrigation (for salt leaching, soil disinfection or other purposes) is useful to prevent possible drip-applied water deficits (Castilla *et al*, 1990 -A).

In soilless culture, irrigation frequency must be several times per day, dividing the daily water requirements, according to the evaporative demand and water storage characteristics of the substrate. In a first approach, the evaluation of solar radiation along the day (that determines primarily the evaporative demand) can be a guide to schedule the irrigation. Various automatic devices have been developed. Some of them are based on maintaining a minimum level of water in the substrate (i.e. using electrodes to activate irrigation) while other automatic devices use a balance to replace the evapotranspirated water. The simplest method is to preprogram the irrigation time or volume; conductivity of the drainage water is used to adjust the irrigation dosage and frequency.

Water Saving and Water Use Efficiency

Normally, the greenhouse grower is not especially interested in water saving. The crude/little knowledge about the irrigation requirements among growers induce them to over irrigate. Proper information on irrigation requirement, spreaded at the farm level, can help to overcome this lack of interest to reduce the water demand.

Different measures to save water and improve its use, at the farm level, include reducing the water requirement, increasing the water availability and raising the yields. Use of mulching (plastic sheet, sand, gravel, etc.) has been widely exploited to limit the evaporation of soil water and reduce ET. Subsurface drip irrigation can reach similar results. Various cultural practices affect the water demand. Use of transplants instead of direct seeding, multiple cropping, varying plant density, electing the cycles, pruning and trellising are effective, when properly managed, to save water or to increase the yield quantity and quality.

An adequate greenhouse environmental management can reduce the water demand, increase the crop yield and therefore, improve the water use efficiency. Manipulating ventilation, misting, shading and carbon dioxide (CO_2) injecting are effective techniques for that purpose. (Boulard *et al.*, 1991; Stanghellini, 1993; Castilla, 1994).

Tapping/Cropping the rainwater from the greenhouse roof is an easy way to increase the water resources, especially relevant for its excellent quality. Another way is to make a better use of water stored in the soil profile ; pulling up the crop immediately after harvesting is over (avoiding the soil water exhaustion with no agronomic interest) or managing rationally, complementary irrigation (as those used for soil disinfection) in the watering schedule. Slight deficit-irrigations have been recommended as they do not affect yield in tomato (Villele, 1984), though they can reduce total biomass.

When soil profile is well-wetted before planting, slight deficit irrigation does not influence yield in greenhouse tomatoes and melons (Castilla *et al*, 1990-C). Condensing water from the saturated air can also help to reduce the fungal disease incidence, but economics are not clear (Boulard *et al,* 1989).

Maximizing the uniformity of water application is one of the easier ways to save water, at the farm level, which is too frequently forgotten. Evaluation of emission uniformity of the drip system should be done periodically. In a study conducted by Orgaz *et al,* (1986) thirty drip systems were evaluated in the Almeria greenhouse industry; the results showed that (A) uniformity coefficient (UC) varied between 51 and 93%, average UC being 76%. And only 4% of the systems had excellent uniformity while UC was unacceptable in 20% of them and (B) applied water was considered excessive in 50% of the farms studied.

Protected cultivation improves the water productivity due to ET reduction and larger outputs of protected growing (Stanghellini, 1993). Drip irrigation also increases the water productivity, relative to conventional irrigation in greenhouses. 20 kg of yield per cubic meter of applied water quantified in open field tomato growing in the Mediterranean area, can be increased to 33 kg per m^3 in unheated plastic house (Castilla *et al*, 1990-C), far from the 65 kg per m^3 obtained in sophisticated greenhouses, with soilless culture and very long cycles, in the Netherlands (Stanghellini, 1994).

Use of recirculating soilless culture can improve the water productivity but poor quality of water, high cost of the equipment (for quantifying the ion concentration and disinfecting the recirculating solution) and the absence of proper information, at the farm level, limits its use. Table-2 summarizes the economic yield per cubic meter of applied water (water productivity) in commercial plastic houses in the

Table 2. Water productivity (economic yield in kg per m^3 of applied water) of various drip irrigated crops, evaluated in commercial unheated greenhouses in Almeria (Spain), in conventional soil and soilless cultures, in the autumn and spring cycle (up) and long season cycle (down)

	Autumn		Spring	
	Soil	Soilless	Soil	Soilless
Short Cycle Crops				
Squash	41.0	-	-	-
Green bean	15.0	18.5	20.5	22.0
Cucumber	27.5	43.5	-	-
Melon	-	-	22.5	19.5
Watermelon	-	-	30.5	
Long Cycle Crops				
Tomato	34.0	29.0		
Pepper	13.5	-		
Eggplant	18.0	-		

Almeria area (Spain). In some cases, the agronomic water use efficiency (water productivity) is lower in soilless culture than in soil-grown crops (Table 2), showing the need of applied research on this subject.

A study was conducted at PFDC, GBPUA&T, Pantnagar, India on tensiometer based drip irrigation and fertigation scheduling on capsicum under polyhouse. In Double Span Naturally Ventilated Polyhouse (DS NVPH), water use efficiency was observed maximum in the treatment irrigated at 50-70kPa soil moisture tension and 75% of normal fertilizer doses (I_3F_1) by 281.31 % higher as compared to the control treatment. Water use efficiency was increased higher in the treatment by 248.05 % as compared to the control treatment in WT NVPH. The water productivity of treatment irrigated at 50-70 kPa soil moisture tension and normal fertilizer doses (I_3F_2) was observed to decrease by -72.43% under DS NVPH and -71.45 % under WT NVPH in treatment irrigated at 50-70kPa soil moisture tension and 75% of normal fertilizer doses (I_3F_1) as compared to control treatment (Table3).

Table 3: Effect of tensiometer based irrigation and fertigation on capsicum yield (kg/m^2), water use efficiency and water productivity in DS NVPH and WT NVPH

Treatments	DS NVPH			WT NVPH		
	Yield (kg/m^2)	WUE (kg/m^3)	Water productivity (liter/kg)	Yield (kg/m^2)	WUE (kg/m^3)	Water productivity (liter/kg)
I_1F_1	12.00	48.18		20.76	12.35	40.73
24.60						
I_1F_2	12.34	49.55	20.33	13.81	45.53	22.01
I_2F_1	10.92	62.74	17.07	11.67	81.37	12.94
I_2F_2	12.28	70.55	14.40	13.49	70.66	14.20
I_3F_1	13.14	100.59	10.15	13.04	90.91	11.00
I_3F_2	12.54	96.00	10.47	11.94	83.19	12.08
C	8.52	26.38	37.98	9.56	26.12	38.53
CD (P=0.05)	NS	20.47	4.56	NS	17.51	4.12

I_1: Drip irrigation at 20-30 kPa soil moisture tension; I_2: Drip irrigation at 30-50 kPa soil moisture tension; I_3: Drip irrigation at 50-70 kPa soil moisture tension; F_1: 75% of recommended fertilizer doses; F_2: 100% of recommended fertilizer doses

Automated Watering

Watering a large number of plants in a well-stocked greenhouse can be a major task, especially if it is to be done by hand. However, there are alternative approaches available, including seep hoses, drip irrigation, misters and sprinkler systems, which make the job much easier. Many of these systems can be automated to varying degrees, using modern timers or controllers, which range from cheap and simple to very sophisticated – with a price tag to match! Water timers, for instance, fit between a irrigation hose and a tap and can be set to allow watering to take place at specific times of the day and for a given period, allowing you to water at night to avoid evaporation, for example, without having to be there yourself. These relatively simple devices are also invaluable if you are out all day or when you go away, keeping the soil nicely moist in your absence. More complex water controllers can take automation even further, measuring soil moisture and watering accordingly to keep the growing conditions at their optimum.

Water Quality Management for Greenhouse Irrigation

A dependable irrigation water supply is a vital component of any greenhouse growing operation. In the past, quality of water source was not a cultural issue considered by growers. Water source may contain essential nutrients such as iron or nitrate in high enough concentrations to justify a reduction in levels applied through a fertility program. Water may also contain harmful impurities which require corrective procedures. Water quality can be a deciding factor when choosing among sites for establishing a new greenhouse business or, where the opportunity exists, to choose among two or more water sources at a particular site. Growers should have their irrigation water tested by a university or reliable private laboratory any time a new water source is established, whether it be from a well, river, pond or municipal system. Afterwards, test the water at least twice a year or often enough to establish how much variability there is in water quality over time. One good approach is to take one test during a wet period and another during a dry period because high rainfall can dilute water impurities and drought can concentrate water impurities. Once a water quality pattern has been established, yearly testing is usually sufficient. The recommended water quality parameters have been presented in Table 4 suitable for protected cultivation.

Other Aspects of Drip Irrigation Management

Well-designed drip irrigation systems must apply water uniformly, in such a way that different emitters discharge almost at the same rate. A good management must maintain a good emission uniformity that is easy to quantify.

Physical, chemical and biological agents can be responsible of emission uniformity decay due to clogging. Physical clogging can be prevented with good filtration of water and regular flushing of the lines and emitters.

Chemical clogging normally originates when soluble salts precipitate on emitters as water evaporates from emitter surfaces between irrigation runs. Injected chemicals (fertilizers, pesticides) are also responsible for frequent clogging. Most cases of chemical clogging can be solved by acid treatment or injection. Acid injection to reduce the pH of water between 1 and 2 should be adequate, while, in severe clogging cases, emitters must be soaked in dilute acid solution (approx. 1%) and even cleaned individually (Fereres, 1981). Acids are highly corrosive and extreme caution must be observed with their use. Surfaces in contact with acid solutions should be of stainless steel or plastic, and must be rinsed well after contact with acid. Biological clogging can be solved with injection of a biocide followed by flushing to clear the system of organic matter.

Table 4: Recommended upper limits of chemical factors in Irrigation water for greenhouse crop production

Quality parameters	Recommended limit of factors	
pH		5.4 to 6.8
Alkalinity		150 ppm $CaCo_3$ (3 meq/L)
	Bicarbonates	122 ppm (2 meq/L)
	Hardness (Ca+Mg)	150ppm $CaCo_3$ (3meq/L)
Electrical Conductivity		
	Plug grown seedlings	0.75 mmhos/cm
	General production	1.5 mmhos/cm
Total Dissolved salts		
	Plug grown seedlings	480ppm
	General production	960 ppm
Sodium Absorption ratio		4 (no unit)
Sodium (Na)		69 ppm (3meq/L)
Chloride (Cl^-)		71 ppm (3meq/L)
Nitrogen (N)		10 ppm (0.72 meq/L)
	Nitrate (NO_3^-)	10 ppm (0.16 meq/L)
	Ammonium (NH_4^+)	10 ppm (0.56 meq/L)
Phosphorus(P)		1 ppm (0.3 meq/L)
	Phosphate ($H_2PO_4^-$)	1 ppm (0.01 meq/L)
Potassium (K)		10 ppm (0.26 meq/L)
Calcium (Ca)		120 ppm (6 meq/L)
Magnesium (Mg)		24 ppm (2 meq/L)
Sulfur (S)		20-30ppm(0.63.94meq/L)
	Sulfate (SO_4^-)	30-45ppm(0.63-0.94meq/L)
Iron (Fe)		0.2-4.0 ppm
Manganese (Mn)		1.0 ppm
Boron (B)		0.5 ppm
Copper (Cu)		0.2 ppm
Zinc (Zn)		0.3 ppm
Fluoride (F)		1.0 ppm
Aluminium (Al)		5.0 ppm

(Castilla, 2000)

Conclusion

A well designed micro irrigation system must be properly managed to attain a high level of application uniformity to obtain fairly good water use efficiency. Use of Class A evaporation pan radiation method and tensiometers may be helpful for the estimation of water requirement and irrigation scheduling. Various other techniques such as use of mulchs can be used to further enhance water and nutrient use efficiency along with improving the quality of produce and precise resource savings. Automation further reduces the manual intervention and increases the input use efficiency.

References

Ayers, R.S. and Westocot, D.W. 1976. Calidad del agua para la agricultura. Estudio FAO: Riego y Drenaje n°29. FAO. Roma. 85 pp.

Boulard, T., Baille, A., Lagier, J. and Mermierm, M. 1989. Watervapor transfer and dehumidification in a inflatable plastic greenhouse. *Acta Hort.* **245**: 462-469.

Boulard, T., Baille, A. and Gall, F.L., 1991. Etude de différentes méthodes de refroidissernent sur le climat et la transpiration de tomates de serre. *Agronomie*, **11**: 543-553.

Castilla, N., 1986. Contribucion al estudio de los cultivos enarenados en Almería: Necesidades hídricas y extracción de nutrientes del cultivo de tomate de crecimiento indetermineando en abrigo de polietileno. Tesis Doctoral. Caja Rural Provincial. Almería. 196 pp.

Castilla, N., 1989. Programation de l'irrigation gouteà-goutte en serre plastique non chauffée. *Plasticulture*, **82**:59-63.

Castilla, N., 1990. Technology transfer on the use of localized irrigation. Proc. XI Int. Congr. "The use of plastics in Agriculture". New Delhi. India. **l-B:** 33-39.

Castilla, N. and Fereres, E. 1990-B. The climate and water requirements of tomatoes in unheated plastic greenhouses. *Agricoltura Mediterranea* **120**: 268-274.

Castilla, N. and Fereres, E., 1990-C. Tomato growth and yield in unheated plastic greenhouse under Mediterranean climate. *Agric. Medit.* **120**(1): 31-40.

Castilla, N., 1994. Greenhouses in the Mediterranean area: Technological level and strategic management. *Acta Hort.* **361**: 44-56.

Castilla, N., 2000. Greenhouse drip irrigation management and water saving. *Cashiers Options Mediterraneennes* , **31**: 189-202.

Doorembos, J. and Pruitt,W.O. 1976. Las necesidades de agua de los cultivos. Estudio FAO: Riego y Drenaje n" 24. FAO. Roma. 194pp.

Fereres, E. 1981. Drip irrigation management. Univ. of California. Berkeley (Calif). 39pp

Hanan, J.J. 1990. The influence of greenhouse on internal climate with special reference to Mediterranean regions. *Acta Horti.* **287**:23-34.

Martinez , A. and Castilla, N.1990. Evapotranspiración del pimiento en invernadero en Almería. ITEA, **85**: 57-62.

Orgaz, F., Bonachela, S. Cuevas, R., Rios, E., Fereres, E., Escalada, S. NMontero, J.I., Lopez, J. and Castilla, N.G 1986. Evaluación de sistemas de riego localizado en cultivos bajo invemd la provincia de Almería. ActaIsI Congreso Nacional de S.E.C.H. C6rdoba, Españaj. **1:** 509-517.

Rana, S. and Sah, R. 2010. Study of the effect of protected environment on pan evaporation and water requirement of different vegetables. U. G. Dissertation, GBPUA&T., Pantnagar

Sirjacobs, M. 1986. Protected cultivation of sweet pepper in arid zone: Evaluation of water requirements and amounts per watering. *Acta Horti,* **191**:199-207.

Stanghellini, C., 1993. Evapotranspiration in greenhouses with special reference to Mediterranean conditions. *Acta Hort.* **335**: 295-304.

Stanghellini, C. 1994. Balance hídrico y manejo de microclima en invernadero. IT[." Tecnología de invernaderos". Alvarez and Parra (Ed.) Junta de Andalucía. *Almería (Spain)*:**49**-62.

Stegman,E., Musick J. and Stwart, J. 1980. Irrigation water management. In: "Design and operation of farm irrigation system". Jensen, M.E. (Ed), ASAE Monograph no 3. Michigan(USA): 763-816.

Vermeiren, I., Jobling., G.A. 1980. Localized irrigation: design, installation, operation, evaluation FAO Irrigation and Drainage paper no" 36F. AO. Roma. 203 pp.
Veschambre, D., P.vaYsse. 1980. Mémento goutte a goutte: Guide pratique de la micro-irrigation par goîìteur et diffuseur. C.T.I.F.L. Paris 2. 04 pp.
Villele, O., 1984. Les besoins en eau des cultures sous serre. Com. Troisième Journee Nationales des Techniques Horticoles dePoints. Hortiforum-84. Orleans. 15 pp.

6

Indigenous Approach Towards Popularization of Greenhouses in Himalayan Regions

Sanjai K. Dwivedi and Deepa H. Dwivedi

Abstract

About 20 percent of the world's total population is residing in the mountains and this proportion is likely to increase in the present century as environmental degradation and pollution are driving the human population for settlement in the clean and serene environment of the mountains. Himalayan region has sizeable area in countries like India, China/Tibet, Nepal, Pakistan, Afghanistan, Bhutan, etc. Remarkable progress has been made in greenhouse research and development in many countries. However, Himalayan regions which reserve unique flora and fauna have attracted little attention so far. Changing climate, fast increasing population and shrinking land and water resources are the major challenges before us to meet the food and nutritional security of the population at global level. Improvement in the production and productivity of crops are the immediate need to boost the availability of food and improve the economic conditions of the farmers.

Greenhouse technology has tremendous potential to improve production and productivity of crops even under uncongenial soil and climatic conditions. The farmers having very small farm holdings in countries like Japan and South Korea have improved their income and living standards by adopting protected cultivation technologies. Similarly, Israel has demonstrated successful crop production under desert and uncongenial climatic conditions by improved water management practices and using greenhouse technology. However, India is yet to trap the potentials of the technology. Modern greenhouses require uninterrupted huge electric power for control of environment (heating/cooling/air circulation etc), which is considered one of the major causes for global warming. High installation and running (energy) cost, high cost of production; poor government support and lack of farmer friendly /cost effective greenhouse technology are some other challenges. Appropriate modifications suiting to our requirements, as per the need of crop, climate and farmers may lead to popularization of greenhouses in the country, especially among the small and marginal farmers. Defence Research & Development Organisation (DRDO) successfully demonstrated the innovative approach to popularize low cost zero energy based solar greenhouse technology in higher Himalayan regions of the country. Such greenhouses are specifically designed using local resources and also trap natural solar radiation/ heat and maintain it for longer period. Similarly, other approaches like using mud/ unbaked bricks, double wall approach, making hot beds, blackening of walls, using water bottles, greenhouse-cum-animal shed, etc. effective to reduce the adverse effects of the environment on crops with little inputs in hilly regions. Need of the hour is to review our approach for the greenhouse technology in farmers' prospective and device simple and cost effective innovations for fast propagation of greenhouses in the country. This will not only improve the economic status of the farmers but also provide nutritive food to their families and enhance production and productivity of various crops in the country without deteriorating the environment.

Keywords: Indigenous, greenhouse, Himalayan regions, vegeatables

1. Introduction

Greenhouse cultivation technology tempts one's mind as it permits enormous modification in microclimate, enabling cultivation of crops in adverse climatic conditions, caring the least for the outside environment. Vrikshayurveda, an epic of 11th century AD by Surapala, states that any plant/tree could be grown anywhere provided king, treasury and destiny are favourable. This is an indication that agricultural experts of that era were aware of protected cultivation methods. The idea of growing plants in environmentally controlled areas also existed during

Roman times. The Roman emperor Tiberius ate a cucumber-like vegetable daily. The Roman gardeners used artificial methods (similar to the greenhouse system) of growing to have it available for his table every day of the year. Cucumbers were planted in wheeled carts which were put in the sun daily, and then taken inside to keep them warm at night. The first modern greenhouses were built in Italy in the thirteenth century to house the exotic plants which explorers brought back from the tropics. They were originally called *Giardini botanici* (botanical gardens). The concept of greenhouses soon spread to the Netherlands and then England. Jules Charles, a French botanist, is often credited with building the first practical modern greenhouse in Leiden, Holland to grow medicinal tropical plants. In the nineteenth century, large greenhouses were built. The conservatory at Kew Gardens in England is a prime example of the Victorian greenhouse. In Japan, the first greenhouse was built in 1880 by Samuel Cocking, a British merchant who exported herbs. In the twentieth century, the geodesic dome was added to many types of greenhouses (Anon, 2008). In a country like India, with its vast and diverse agro-climatic conditions, protected cultivation technology in the last three decades has made good progress and proved as a boon for production of high value, low-volume crops with better productivity and quality.

2. Greenhouse Technology

Greenhouse technology implies production of plants for economic use in covered structure that allows rapid harvesting of solar radiation and modification of agro-climatic conditions conducive for plant growth and development. The technology embraces infrastructure modelling, selection of plants for adaptation, production economics, agronomic management and commercial potential, etc.

"A greenhouse" is a framed or an inflated structure covered with a transparent or translucent material which permits at least partial control of plant environment and which are large enough to permit a person to carry out cultural operations" (Chandra and Panwar, 1987). While full advantage is taken of the available sunshine for crop production by way of selecting proper covering materials, the enclosure as well provides an opportunity to control the other environmental parameters. As a result, greenhouse crop productivity is largely independent of outdoor environmental conditions. However, environmentally controlled greenhouses are not popular in hilly regions simply because of acute shortage of electric power in the region. Hence, greenhouse technology is discussed with special emphasis on crop production in hilly regions. Types and design of protected structures mainly depend on the availability of materials, purpose of protection, climate of the region and market for quality produce. Environmental conditions in general include light, temperature, air composition and nature of root medium. Depending upon the local climate and crop requirements, adequate environmental control is attempted rather than full control of all components to maximize the profit.

Principle

A greenhouse is generally covered with transparent material such as polyethylene, glass or polycarbonate. Depending upon the clading material and its transparency, major fractions of sunlight are absorbed by the crops and other objects. These objects in the greenhouse in turn emit long wave thermal radiation for which cladding material has lower transparency with the result that solar energy is trapped and resist the temperature inside the greenhouse (Nimje and Shyam, 1991). This is known as greenhouse effect. This rise in temperature in greenhouse is responsible for crop forcing in adverse climate. During winter, loss of trapped temperature should be minimised whereas during summer, air temperature in greenhouse has to be brought down by providing cooling device or ventilation.

Uses

Greenhouses are often used for growing flowers, vegetables, fruits and other plants. Many vegetables and flowers are grown in greenhouses in late winter and early spring, and then transplanted outside as the weather warms. Started plants are usually available for gardeners in farmers' markets at transplanting time. The closed environment of a greenhouse has its own unique requirements as compared to outdoor production. Pests and diseases and extremes of heat and humidity have to be controlled and irrigation is necessary to provide water. Significant inputs of heat and light may be required, particularly with winter production of warm-weather crops. Special greenhouse varieties of certain crops, like tomatoes, are generally used for commercial production. Greenhouses are increasingly important in the food supply of high latitude countries. One of the largest greenhouse complexes in the world is in Almeria, Spain where greenhouses cover almost 50,000 acres (200 km^2) and where almost 5% of Spain's salad vegetables are grown. Greenhouses protect crops from too much heat or cold, shield plants from dust storms and blizzards, and help to keep out pests. Light and temperature control allow greenhouses to turn inarable land into arable land. Greenhouses can feed starving nations where crops can't survive in the harsh deserts and Arctic wastes. Hydroponics can be used in greenhouses as well to make the most use of the interior space. Biologist John Todd invented a greenhouse that turns sewage into water, through the natural processes of bacteria, plants, and animals (Anon, 2008).

Agro-climate

There are several climatic factors which need care in the management of a greenhouse. The following factors are important for greenhouse cultivation:

a) Sun Light

Direction of the greenhouse should be such that it receives sunlight for maximum hours. In hilly region, greenhouse should face south-east direction and its location should not be under shade. However, in tropical conditions, temperature retention can be reduced by choosing a less sunny site. Sunlight is essential for augmenting the process of photosynthesis and minimum sun requirement of crop should be kept in mind. In addition, sunlight also provides requisite thermal environment for various other physiological processes including photosynthesis. Hilly regions receive clear sunny days in winter so sunlight as such may not be a limiting factor but its trapping and conservation could be a major concern for further investigation.

b) Temperature

Sunlight is the major rather sole source of temperature. Examples of providing external heat source other than sunlight are rare in Hilly region. Temperature during night hours may drop down to below zero degree Celsius. Heat obtained from trapped solar radiation is not properly conserved. Following steps are advantageous to maintain thermal environment of greenhouse:

i. Blackening of internal surface of brick wall and soil bed surface.

ii. Providing insulation by doubling brick wall.

iii. Covering polyethylene or glass external surface with multi layer covering during night hours.

iv. Mud plastering in case of stone walls.

v. Placing of black coloured stones in the greenhouses.

vi. Keeping containers filled with water.

c) Relative Humidity

The minimum limit of relative humidity for normal physiological functioning of the plants is 50%. Extreme xeric conditions hardly allow relative humidity to rise to this level in outer environment but inside greenhouse, sufficient water application and restricted air circulation gear up and maintain relative humidity at higher level. It may be increased by keeping pan evaporation, a mechanical device which automatically controls relative humidity by periodical sprinkling of water. Cooling pads are usually installed in low humidity and tropical areas.

d) CO_2 Enrichment or Maintenance

Carbon dioxide (CO_2) is the basic ingredient required for manufacturing of organic matter by process of photosynthesis. In cold arid regions, CO_2 concentration is below normal (NAEDB, 1992) and in closed structures like greenhouses, it may drop down further because of its consumption by the plants. At relatively high sunlight intensities and temperature, photosynthetic rate is directly proportional to CO_2 concentration upto 2400 ppm. Hence, supplementary application of CO_2 will boost plant productivity inside the greenhouse. Under hilly region situations, appropriate air circulation is better for conservation of CO_2 produced at night, a tight closed greenhouse is a pre-requisite.

e) Soil

Soil serves as growth medium for plants. Sandy loam soil is the best choice. It should be supplemented with well rotten FYM, compost or biofertilizers. Inorganic fertilizers need to be used as per recommendations. Continuous cultivation inside the greenhouse may give rise to pathogens. To destroy pathogens, pasteurization of soil is recommended (steam aeration) before every crop and soil replacement is recommenced after every 3 years. For steam aeration temperature of 60°C for 30 minute is good enough to accomplish the task. Chemical fumigation can also be used but steam is the best as it is not selective thus, attacking all sort of pathogens.

Benefits of Greenhouses

Major benefits of greenhouse are given below:

i. Cultivation of crops during off season when it is not possible to grow in open fields. Table 1 gives yield of various leaf vegetables grown under polyhouse in Ladhak in peak winter season.

ii. Tropical vegetables like cucurbits, capsicum, brinjal, okra, etc. are rare in hilly region but these crops can be grown in greenhouses.

iii. Greenhouse provides an excellent opportunity to produce quality crops for export.

iv. Raising nursery earlier and advancing the availability of crops is also possible by use of greenhouse. Table 2 shows earliness achieved in some crops under polyhouse conditions

v. Productivity in greenhouse is increased manifold in comparison to open field.

vi. Conservation of valuable planting materials and its cultivation are also possible under greenhouse.

vii. Greenhouse can also be used for growing flower plants, strawberries, grapes and propagation of quality fruit plants.

viii. Insect pest and weed management are easier in greenhouse than open.

ix. Greenhouse conserves the moisture. Hence, frequency of watering plants gets reduced.

x. Greenhouse is ideally suited for Indian farmers having small holdings.

xi. Productivity per unit area and time can be increased by adopting suitable crop sequences.

Table 1. Performance of Various Leaf Vegetables Grown Under Polyhouse in Ladakh during Peak Winter (Nov.-March).

Crop/Variety	Av. Plant height (cm)	Yield /m^2 (kg)
Beet leaf (All Green)	21.36	1.25
Beet leaf (Mongol)	34.44	2.50
Coriander (Local)	39.8	2.25
Fenugreek	39.92	2.50
Lettuce (Great Lake)	23.84	4.50
Veg Mustard (ARU Black)	46.56	4.50

Table 2. Effect of greenhouse in achieving earliness in some crops

Crop	Days to first picking		Harvesting period (days)	
	FRP	Open	Polycarbonate	Open
Tomato	70	105	100	50
Capsicum	90	127	90	45
Brinjal	92	128	-	-

3. Infrastructure of Greenhouse

Types of Glazing Materials

Glass

Glass is the preferred covering material for greenhouse world wide because of its light transmissivity characteristics. Transmissivity of 40-50 years old glass differs a little from that of new glass. Temperature retension in glasshouse is pretty good. However, high installation cost is a major limitation. Moreover, transportation of materials and frequent damage of glasses due to high wind velocity, glasshouse is discouraged.

Polyethylene

Plastic polyethylene is the most widely used greenhouse film around the globe. It is produced by mixing homopolymers of ethylene with or without an ultra-violet (UV) inhibitor package but only UV stabilized polyethylene sheet is recommended because non UV inhibitor package added polyethylene will break down after 3-5 months due to photochemical reactions. Polyethylene film is tough, flexible and relatively inexpensive. It can withstand as low as -50°C but 80-90°C temperature will cause it to melt. Above 60°C, it loses much of its strength and may stretch markedly. The life of UV stablized polyethylene film is 1-3 years. However, regidex polyethylene sheet may be used for 5-6 years at a stretch.

Fibreglass Reinforced Panel (FRP)

These panels consist of glassfibre reinforced polyester. The panels have been very popular in areas of high light intensity such as Southern USA but about 10-15% heat loss due to increased exposed area makes it unfit for most of Indian region. Further, with its age under use, the transparency gets reduced due to yellowing.

Polycarbonate Panels

The panels are available in double skinned sheets. SDP polycarbonate sheets are similar in physical dimension to the SDP acrylite but are stronger and have a lower light transmissivity. Its price is approximately 25% higher than SDP acrylite.

Construction Materials

Besides glazing materials (Polyethylene/glass) several other materials are also required for construction of a greenhouse. It is always better to prefer use of locally available materials for the construction of a greenhouse. It not only reduces the installation cost but also helps to popularize the technology. The major construction material required for greenhouse is frame structure which could be either of a metal or wood. Iron angle, GI pipes or aluminum frame are the most common metals used in greenhouses. Wooden frames can also be used which are locally available in the region. Such frames are cheaper than those of metal frames and also easy to use. Wooden frame should also be preferred in hilly regions since metal is a good conductor of heat thus, its temperature gets reduced during winter months and during summer, it gets hot and polyethylene sheet gets stuck to metal which tears off the sheet. Sometimes supporting wares are also used in greenhouse to hold the polyethylene sheet. It is always advisable to avoid direct contact of metal with polyethylene sheet. If at all it is essential, gunny bags or cloth should be wrapped over the metal. Other precautions include use of smooth surfaced frame and removal of sharp edges of frame. Strong foundation of frame is essential in cold arid region to resist the strong winds. Stability is further ensured by constructing earthen brick wall at the back and on either side.

Types of Greenhouses

Various types of greenhouses are seen in India. The brief description of each type is being discussed below:

Glasshouse

As its name indicates, glass is used as glazing material in this greenhouse. Perhaps this is the first and oldest among all types of greenhouse structures. Glass panels are fitted with the help of wooden or metal frame. It can be of any shape and size and it is pretty effective for winter cultivation but due to increase in day temperature in summer, it becomes unfit for cultivation during summer. High initial cost, difficulty in construction and frequent damage of glass panels by strong winds are other discouraging factors. In Hilly regions, only a few such structures are found that too for research. One oldest glasshouse constructed in 1964 is available with FRL-Leh.

Polyhouse

Ideal features of polyethylene have increased the use of polyhouses in place of glasshouse throughout the globe. It has not only reduced the initial cost but also increased the popularity of greenhouse by simplifying the installation technology. Generally, two type of polyhouses are there in hilly region:

Pre-fabricated polyhouses

A pre-fabricated metal structure of convenient size is installed and polyethylene film is covered over it. This structure is not recommended at all for the region because of poor temperature retention, low crop yield and high installation cost (Singh *et al,* 1998).

Ladakhi polyhouse

This is one of the innovative and low cost greenhouses of Ladakh region. It is similar to normal greenhouse but the only difference is its surrounding mud brick wall in place of polyethylene sheets which not only cut down the installation cost but also reduce the adverse effects of strong winds and also increase temperature retention and ultimately net profit get increased. This is generally lean type, have three sides made of mud bricks. The back wall is 7 ft in height, while the front has no well. The length is 32 ft with a width of 16 ft (Singh *et al,* 2000). The polyethylene is supported on wooden poles and side walls. The two side walls descend towards the front. J&K State Deptt. of Agriculture/ Horticulture provides cash subsidy besides 32 x 16 ft free polyethylene sheet. This has boosted greenhouse revolution in Ladakh.

Trench (Underground greenhouse)

This is a unique, innovative, very simple, cheap and useful underground greenhouse structure for hilly region and thus, has unlimited potential in the region. This may be of any convenient dimension. However, a trench of 30 × 10 × 3 ft size is ideal. In this pit type of structure, wooden poles are used to hold UV stabilized polyethylene film. The polyethylene is also covered by an additional woolen or cotton sheet during night to reduce the heat loss during extreme winter. The damage of blowing off of polyethylene film by strong winds is minimised by putting stones along the sides. Cultural practices and other operations are done by removing polyethylene sheet from top of the trench. The structure does not require much skill in its construction and management. Its cost is the lowest among all other greenhouses and being an underground structure heat loss is minimal and temperature retention is high (Singh and Dhaulakhandi, 1998) and thus, yields good crop. Strong winds do not affect polyethylene cover much and hence, it is long lasting. This structure is therefore, being recommended as most suitable greenhouse for the region (Singh *et al.,* 1998 &2000).

Plastic Low Tunnel

It is a small semi spherical structure frame made of metal, wood or plastic and covered with polyethylene or fiber reinforced plastics to create protected environment. Plastic low tunnels are flexible transparent coverings that are installed over rows of individual beds of transplanted vegetables to enhance plant growth by warming the air around plants (micro-climate) in open field during winters. This helps in warming the soil and protects the crop against the hails and snow and advances the crop by 30-60 days than their normal season. This low-cost technology is very simple and highly profitable for off-season cultivation and catching the early market. Generally, galvanized iron arches are fixed manually 1.5-2.0 m apart to support the plastic tunnel. Width and height are kept about 45-60cm to cover the plant as per the crop requirement. Transparent non-perforated plastic of 30-50 micron thickness is sufficient. It partly reflects infrared radiation to keep the temperature of low tunnel higher than outside. Small vents can be made at the side of tunnel to facilitate proper aeration in tunnel. These structures serve as mini greenhouse. This is a temporary structure which can easily be shifted to any place. However, major problem in tunnels is faced in watering, weeding and harvesting which involves removal of cover and again putting them back.

Double Wall Polyench

A structure designed by FRL, Leh, to harness soil and solar heat for growing vegetables especially during winter months. Polyench refers to a polyhouse erected over a trench where back and sidewalls are made up of mud bricks. The sidewalls are provided with double walls filled with insulating material available locally i.e. sawdust, for better retention of heat inside the polyench. The inner walls are painted black to absorb more solar radiation during morning hours. The structure is also like gable uneven span with modified roof having polythene sheet towards sun facing side and grass thatch support over wooden frame opposite to sun facing side with provision of ventilators in the roof. Since locally available material has been used in this greenhouse, the cost is reduced drastically compared to normal greenhouse with better efficiency of temperature retention than trench and polyhouse. Considering the wear and tear due to high wind velocity in the Hilly regions, the polythene sheet has been replaced with FRL sheets to increase life span of the structure. The structure has been accepted by the Ladakh Autonomous Hill Development Council (LAHDC) and State Govt. is providing subsidy of 50 per cent to the farmers for promoting the protected cultivation in Ladakh region.

Multipurpose Nets

Shade nets are used to reduce adverse effect of scorching sun and heavy rain. Shade houses are becoming popular for growing crops and nursery during summer. Net houses are used for raising vegetables/fruits/flowers/medicinal plants in high rainfall regions. Roof of the structure is covered with suitable cladding material mostly HDPE, which does not absorb moisture. Slides are made of wire mesh of different gauges, 25-90 percent shade depending upon requirements. Such structures are popular in North-Eastern region of the country.

Similarly, weed nets are made of HDPE and covered on ground to control the weeds by reducing light on the soil. Insect proof nets are effective to reduce the incidence of a number of pests and viral diseases in crops. These nets are used like mosquito net around the crops, having 40-50 mesh size. Other similar nets include bird protection nets to control the bird damage of plats/crops, hail protection nets and reflector nets, etc.

Life span and cost of construction of different types of greenhouse are given in Table 3.

Table 3. Cost of Different Greenhouses in Ladakh Per m^2 Per Year (Rs)

Structure	Total Cost	Life span (Yr)	Area (m^2)	Cost(m^2/Yr)
Glasshouse	115200	10	40	288.00
Ladakhi Polyhouse	5240	3	50	34.94
Polyhouse	52000	2	70	125.00
Trench	1960	6	24	13.59
TunnelUnit Polyhouse	62405300	22	14070	22.2937.86

Features of an Ideal Greenhouse

Peculiarity of region's weather i.e. very high day-night temperature fluctuations, strong winds and high light intensity coupled with UV radiation etc must be kept in mind and design should be done as per needs of the region. Since hilly regions are remotely located and isolated from main land hence, all the required materials for greenhouse must be locally available and cheap in cost so that poor farmers can afford the expenditure. The greenhouse must be capable enough to trap the solar energy and produce crops even at subzero temperatures (Tewari and Dhiman, 1986). It should also be strong enough to resist the snow load, strong cold waves and extremely low temperature. The structure in hills should also be able to retain the temperature for longer period, besides being simple and easily manageable so that even a layman can built and manage it without any difficulty. All the features may not be accommodated in one type of greenhouse (Singh *et al,* 1998).

Selection of Suitable Crops and Production Technology for Greenhouse

Selection of crops for production in greenhouse is the most important factor in success. The crop should not only be able to tolerate low winter temperature but also grow under the cover where very little and diffused sun light is available. Fast vertical growing and ratoon yielding crops should be preferred so that horizontal and vertical space are fully utilized, more number of cutting/harvesting can be taken. Root crops like radish, beetroot, carrot and turnip can also be grown (Singh *et al,* 1997). A list of vegetables suitable for winter, their recommended varieties, sowing time and planting distance is given in Table 4 (Singh, 1995 b). Among all, most important factor is sowing time. Even one day delay in sowing causes significant reduction in yield. Thus, crop should be sown before onset of winter which enables the plants to resist low temperature.

There are certain other managerial tips required to be adopted for greenhouse cultivation which are listed below :-

i. Use of well rotten FYM @ 1.5-2.0 kg per m^2 atleast 30 days before sowing, improves the growth and yield of crop. Use of chemical fertilizer is generally avoided. However, 10-12 g of DAP per m^2 before sowing and 5-7 g per m^2 urea, enhance the growth of plants.

ii. Since every bit of soil under cover is highly precious, hence it should be utilized properly. Even ridges of beds and channels should be used for growing crops.

iii. Water soaking of seeds for 36-48 hours before sowing results in faster and higher germination percentage. Soon after sowing, light irrigation is beneficial.

iv. Frequent light irrigation has been recommended for winter cultivation. It not only enhances the crop growth but also improves low temperature tolerance of the plants.

v. Since greenhouse temperature falls down below zero degree Celcius during peak winter, cold protection measures are essentially required for crops growing under the cover. For this, one additional covering of polyethylene just two feet above the plants (inside greenhouse) has given encouraging results. This improves microclimate around the plants and hence, crop growth gets increased. This covering is required especially during nights in the months of December and January. Depending upon environmental temperature and sunlight, the polyethylene film may be removed during day time.

List of vegetables, their varieties, sowing time and planting distance for winter cultivation in Ladakh is given in Table 4.

Table 4. List of Vegetables, their Varieties, Sowing time and Planting Distance for Winter Cultication in Ladakh

Crops	Variety	Sowing Time	Planting distance (cm)
Beetleaf	All green Mongol	First fortnight of Sep.	20
Beetroot (leaf)	Detroit Dark Red	-do-	20
Celery	—	-do-	25
Chenopodium (Bathua)	—	First fortnight of Sep.	10
Coriander	RIL-195	End of Sep.	15
Dums	Local	End of August	20
Fenugreek	Kasuri	Second fortnight of Sep.	15
Karam sag	Khanyari	Second fortnight of August	20
Lettuce	Great Lake	End of Sep.	25
Mint	—	-do-	20
Parsely	Curld leaf	-do-	25
Radish (leaf)	Pusa Himani/Local	-do-	20
Vegetable Mustard	ARU Black	End of Sep.	20

Future Thrust

Greenhouse technology development has made good progress in India during last two decades and steps necessary for promoting greenhouse cultivation of vegetable crops have been started. Energy efficient greenhouse cultivation continues to be an area of active research and development and this is sought to be achieved through precision equipment and protocols. While the gap between demand and supply of most horticultural crops remains wide and country plans to double the production of horticultural crops by 2015, protected cultivation technology holds the key to meet the targets. It clearly emerges that greenhouse technology has multipurpose application for sustainable development of hilly zone. Harsh winters which otherwise threaten survival of human being in these remote hilly areas, popularization and further improvements in greenhouse management could provide a sign of relief in solitude of white snow cover. Thrust areas for future research are enumerated below:

- Conservation of more thermal radiation in winter.
- Durable infrastructure modelling

- Development of successful commercial models
- Modernization of poly/greenhouse to permit agro-climatic manipulations of higher magnitude
- Identification of new crops / varieties suitable for cultivation inside the greenhouse
- Socio-economic impacts of poly/greenhouse technology in Hilly region / cold arid zone
- Plastic films of suitable grades need to be made available in adequate quan tity
- Development of area specific suitable greenhouse designs
- Standardization of package of practices for growing various crops in green-house for specific regions
- Varieties suitable for greenhouse cultivation need to be made available to enhance the productivity
- Vocational training to the unemployed youth and greenhouse practitioners in remote areas to update their skills in this technology.
- Financial assistance as low interest loans should be made available for poor farmers to adopt this technology
- Insurance of greenhouse structure and crop to protect farmers against the loss caused by natural calamities
- Rigorous extension programme to disseminate technology information to the poorest of the society.

References

Anonymous 2008.www.wikipedia.org/green house.

Chandra, P. 1991. The features of solar greenhouse for hill regions of India. Workshop on renewable energy option for Ladakh. SESI, H.P. Chapter, Forestry, Solan, May 3-4, 1991

Chandra, P. 1996. Protected cultivation for horticulture development. Strategies for horticulture development in India, Dept of Agri. & Co-operation, GoI. 321-8.

Chandra, P. and Panwar, J.S. 1987. Greenhouse technology and its scope in India. Paper presented in National symposium on use of plastics in Agriculture. Feb. 6, 1987 New Delhi. Proceedings pp 62-66.

Dwivedi, S.K.; Sareen, S.K. and Paljor Eli 1999. Trench for winter vegetable production in Ladakh (In Hindi) *Phal-Phool.* **21**(4) : 23-24.

Dwivedi, S.K. and Singh, N. 1999. Roots of green revolution in Ladakh. *Indian Farming.* **49**(I) : 20-21.

Dwivedi S K and Eli Paljor 2010. Green House Technology for higher Himalays.In: Agro Animal Resources of Higher Himalayas. Satish Serial & Publishing House. Delhi

NAEDB, (1992). Action plan on cold deserts : An integrated approach for sustainable development. NAEDB, UHF, Solan - 173230 (H.P.)

Nimje, P.M. and Shyam, M. 1991. Greenhouse as an alternative technology for commercial vegetable crop production. *Indian J. of Agri. Sc.* **61**(3): 85 -89.s

Sharma, J.L. (1997). Present status and scope of greenhouse in Ladakh. Office of chief Horticulture Officer, Leh.

Singh, B. 1995 a. Vegetable Production in Ladakh. Field Research Laboratory, C/O 56 APO.

Singh, B. 1995 b. Greenhouse Development in Leh. Paper presented in National workshop on commercialization of tissue culture and greenhouse technology, Banglore, India 5-6 May 1995.

Singh, B and Dhaulakhandi, A.B. 1998. Application of Solar greenhouse for vegetable production in cold desert. Renewable Energy : Energy efficiency, Policy and the environment. Proceedings of World Renewable Energy Congress, V, 20-25 September 1998, Florence, Itali 2311-2314.

Singh, B.; Jadhav, K.L. and Paljor, Eli 1997. In high altitude cold desert of Ladakh. Protected vegetable farming potentials. *Indian Horticulture*, **42**(3):16-17.

Singh, B.; Dwivedi, S.K. and Paljor, Eli. 1998 a. Studies on suitability of various structures for winter vegetable production at subzero temperatures. Paper presented in 25th IHC - Belgium, 2-7 August 1998. Abstract, pp -290.

Singh, B., Dwivedi, S.K., Singh, N. and Paljor, Eli. 1998b. Sustainable horticultural practices for cold arid areas. Paper presented in Internation symposium on sustainable agriculture in hill areas. 29-31 Oct 1998. H.P.K.V. - Palampur, Absctact pp 35-36.

Singh B.; Dwivedi, S.K. and Sharma, J.P. 2000. Greenhouse technology and winter vegetable production in cold arid zone. In: Dynamics of cold and agriculture (Eds J P Sharma and AA Mir), pp. 279-93, Kalyani Publishers, New Delhi.

Sood, Deepak 1987. Greenhouse in India - an overview. Paper presented in National seminar in use of plastice in Agricultrue. Feb.-6, 1987, New Delhi. Proceedings pp 62-66.

Tewari, G.N. and Dhiman, N.K. 1986. Design and optimization of a winter greenhouse for Leh - Type Climate. *Energy covers. Mgmt.* **26**(1):71-78.

7

Protected Cultivation Technologies for Diversification and Livelihood Security for Indian Farmers

Balraj Singh

Abstract

Keeping in view the increasing demand of off-season and high value vegetables in several big cities of the country, there is an urgent need for diversification from the traditional agriculture by production of high value vegetable crops under different protected conditions for increasing their productivity and quality not only for getting high returns but in a Horti-Business / Veggie-business model. Nursery raising under protected cultivation is highly suitable for such Horti-business model or can be adopted as an agri entrepreneur business in major vegetable growing areas of the country by unemployed youths who are graduates in agriculture or post graduates in horticulture. Low cost protected technologies like plastic low tunnels or walk in tunnels, shade net houses can be used for off season vegetable cultivation for getting high returns. Similarly, insect proof net houses can be used on a large scale for safe vegetable cultivation by way of minimizing the use of pesticides in vegetable cultivation and virus free quality seed production in large number of vegetables in a horti-business model. Some of the technologies like low pressure drip irrigation and low cost nursery raising technology can also be replicated in other parts of the country even among the resource poor and tribal farmers.

Keywords: Protected cultivation, production technology, vegetables

Introduction

Several biotic and abiotic stresses are the major factors responsible for low productivity and poor quality in large number of vegetable crops under open field cultivation more specifically during rainy and post rainy season. The increasing demand of off-season and high quality vegetables in various markets of the big cities have called the attention of vegetable growers for diversification from traditional ways of vegetable cultivation to modern methods of vegetables cultivation in an agri-business entrepreneurial model. Under the new era of FDI (Foreign Direct Investment) in retail, these kinds of models are having high potential for betterment of the farmers opting for quality and off-season vegetable cultivation through protected cultivation (Singh *et. al*, 2012).

Protected cultivation structures are useful for combating both biotic and abiotic stresses which limits the productivity and quality of vegetable crops. This requires careful planning, attention and details about timing of production and moreover, harvest time to coincide with high market prices, choice of varieties adopted for off season environments and able to produce economical yields of high quality produce, etc.

Protected cultivation of vegetable crops offers the best choice for diversification from traditional production systems to this modern technology for a number of reasons. Production of crops under protected conditions have great potential in augmenting production and quality of vegetables, in main and also during off season for maximizing water and nutrient use efficiency under varied agro-climatic conditions of the country. This technology has very good potential, especially in rural and peri-urban areas adjoining the major cities of the country (Singh *et. al.*, 2012), since it can be profitably used for growing high value vegetable crops like tomato, cherry tomato, coloured peppers, parthenocarpic cucumber, healthy and virus free seedlings in agri- entrepreneurial models.

Large number of farmers, who are following traditional agriculture in rural and peri- urban areas of the country can successfully diversify their traditional agriculture by using various levels of protected cultivation technologies for production of high value vegetable crops looking to availability of resources, availability of emerging markets of usual and unusual off season vegetable crops, year round demand of high value vegetables like standard tomatoes, coloured peppers and parthenocarpic cucumbers, etc. High quality nursery raising in vegetables is the other area where improvement over the traditional system of nursery raising is necessarily required (Singh and Kalia, 2005). All kinds of protected technologies may not be economical and suitable to various groups of farmers in India because of their very high initial, running and maintenance cost, but some protected technologies are low cost, simple and highly profitable under Indian conditions and more specifically for rural and peri-urban areas, which can be adopted by Indian farmers for production of diverse high value vegetable crops and nursery raising in profitable agri-business models. Some of the technologies like low pressure drip irrigation and low cost nursery raising technology are highly suitable for livelihood security even in tribal areas of the country. These two technologies have already proved its suitability, efficiency and profitability in tribal areas of Jharkhand and Madhya Pradesh in the last two years (Singh and Hasan, 2011). Similarly, these technologies can also be replicated in other parts of the country among the resource poor and tribal farmers.

Modern Nursery raising Technology

To ensure high productivity and high quality in vegetables, raising of high quality seedlings through use of good quality seeds at right time and at an appropriate place is one of the cheapest but most important way. Most of the Indian farmers are raising their vegetable seedlings under open field conditions which are always inferior in quality, as the seedlings are infected with virus when raised in open during rainy and post rainy season. On one side, soil borne fungus and nematodes create severe problems for raising the seedlings in soil media in open fields during hot summers and rainy season but on the other hand the very high cost of hybrid seeds in vegetables has also warranted the farmers to improve or change their traditional nursery raising method to increase the productivity and quality of vegetables. Plug tray nursery raising in vegetable crops has already working as a well established industry in several countries. Under this system, seedlings are raised in plastic multi-celled plug-trays in artificial soil-less media in especially designed greenhouses or other protected structures. A large number of virus free healthy seedlings of different vegetables can be raised in a small area of green house in plastic plug-trays by using soil-less media for growing vegetables either for main season or for their off season cultivation. With the use of this technology, it is now almost possible to raise healthy vigorous seedlings of different cucurbits which otherwise was not possible in the traditional system of nursery production. The farmers or unemployed agriculture graduate youths of our country can very successfully start nursery production as an agri entrepreneur in major vegetable growing pockets of the country. By this way, vegetable growers will get the virus free or off-season healthy nursery as per their requirement and it will also generate additional employment in agriculture sector. Therefore, this is the first and most important step for enhancing productivity, quality and profitability in vegetables through protected cultivation (Singh *et. al.*, 2005).

Insect Proof Nethouses for Safe Vegetable Cultivation

Usually, the farmers grow their vegetable crops like tomato, chilli, sweet pepper, okra, etc. under open fields. But during rainy and post rainy season, it is very difficult to grow these crops successfully due to leaf curl and yellow vein mosaic and other viruses, respectively. These viruses are mainly spread by insect vectors like whiteflies, aphids, thrips, jassids and sometimes also by hoppers. The population of white fly after on start of monsoons is very high and it remains in the open environment up to end of October depending upon the temperature (Singh *et. al,* 2005). The farmers are using several sprays of several insecticides to control these vectors, even then they could not control these vectors and their tomatoes, chilli or okra crops are highly infected with viruses. The second most common and most severe problem in tomato, brinjal and okra is the fruit borer against which the growers are using huge amount of insecticides even then they are unable to control this insect. The only way to control the virus and fruit borer is to put a mechanical barrier between the crops and open environment and this is possible with the use of insect proof net of 40 or 50 meshes in form of net houses or insect proof net covered walk-in- tunnels (Singh and Tomar, 2007). By this way, the growers can directly reduce the use of insecticides and they can grow virus free crops of tomato, chilli, sweet pepper and okra during rainy or post rainy season. But for growing these crops under insect proof net houses, it is a pre-requisite to raise virus free healthy seedlings of these crops either in the greenhouse or by covering the nursery beds with insect proof net. The farmers can erect these insect proof net houses by using half inch size GI pipes after bending them in half circle shape. Other insect proof net houses can also be made by covering all sides and top with insect proof net of 40 or 50 meshes, but the net should be UV stabilized. Under these net houses crops like sweet pepper, tomato, chilli or okra can be grown successfully without infestation of viruses or other insects like fruit borer, etc. and the growers can save the huge amount spent on pesticides. Insect proof net houses can also be fabricated in greenhouse design for maximum utilization of the space through vertical growing of high value vegetables. These structures can be fabricated with a cost of Rs.350-400/m^2. Such structures can be covered with 40-50% shading nets during critical summer months (April – June) and with plastic during critical winter months (Dec- Feb) and with transparent plastic under arid and semi-arid climatic conditions. High value vegetables like tomato, cherry tomato (crop duration 7-8months), two crops of parthenocarpic cucumber (summer and post rainy season) and capsicum (crop duration 7-8 months) can be successfully grown in these net houses. The basic objective of Insect proof net house vegetable cultivation is to minimize the use of pesticides in fresh vegetable cultivation for producing safe vegetables (Singh, 2011).

Zero Energy Naturally Ventilated Greenhouses for High value Vegetable Cultivation

Naturally ventilated greenhouses are the protected structures where no heating or cooling devices are provided for climate control. These are simple and medium cost greenhouses which can be erected with a cost of Rs.650-700 sq. meter and these greenhouses can be used successfully and efficiently for growing year round parthenocarpic slicing cucumber, off season muskmelon, tomato and sweet pepper crops for 8-9 months duration. These structures are having a manually operated cross ventilation system as and when required. Looking to the year round, increasing demand of high quality parthenocarpic slicing cucumber in up markets of metros and other big cities of the country, this is one of the most suitable and profitable crop for cultivation under naturally ventilated greenhouses in peri-urban areas of the country. Three successful crops of cucumber can be grown in a naturally ventilated greenhouse in a period of one year. Muskmelon is the second crop, which can be successfully cultivated for its complete off-season availability, which can fetch very high price as off-season produce in the up markets of metros and other big cities of northern parts of the country (Singh and Kumar, 2006).

Similarly, high value vegetables like standard tomatoes, cherry tomatoes and colored peppers are three crops which can be grown for long duration (8-10 months period) under naturally ventilated greenhouse conditions. In metro cities like Delhi, there is year round demand of these high quality vegetables in the up markets viz. five star hotels, shops of embassies or high commissions of various countries situated in Delhi. They are ready to pay very high price for the high quality produce; therefore, it may be a profitable venture only if this technology is adopted around metro cities of the country. Greenhouse vegetable production is a highly intensive enterprise requiring substantial labor and 24 hour-a-day commitment, which restrict the adoption of this technology. But now the time has come when the vegetable growers in various parts of the country can use the naturally ventilated greenhouse technology for cultivation of high value vegetables for high profits.

Off-season Vegetable Cultivation Under Shade Nethouses

Shade nets are perforated plastic materials used to cut down the solar radiation and prevent scorching or wilting of leaves caused by marked temperature increases within the leaf tissue from strong sunlight. These nets are available in different shading intensities ranging from 25% to 75%. Leafy vegetables and ornamental greens are recommended to be grown under shade nets whose growth rates are significantly enhanced compared to un-shaded plants when sunlight is strong. The basic objective of shade net is to reduce radiation and temperature up to some extent during critical summer months (May-Sept.). Black color shade nets are most efficient in reduction of temperature compared to other colors like green, white or silver, etc. as the black color is the maximum absorbent of heat. Mostly leafy vegetables like beet leaf and green coriander are preferred to be grown under shade nets, but it is also suitable for growing early cauliflower and radish cultivation during June to September months.

Off-season vegetable Cultivation Under Walk in Tunnels

The demand for off-season vegetables is increasing day by day in several big and medium sized cities of the country, which provide wide scope of vegetable production through off-season cultivation of some vegetables, mainly the Cucurbits in peri-urban areas of the country. Walk-in- tunnel technology is a simple and profitable technology for off-season cultivation of cucurbits during the winter season in northern plains of our country. Crops like summer squash can be grown as a complete off-season crop whereas, other cucurbits like muskmelon, round melon, bottle gourd, cucumber, bitter gourd, watermelon are mainly grown during off season (Singh *et. al,* 2010).

Walk in tunnels are the temporary structures erected by using G.I. pipes & transparent plastic. Walk in tunnels are used for complete off season cultivation of vegetables like bottle gourd, summer squash, cucumber etc. during winter season (Dec.–mid February) the basic objective & utility of walk in tunnels is to fetch high price of the complete off season produce to earn more profit per unit area. The ideal size of a walk in tunnel can be of 4.0 m width and 30m length ($120m^2$) and total cost of fabrication may be Rs.16000-18000/-.

Off-season Cultivation of Vegetable Crops Under Plastic Low Tunnels

The demand of off-season vegetables is increasing day by day in several big and medium sized cities of the country which provide wide scope of vegetable production for off-season cultivation of some vegetables mainly the Cucurbits in peri-urban areas of the country. Plastic low tunnel technology is a simple and profitable technology for off-season cultivation of Cucurbits during the winter season in northern plains of our country. Crops like summer squash can be grown as a complete off-season crop whereas, other cucurbits like muskmelon, round melon, bottle gourd, cucumber, bitter gourd, watermelon can be advanced by 30-40 days over their normal growing season (Singh *et al*, 2004; Singh and Kumar, 2009).

Plastic low tunnels are flexible transparent coverings that are installed over single or multiple rows of vegetables to enhance the plant growth by warming the air around plants in the open field during winter season when the temperature is below $8^{0}C$. Plastic low tunnels are often used to promote the growth of plants during the period of winter season. Low tunnels are supported above the plants by using hoops of GI wire and a clear or transparent plastic of 20-30 micron is covered/ stretched over the hoops and the sides are secured by placing in soil.

The plastic is vented or slitted during the growing season as the temperature increase within the tunnels. The farmers can grow different varieties of summer squash (round fruited, long fruited) which is an emerging crop along with cultivation of netted muskmelon varieties in place of traditional varieties. Bitter gourd and round melon are two other crops with increasing demand and which usually fetches very high price during off-season and can be grown successfully by using the plastic low tunnel technology. This technology is highly suitable and profitable for the farmers living in northern plains of India.

Quality Seed Production of Vegetables Under Protected Conditions

Recently there has been a technical shift towards the production of high quality hybrid seed of important high value vegetables under protected conditions. Selection criteria of protected structures to be used for vegetable seed production certainty depends upon a few important factors viz., climatic conditions of the area selected for seed production, vegetable crop, type of seed crop, season of seed production and targeted quantity of seed production. Generally, semi-climate controlled greenhouses are only suitable for hybrid seed production of determinate type varieties/hybrids of standard tomato, cherry tomato, sweet pepper and parthenocarpic cucumber varieties, which are suitable for protected cultivation. Seed yield of such crops can be 3-4 times more compared to their open field cultivation along with high quality of seed. Similarly, zero energy naturally ventilated greenhouses are also equally suitable for hybrid seed production of these high value vegetable crops, where the seed yield is usually 2-3 times more than open fields, but the cost of seed production is only one third (1/3) of the seed produced under semi-climate controlled greenhouse conditions (Singh *et. al.,* 2010).

Insect proof net houses are the most suitable and low cost protected structures suitable for quality hybrid or seed production of OP varieties in large number of vegetables viz., tomato, sweet pepper, chilli, okra, brinjal and several Cucurbitaceous vegetables. The major objectives of seed production under insect proof net houses are to grow virus free seed crops, to protect the seed crops against other major insects/pests like shoot and fruit borer in brinjal, fruit borer in okra and tomato, fruit fly in cucurbits, leaf minor in cucurbits and solanaceous vegetables and red pumpkin beetle in cucurbits, . Not only this, insect proof net house also provides protection to the seed crops against mild frost conditions. Some other low cost protected structure like walk- in- tunnels and plastic low tunnels are used for off season seed production or advancing the season of seed production, but these structures are technically suitable only for northern plains of India during winter season. Massive work on quality seed production of vegetables is being carried out at the Centre for Protected Cultivation Technology, IARI, New Delhi and excellent results in hybrid seed production of brinjal, pumpkin, bitter gourd, cucumber and summer squash were recorded as compared to open field seed production of these crops. Hybrid seed production by using insect proof net houses can be highly profitable venture for progressive farmers in various regions of the country (Singh, 2012).

Strategies for Accelerating Production of Vegetables Under Protected Conditions in India

- Large scale motivation and training of educated unemployed youths in the field of protected cultivation.
- Government support must be extended for self fabrication mode of temporary low cost structures like insect proof net houses, shade net houses, walk-in-tunnels and plastic low tunnels for production of vegetables.
- Govt. support may also be extended to farmers for self-fabricated low cost protected structures.
- Government should support and promote protected cultivation in cluster approach especially in peri-urban areas of the country.
- Government should promote to develop input hubs for protected cultivation in multi-locations in PPP mode in various regions of the country.
- Marketing is the key for success of protected cultivation. Electronic auctioning of produce through cluster representatives may be initiated.
- Protected cultivation has hitherto been promoted from the view point of more and more construction of greenhouses by providing subsidy. However, there is need to link such subsidies with production system i.e. when the protected cultivation produce is sold/auctioned by the grower some of the subsidy may be realized to him at this level as incentive. A special cluster club of protected cultivation growers may be established along with protected cultivation blocks at state level.
- All the protected cultivation clusters must be mandatorily clubbed with rain water harvesting infrastructure and facilities.
- Practical demonstration units at each state level may be established for ensuring demonstrations of all low cost protected structures along with the production and pest management strategies.
- Suggest most suitable crop sequences for different protected structures and seasons based on research data.
- Large scale promotion of low pressure drip irrigation system for low cost small scale protected cultivation in various parts of the country.
- Selection and promotion of suitable designs of protected structures for different regions of the country for high quality long duration vegetable cultivation.

- Large scale use of different color plastic mulches for different seasons clubbed with raised beds and drip fertigation system for vegetable and flower production under open field conditions and also to discourage expensive surface irrigation in vegetable crops.
- Promotion of large scale mechanization in vegetable and flower cultivation by using raised bed makers, plastic laying machines, plastic low tunnel making machines, pipe bending machines for making walk- in -tunnels, drip lateral laying and binding machines
- To establish convergence and synergy among various ongoing and planned government programmes in the field of protected cultivation development in the country.
- To ensure adequate, appropriate, time bound and concurrent attention to all links in production under protected conditions, post production on farm value addition, processing and consumption chain.
- Major ITIs in the country should include a diploma course on fabrication of protected structures, repair and their maintenance.
- Use of self driven trolleys for training and pruning in greenhouse vegetables along with the equipments related to aided pollination in vegetables under protected cultivation.
- Use of solar energy for running drip system and up to some extent for running heating and cooling devices of the protected structures.
- Under the new era of FDI (Foreign Direct Investment) in retail, these kinds of models are having high potential for betterment of the farmers opting for quality and offseason vegetable cultivation through protected cultivation

References

Singh B., Tomar B.S. and Kumar M. 2004. Plastic low tunnel: A profitable and sustainable technology for peri-urban areas. *Intensive Agriculture*, **42**(7-8):18-20

Singh B and Kalia P. 2005. Protected cultivation of vegetable crops: Problem, potential and prospects in India. *Indian Journal of Fertilizers*, **1**(4): 93-97.

Singh B, Kumar M and Yadav H.L. 2005. Plug-tray nursery raising technology for vegetables. *Indian Horticulture*, **49**(4): 10-12.

Singh B., Kumar M and Singh V. 2005. Nylon Mesh Screens Reduce Incidence of Leaf Curl Virus and Improves Yield in Sweet Pepper. Journal of Vegetable Science (USA). **12**(1): 65-70.

Singh B. and Kumar M. 2006. Techno-economic feasibility of Israeli and indigenously designed naturally ventilated greenhouses for year round cucumber cultivation.*Acta Horticulturae*.**710**:535-538.

Singh B. and Tomar B.S. 2007. Keet avrodhi nylon net dwara sabjion kee vishanu rog rahit swasth paudh utpadan.Phal Phool **29**(5): cover page 1 &3.

Singh B. and Kumar M. 2009. Evaluation of Summer Squash varieties under plastic low tunnels during their off season cultivation. *Indian J. Hort.* **66**(1):135-136.

Singh B., Singh A.K. and Tomar B.S. 2010. In Peri- Urban: Protected cultivation technology to bring prosperity. *Indian Horticulture.* **55**(4): 31-32.

Singh B., Tomar B. S. and Thakur S. 2010. Quality seed Production of Parental lines of Pumpkin (C. moschata) under Insect Proof net house. *Acta Horticulturae* (ISHS) **871**: 275-278.

Singh B. 2011. Insect proof net house for cultivation for high value vegetables. *ICAR News.* **17**(3) : 6.

Singh B and Hasan M. 2011. Low pressure drip irrigation system: A success story *(In Hindi)*. Agriculture Extension Review. **21**(2):11-12.

Singh B, Singh A and Kumar M. 2012. *Pari nagriya shetroo me: Aiise hogi sabjiyo kee sanrakshit kheti.* Phal Phul. **33**(3): 3-6.

Singh B. 2012. Seed Production of summer squash in North Indian plains. ICAR News **18**(2), 14.

8

Polyhouse Vegetable Crop Breeding

D.K. Singh

Abstract

Space, water and labour have become limiting factors in vegetable production. Moreover, production of off-season vegetables during period of demand fetches better returns. Vegetable forcing for year round production has become a necessity in urban and peri-urban olericulture. With use of pesticides, fungicides, weedicides and chemical fertilizers, health hazards have become visible. Being treated as a field crop, vegetables are exposed to vagaries of weather and floods and droughts take heavy toll. Research on protected cultivation in India is limited but for attempts by private seed companies and enterprising farmers in Karnataka, Maharashtra, Haryana and Punjab. More emphasis is given to tomato, cucumber, capsicum, lettuce and beans. Breeding for varieties suited for protected cultivation is a recent innovation in public sector laboratories. Now there are parthenocarpic cucumbers, tomatoes and capsicum especially to suit protected cultivation. The first four vegetable varieties were developed and released by Uttarakhand State Varieties Release Committee for Horticultural Crops during 2011. These varieties namely Pant Polyhouse Hybrid Tomato -1 , Pant Polyhouse Tomato -2 in tomato and Pant Parthenocarpic cucumber -2 and Pant Parthenocarpic Cucumber -3 in cucumber. These varieties were delivered to ployhouse farmers cross the country.

Keywords: Vegetable forcing, protected cultivation, polyhouses, parthenocarpic cucumber, tomatoes, beans, capsicums, crop breeding

Introduction

About 70 per cent of Indian population live in rural areas and the overwhelming majority of them depend upon agriculture as their primary source of income. Agriculture and allied sectors contribute 14.6% of the total GDP and provide around 65% employment to Indian population. Indian farmers face several challenges such as small land holding, poor yields due to reliance on inefficient methods of farming, too much reliance on natural phenomena such as rainfall and lack of knowledge regarding modern techniques of agriculture. Polyhouse farming is an alternative new technique in agriculture which is gaining foothold in rural India. Though, polyhouse vegetable production and breeding in the country are very ancient but it is still in infancy stage. Very little efforts are made to exploit vast potential of polyhouse vegetable production. It reduces dependency on rainfall and makes optimum use of land, fertilizers and water resources. A typical, traditional farm of 500 square meters would generate an estimated annual income of Rs. 10,000 to 20,000 while the estimated annual income from similar sized polyhouse goes upto Rs. 1,00,000 to 1,50,000. Potentially, polyhouse farming can help farmers to generate income round the year by growing multiple crops. It enables cultivation of crops that can give maximum yield on specific days and exotic crops which cannot be normally grown in Indian conditions (e.g. tomato, coloured capsicum, parthenocarpic cucumber, etc.). It also enables cultivation of off-season crops regularly, thus, fetching a higher price. Therefore, polyhouse cultivation with engineered vegetable plants, especially for polyhouse has to be utilized to enhance the yield manifolds in different crops by using their economically important characters and thus, increasing the overall productivity.

Introduction of protected cultivation technology in India has a time lag in relation to many countries. China started protected cultivation technology almost at the same time when India made its beginning. But now China, Japan and USA are the leading countries having large area under polyhouse for horticultural crops in the world and have also developed a number of tomato and parthenocarpic cucumber varieties, especially suitable for polyhouse. Despite of all the efforts, India is still lagging behind in developing varieties for polyhouses. Reasons behind this are less attention among the researchers, lack of suitable varieties and other limiting aspects like cost, infrastructures, etc. In India, the area under green/polyhouse cultivation crop is 2000 hectares, which is mainly in Maharashtra, Uttarakhand, Karnataka, Himachal Pradesh and Jammu & Kashmir. There appears ample potential and scope for increasing area under low cost polyhouse in periurban areas for production of high value vegetables during off-season for taking advantage of high price of the available market in nearby cities.

Global Scenario

The area under greenhouses in Spain has been estimated to be around 25,000 ha and in Italy 18,500 ha, used mostly for growing vegetable crops like watermelon, capsicum, strawberries, beans, cucumbers and tomatoes (Table.1). In Spain, simple tunnel type greenhouses are generally used without any elaborate environmental control equipments. There is mostly use of UV stabilized polyethylene film as cladding material.

Table 1. Area under protected cultivation of vegetables

Country	Area (ha)	Country	Area (ha)
Japan	54000	Turkey	10000
China	48000	Holland	9600
Spain	25000	USA	4000
South Korea	21000	Israel	1500
Itlay	18500	India	2000

In Canada, the greenhouse industry caters both to the flower and off-season vegetable markets. The main vegetable crops grown in Canadian greenhouses are tomato, cucumbers and capsicum. Hydroponically grown greenhouse vegetables in Canada find greater preference with the consumers and could be priced as much as twice the regular greenhouse produce.

The Netherlands is the traditional exporter of greenhouse grown flowers and vegetables all over the world. With about 89,600 ha under cover, the Dutch greenhouse industry is probably the most advanced in the world. Dutch greenhouse industry however, relies heavily on glass framed greenhouses, to cope up with very cloudy conditions prevalent all the year round. A very strong research and development component has kept the Dutch industry in the forefront.

The development of greenhouses in Gulf countries is primarily due to the extremity in the prevailing climatic conditions. Israel is the largest exporter of cut flowers and has wide range of crops under greenhouses (15,000 ha) whereas Turkey has an area of 10,000 ha under cover for cultivation of cut flowers and vegetables. In Saudi Arabia, cucumbers and tomatoes are the most important crops contributing more than 94% of the total production. The most common cooling method employed in these areas is evaporative cooling.

Egypt has about 1000 ha greenhouses, consisting mainly of plastic covered tunnel type structures. Arrangements for natural ventilation are made for regulation of temperature and humidity conditions. The main crops grown in these greenhouses are tomatoes, cucumbers, peppers, melons and nursery plant materials.

In Asia, China and Japan are the largest users of greenhouses. The development of greenhouse technology in China has been faster than in any other country in the world. With a modest beginning in late seventies, the area under greenhouses in China has increased to 48,000 ha in recent years. Out of this, 11,000 ha are under fruits like grapes, cherry, Japanese persimmon, fig, loquat, lemon and mango. The majority of greenhouses use local materials for the frame and flexible plastic films for glazing. Most of the greenhouses in China are reported to be unheated and use straw mats to improve the heat retention characteristics.

Japan has more than 40,000 ha under greenhouse cultivation of which nearly 7,500 ha are devoted to only fruit orchards. Greenhouses in Japan are used to grow wide range of vegetables and flowers with a considerable share of vegetable demand being met from greenhouse production. Even a country like South Korea has more than 21,000 ha under greenhouses for production of flowers and fruits. Greenhouses permit crop production in areas where winters are severe and extremely cold as in Canada and USSR, and also make production possible even in areas where summers are extremely intolerable as in Israel, UAE, and Kuwait, Greenhouses in Philippines make it possible to grow crops inspite of excessive rains and also in moderate climates of several other countries. Thus, in essence greenhouse cultivation is being practiced and is possible in all types of climatic conditions.

Indian Scenario

While greenhouses have existed for more than one and a half centuries in various parts of the world, in India, use of greenhouse technology started only during 1980's and it was mainly used for agronomical and structural design research activities. This may be because our emphasis, so far had been on achieving self-sufficiency in vegetable production. However, in recent years in view of the globalization of international market and tremendous boost and fillip being given for export of horticultural produces, there has been a spurt in the demand for greenhouse technology. The area under greenhouse in India is 2000 ha. The National Committee on the use of Plastics in Agriculture (NCPA-1982) recommended location specific trials of greenhouse technology for adoption in various regions of the country.

Greenhouses are being built in Ladakh region for extending the growing season of vegetables from 3 to 8 months. In the North-East, greenhouses are being constructed essentially as rain shelters to permit off-season vegetable production. In the Northern plains, seedlings of vegetables and flowers are being raised in the greenhouses either for capturing the early markets or to improve the quality of seedlings. Propagation of difficult-to-root tree species has also been found to be very encouraging. Several commercial floriculture ventures are coming up in Maharashtra, Tamil Nadu and Karnataka to meet the demands of both domestic and export markets.

The commercial utilization of greenhouses started from 1988 onwards and now with introduction of Government's Liberalization policies and developmental initiatives, several corporate houses have entered to set up 100% export oriented units in just four years. Since implementation of the new policies in 1991, 103 projects with foreign investment of more than Rs. 80 crores were approved to be set up in the country at an estimated cost of more than Rs. 1000 crores around Pune, Bangaluru, Hyderabad and Delhi. Thus the area under climatically controlled greenhouses of these projects is estimated around 300 ha. Out of which many have already commenced exports and have received very encouraging results in terms of the acceptance of the quality in major markets abroad and the price obtained.

Principle of Greenhouse

A greenhouse is generally covered with a transparent material like polythene or glass. Depending upon the cladding material and its transparency, major fraction of sunlight is absorbed by vegetable crops and other objects. These objects in the greenhouse emit long wave thermal radiations for which cladding material has lower transparency. With the result, solar energy is trapped and raises the temperature inside the greenhouse. This is popularly known as greenhouse effect. This rise in temperature in greenhouse is responsible for growing vegetables in cold climates. During summer months, air temperature in greenhouse has to be brought down by providing cooling device. In commercial greenhouses, besides temperature controlled humidity, carbon dioxide, photoperiod, soil temperature, plant nutrients etc. facilitate round the year production of desired vegetable crops. Controlled climatic and soil conditions provide an opportunity to the vegetable crops to express their yield potentials.

Grafting

Grafting is an emerging technique and is becoming popular throughout the world, especially for the greenhouse cultivation of vegetable crops, mainly Solanaceous and Cucurbitaceous ones, from the point of resistance against soil borne pathogens including the root knot nematode Meloidogyne spp. Apart from obtaining better yield and quality, it is especially successful in cucurbits like melons for resistance against diseases. Grafting involves the union of suitable varieties of green house crops used as scions over the resistant root stocks usually from the same family. The grafted seedlings are conditioned in the dark and cool shed nets are prepared specially for the purpose before planting in seedling trays and finally, they are transplanted in the green houses.

Salinity is one of the major abiotic stresses that reduces plant growth and crop productivity in many vegetable production areas of the world. Grafting can represent an interesting tool to avoid or reduce yield losses caused by salinity stress in high-yielding genotypes belonging to Solanaceae and Cucurbitaceae families. Grafting of vegetable seedlings is a unique horticultural technology to overcome soil-borne diseases and pests and to add extra vigour to the plants under various environmental stress conditions. The technique was practiced for many years in East Asia, and more recently it has become a common pest management technique worldwide. In 1990s, nearly 60% of open fields and greenhouses for producing muskmelon, watermelon, cucumber, tomato and eggplant were reportedly planted with grafted seedlings in Japan and 81% in Korea.

There are three primary techniques used for grafting: tongue, approach grafting, cleft grafting and tube grafting. Cleft grafting and tube grafting are quite similar in that that the shoot of the fruit producing variety is completely cut off from its own roots and attached to the severed stem of the rootstock plant (the name 'Tube Grafting' originated because when the technique was first developed, a tube was used to attach the shoot to the root. Clips are now used to make this graft.) Tube grafting is quicker and less complicated to do than cleft grafting because it only requires a single straight cut on both the root and shoot portions of the graft. Also, because a fewer intricate cuts are involved, this technique can be used on very small seedlings. With both cleft and tube grafting, the newly grafted plants must be protected from drying out until the graft union gets healed. This usually involves covering the plants with a plastic cover to reduce light and retain moisture, and then misting the grafted plants periodically during the day. Tomato grafts heal quickly and the plants begin to be acclimated back into the greenhouse after 4-5 days, and usually plants are fully set in the greenhouse 7-8 days after the initial grafting.

Grafting was used for millennia to increase uniformity, vigour and resistance to biotic and abiotic stresses of vegetatively propagated plants. In vegetable crops propagated by hybrid or homozygote seeds, stand uniformity and vigour are nearly always assured by the genotype. The main reason for vegetable grafting is to improve adaptation to stresses. For example, rootstocks for Solanaceous and Cucurbitaceous species can tolerate, to a different degree, various biotic stresses. Hence, on the whole, grafting can allow for a quicker use of genetic resistance developed in breeding programmes, a reduction in the use of geo-chemicals and planting valuable cultivars which are not resistant to specific soil-borne pest and diseases.

Vegetable grafting was first adopted to enhance ability to cope up with biotic stresses. Lately, this technique has also been proposed as a means to enhance vegetable tolerance to abiotic soil stresses, like low soil temperature, drought, and salinity and flooding. In general, vegetable grafting is motivated by the need to increase plant yields.

Vegetable Crop Breeding for Polyhouse

A great diversity of temperature i.e., 10-45^{0}C exists in the greenhouses during winter to summer season. In each location, conditions vary enormously during growing season and changes can occur very rapidly. Therefore, it is necessary to breed varieties not having too narrow adaptation to climatic conditions and also having wide temperature adaptability for cultivation under polyhouse conditions round the year.

Breeders are considering 3 main aspects: poor light conditions, low temperatures in the early stage of growing season and high temperatures in the late stage of the production. Genotypes are usually chosen after several trials in greenhouse. Efficiency of selection would be increased if the breeders could use simple methods for measuring different responses of the plant to different climatic conditions. Tests must be easy to practice rapidly on a large number of genotypes, without destruction of the plant, if possible at an early stage.

Vegetable improvement shows many characters which require consideration by the breeders to get varieties which in turn can give the highest returns for all the investments made in the greenhouse cultivation.

The genes required for crop improvement are present in different lines, varieties, strains or populations of the crop species and their relatives which could be wild forms or wild relatives. Some useful genes, often critical genes, are contributed by the wild relatives of the crops like the genes for insect and disease resistance, tolerance to abiotic stresses like drought, salinity, cold etc., and even for quality traits and yield. Various lines, strains, populations of a crop species and their related wild species constitute the germplasm of the crop. The genes useful for the improvement of existing varieties are needed to be explored and incorporated for their further improvement. Backcross method, whose objective is to improve one or two specific defects of a high yielding variety, which is well adapted to the area and has other desirable characters can be utilized for the purpose. The characters lacking in the existing variety are transferred into it from a donor parent without changing the genotype of the existing variety, which is well adapted to the area and has other desirable characters. In the backcross method, the hybrid and the progenies in the subsequent generations are repeatedly crossed to one of the parents of the F_1. As a result, the genotype of the backcross progeny becomes increasingly similar to that of the parent to which the backcrosses are made. At the end of 6-8 backcrosses, the progeny would be almost identical with the parent used for backcrossing. Thus, the end result of the backcross programme is a well adopted variety with one or two improved characters.

Requirements of a backcross programme

For the successful development of a new variety through backcross method, the following requirements must be fulfilled:

1. A suitable recurrent parent must be available, which lacks in one or two characteristics.
2. A suitable donor parent must be available. The donor parent should have the character or the characters to be transferred in an intense form.
3. The characters to be transferred must have high heritability and, preferably, it should be determined by one or a few genes.
4. A sufficient number of backcrosses should be made so that the genotype of the recurrent parent is recovered completely. Ordinarily, 6-7 backcrosses are sufficient for the purpose. In practice, 4-6 backcrosses are usually made.

Populations of cross pollinated crops are highly heterozygous as well as heterogeneous. Their genetic makeup is such that they show variable inbreeding depression, which in some cases may be severe. Therefore, population improvement becomes a vital step to overcome the problem. In population improvement, mass selection or its modification is used to increase the frequency of desirable alleles, thus improving the characteristics of populations.

The recurrent selection schemes were devised in relation to heterosis breeding. The idea is to ensure the isolation of superior inbreds from the populations subjected to recurrent selection for their ultimate utilization in the production of hybrid and synthetic varieties. This is a population improvement method followed in cross pollinated vegetable crops with the objective to concentrate favourable genes distributed among individuals in a best population. This method is effective in improving yield and other quantitative characters by the accumulation of additive genetic variance, and is highly efficient for creating superior germplasm. When a population is attained off its peak potential, recurrent selection may be useful in breaking out new types. The source population for recurrent selection may be an open pollinated variety, F_1 hybrid, synthetic variety, composite variety or inter crossed population of selected inbred lines.

Cucumber

Crop Introduction

Cucumber is a member of Cucurbitaceous family and native of Asia and Africa. It is an annual, deep rooted plant with tendrils and hairy leaves. The plant may have indeterminate, determinate or compact plant habit. Its optimum growth occurs between 20 -25^{0}C. It's a warm season fruit vegetable used as salad and for pickling purpose. In developed countries, it is grown as a greenhouse vegetable and in developing countries as an open field vegetable. The average productivity of cucumber in India is 10 t/ha. But under polyhouse, the productivity goes upto 130 t/ha, which is almost 13 times higher. Considering the importance of Parthenocarpic cucumber in terms of yield advantage and suitability of cultivation under polyhouse condition, Parthenocarpic cucumber breeding programme was started at Pantnagar with the following objectives and the following genes were exploited:

Breeding objectives

1. High productivity
2. Development of gynoecious line
3. Incorporation of Pc gene.
4. Short internodal length
5. Fruit dark green with smooth surface
6. Resistance against PM, DM, anthracnose, mosaic and fruit fly

Some important genes of cucumber

Gene Symbol	Phenotype
A	Androecious, produces only male flower if recessive at Acr/acr locus
Acr (F, acr^f, D)	Female, high degree of female sex expression
In-F	Intensifier of female sex expression
cp	Compact, reduced internode length, poorly developed tendrils, small flowers
de	Determinate habit. Short vine with stem terminating in flowers
In-de	Intensifier of *de.* Reduces internode length and branching
gy	Gynoecious, recessive gene for high degree of female sex expression
	Pathenocarpy, fruit setting without pollination
Pc (P)	Fruit length (expressed in additive fashion)
gn	Green mature fruit
Ar	Anthracnose *(Colletotrichum lagenarium),* PI 175111, PI 175120, PI 179676, PI 182445
Ccu	Scab *(Cladosporium cucumerinum),* Davis Perfect
dm	Downy mildew *(Pseudoperonospora cubensis),*
pm-1, pm-2, pm-3	Poinsette Powdery mildew (*Sphaerothe fuliginia*), Natsufushinari, PI 200815, PI 200818
Foc	Fusarium wilt (*Fusarium oxysporum* f. sp. *cucumerinum*), Wisc. 248
Cmv	Cucumber mosaic virus, Wisc. SMR-12, Wisc. SMR-15, Wisc. SMR-18

Parthenocarpic Cucumber Breeding

The crop shows several flowering habits. Most of the cultivars are monoecious with separate male and female flowers. Gynoecious cultivars produce only female flowers.

The biological processes leading to sex expression in plants are of tremendous practical significance for fruit production of many agricultural and horticultural crops. Sex-expression studies in cucumber showed that the different sex types are determined by three major genes: *M/m*, *F/f* and *A/a*. The *M/m* gene in the dominant condition suppresses stamen development and thus leads to female flowers. The *F/f* gene in the dominant condition shifts the monoecious sex pattern downwards and promotes femaleness by causing a higher level of ethylene in the plant.

The cucumber, with its diverse sex types, represents a perfect model plant for studying the molecular and physiological basis of sex expression. All floral buds contain staminate and pistillate primordial at early stages of their development; however, abortion of either staminate or pistillate parts results in female or male flower, respectively, while in hermaphrodite flowers the staminate and pistillate primordial get matured. The distribution of these three different flower types (staminate, pistillate and hermaphrodite) on the plant results in different sex types in cucumber. These three sex types are mainly determined by three major genes *F/f, M/m* and *A/a.*

The gene *F* promotes femaleness on the whole plant level while the gene *M* determines whether flowers are unisexual *(M_)* or bisexual *(mm)*. Thus the genotype *M_FF* is gynoecious; the genotype *M_Ff* sub gynoecious due to incompatible dominance of *F*; and the genotype *mmF_* is hermaphrodite. If *F/f* is homozygous recessive, the sex expression is influenced by the gene *A/a; MMffA_* genotypes are monoecious; *MMffaa* genotypes are androecious. Although, the three major genes determine the basic regulation of sex types, the sex expression is influenced by additional modifying genes and environmental factors such as photoperiod, temperature and irradiation.

Parthenocarpy is the ability to develop fruits without pollination. The inheritance of pathenocarpy in cucumber is conditioned by an incomplete dominant gene Pc. In the homozygous condition PP produces parthenocarpic fruit early, with the first developing generally by 5th node. Heterozygous Pp plant produces parthenocarpic fruit later than homozygous plant and a fewer in number. The homozygous recessive pp produces no parthenocarpic fruits. Trapping of pistilate flowers was effective in identifying homozygous pp plant but failed to identify heterozygous plants. Screening cages and field isolation of gynoecious lines are effective means for accurately identifying all parthenocarpic plants. It is being explained that the inheritance of parthenocarpy is by three independent, isomeric major genes with additive action, together with a non-allelic interaction of the homozygote-heterozygote type. Indications are for linkages between genes which govern parthenocarpy femaleness and the spined/hairy fruit character. Cucumber is a cross pollinated crop and bees are the main pollinating agents. The female flowers develop later than the more numerous male flowers and have a ring of nectiferous tissue surrounding the style. The main sex forms are gynoecious, hermaphrodite and monoecious, which have been greatly used in developing single and triple F_1 hybrids. Since, the emergence of female flowers takes much time and hand pollination seems laborious process for hybrid seed development programme, the concept of polyouse breeding with the utilization of *F* gene which gives the gynoecious line is a mile stone in the process.

Greenhouse or parthenocarpic cucumber production is very popular in many areas of the world. The fruits of parthenocarpic cucumber are mild in flavour, seedless and have thin edible skin that requires no peeling. Fruits are generally 25-35 cm in length and 500g in weight. The parthenocarpic varieties require no pollination for fruit setting. If pollination occur, the fruit will form seeds, its shape will be destroyed and bitter tasting fruit will develop. It is therefore essential to prevent bees and other pollinators from entering the greenhouses and carry pollen from field planted cucumber.

Cucumber cultivars are selected for commercial use on the basis of quality and yield characteristics. An important quality factor is the shape of cucumber referred to as the length to diameter (L/D) ratio. The presently available parthenocarpic cucumber cultivars are bred and developed in Europe under similar climatic conditions in Puget Sound and produce fruit with desirable L/D ratios. The unique characteristics of parthenocarpic cultivars are their ability to set fruit without pollination and fertilization, resulting in no seed produced in the fruit's interior. The parthenocarpic cultivars have potential for improvement of both L/D ratios and interior quality of fruit.

Almost all slicing cucumber varieties cultivated in greenhouse in Western Europe are parthenocarpic. Parthenocarpic pickling cucumber varieties are also available in the market. In Germany, the use of parthenocarpic varieties in production for processing industry is a major part of the total cultivable area of pickling cucumber. Parthenocarpic fruit setting results in an earlier and much more regular production rhythm. Parthenocarpy circumvents the inhibitory effect of seed formation on subsequent fruit development. It must be combined with constant femaleness, because the fruits formed after fertilization in genetic parthenocarpic varieties become misshapen and are of no economic value and ultimately lead to loss of production. It is possible to induce artificially parthenocarpic fruit set by treatment with morphactin (Chlorfluorenol).

Parthenocarpic varieties do not require pollination to set the fruit and the fruits are seedless. So maintenance is a major problem in these lines. The parthenocarpic lines are usually maintained by the use of plant growth regulators and chemicals e.g. Silver nitrate and GA_3. Kasrawi(1988) investigated the effect of $AgNO_3$ concentration, number of sprays and time of application on sex expression and pollen variability of parthenocarpic cucumber. He found that greater number of staminate flowers are produced on plants when $AgNO_3$ was sprayed twice with 300 ppm concentration at one week interval with initial spray at first true leaf stage. He also found that 2-3 sprays of 200-500 ppm $AgNO_3$ starting at first true leaf stage produced the maximum number of staminate nodes. When the plants were sprayed at first true leaf stage, production of staminate flowers began with the 1st node and at the time of onset of flowering. Experiments reported the effect of silver nitrate on induction of staminate flowers of gynoecious parthenocarpic flowers and found that $AgNO_3$ concentration and numbers of sprays affect the induction of staminate flowers. All the treatments using one spray were ineffective in producing staminate flowers. The treatment with 100 ppm $AgNO_3$ failed to induce staminate flowers The greatest number of staminate nodes was produced on plants by 2-3 sprays of 400-500 ppm $AgNO_3$. The initial spray was applied at first true leaf stage and subsequent treatment was applied at one week interval. The plants showed injury a few days after spray of 400-500 ppm $AgNO_3$ and recovered within 7-10 days. At Pantnagar, staminate flowers were induced in parthenocarpic line through use of plant growth regulator GA_3 and silver nitrate by four sprays at 4 days interval.

Varieties

There are many cultivars available for greenhouse cucumber production from private sector. Cultivar selection is one of the most important decisions during crop production process. Several factors should be considered when selecting cultivars including disease resistance, plant vigour, earliness, yield, fruit size, quality and fruit colour. The available parthenocarpic cultivars are generally produced by European seed companies, however, a few number of companies have sales and technical support representative in other parts of the world. Some of the popular parthenocarpic varieties are Kalunga, Bellissma, Millagon, Discover, Marianna, Fitness, Aramon, Fidelio 90-0048, E 1828, B 1157, Sweet Success (AAs), County Fair, Bush Crop, Space Master, Patio Pick, Bush Whopper, Bush Champion, Bush Pickler, Euro-American, Adrian, Al Rashid F_1, Mustang, Bronco, Sandra, Boneva, Daleva, Padex, Fertila, Factum, Fanspot, Femfrance, Toska 70, Farbio, Corona, Sweet Slice, Radja, Bella, La Reine, etc. Knowing the fact that there is no such variety which is exactly suitable for polyhouse cultivation in Indian conditions, research work was initiated to develop parthenocarpic cucumber varieties at Pantnagar and two varieties were developed for the first time in India among public sector.

Development of Parthenocarpic Cucumber Varieties at Pantnagar

The parthenocarpic cucumber lines, PCUCP-2 and PCUCP-3 were isolated from more than 300 accessions collected from India and abroad. Gene 'pc' responsible for parthenocarpy, 'F' responsible for gynoecy and 'cp' responsible for short internodal length were utilized through pure line, back cross and recurrent breeding methods to establish these genes in plant population. For commercialization, first time in India among public sector two varieties were developed and released during 2011. The details of morphological characters are as follows:

1. Pant Parthenocarpic Cucumber-2

1. Parthenocarpic cucumber variety suitable for polyhouse cultivation in India
2. Plants bear only female flowers (gynoecious)

3. The single fruit weight is 630 gm.
4. The average yield is 1755 q/ha.
5. Plant produces seedless fruits (Parthenocarpic in nature)

2. Pant Parthenocarpic Cucumber-3

1. Parthenocarpic Cucumber variety suitable for Polyhouse cultivation in India
2. Plant bears only female flowers (gynoecious)
3. The single fruit weight is 415 gm.
4. The average yield is 1605 q/ha.
5. Plant produces seedless fruits (Parthenocarpic in nature)

Grafting in Cucumber

Cucumber can be successfully grafted onto *Cucumis* spp., *Cucurbita* spp., Cucurbita interspecific hybrids, bottle gourd, wax gourd, fig leaf gourd (*C. ficifolia* Bouché), and luffa (Sakata *et al.*, 2008). The primary reasons for grafting cucumber are to increase cold tolerance and resistance to *Fusarium* wilt. The most common rootstock for cucumber has been fig leaf gourd since it provides superior cold tolerance, resistance to *Fusarium* wilt and has a high affinity for cucumber (Marukawa and Takatsu, 1969). The *Cucurbita* interspecific hybrids are also popular on summer grown cucumbers since, they provide better heat tolerance along with durable *Fusarium* resistance. Additionally, *Cucurbita,* interspecific hybrids may also give some protection from cold. In Japan, there has been a recent shift in cucumber rootstocks to specific genotypes of *C. moschata* which cause bloomless (waxfree) cucumber fruit. Bloomless fruits produced on these rootstocks have a distinct appearance, a longer shelf-life and preferred by local consumers (Sakata *et al.,* 2008). The first generation of bloomless rootstocks had several deficiencies, including an increased susceptibility to powdery mildew and *Corynespora* leaf spot, a tougher skin and softer flesh. It is reported that the third generation of bloomless rootstocks are now available where these deficiencies have been corrected (Sakata *et al.,* 2008), which indicate that there has been recent breeding activity for improvement of cucumber rootstocks.

Tomato

Crop Introduction

Tomato (*Lycopersicon esculentum* L. 2n=2x=24) is being extensively grown as warm season fruit vegetable all over the world. In India, tomato has wider coverage in comparison to other vegetables. The leading tomato growing states are

Uttar Pradesh, Karnataka, Maharashtra, Haryana, Punjab and Bihar. Tomato is a very good source of income to small and marginal farmers and contributes to the nutrition of the consumers. It is a rich source of vitamins and organic acids. Tomato is a warm season crop and thrives in temperatures between 1-30^{0}C and is neither tolerant to frost nor to waterlogged conditions. It is well suited in different cropping systems which include cereals, grains, pulses and oilseeds. The differences are observed in production and productivity in polyhouse bred tomato (350-450 tonnes/ ha) and field bred tomato (150-200 tonnes) under polyhouse condition. The average productivity of open field bred variety in field condition is 250 q/ha whereas under polyhouse condition, is about 800q/ha (AICVIP Annual Report 2002-2004.)

Considering the importance and high yield potential of Israeli polyhouse bred tomato variety, breeding of polyhouse tomato was initiated during 2003 for the first time at G.B. Pant University of Agriculture & Technology, Pantnagar with the following objectives and exploitation of following genes:

Breeding Objectives

1. Large fruit size (150-200g)
2. Thick pericarp (0.8-1cm)
3. 4-5 fruits per cluster
4. Long fruiting period (8-9 months)
5. Long plant height (6-8 m)
6. Potato leaf type
7. Ability to set fruits in wide temperature range (10-45^{0}C)
8. Resistant to fruit and shoot borer
9. Long storage life under room temperature
10. Indeterminate, single stemmed plants
11. Medium sized bright colored fruits
12. Short internode
13. More number of fruits per plant
14. High TSS and acidity
15. Thermo-insensitivity
16. Resistance to white fly and other important insects and pests.
17. Exploration of male sterility having specific marker trait (green cotyledon)

Gene useful in tomato improvement

Gene Symbol	Gene designation
Sp	Self- pruning/indeterminate
c	Potato leaf
Cf-1, Cf-2, Cf-3, Cf-4	Leaf mold resistance
	Fusarium immunity
I-1	race-1
I-2	race-2
Ve	*Verticillium* resistance
Se	*Septoria* resistance
Ph-1	Late blight resistance
Ad	*Altemaria* resistance
Sm	*Stemphylium* resistance
Tm	Tobacco mosaic resistance
u	Uniform ripening
hp	High pigment
B	High beta carotene

Polyhouse tomato breeding

The genus *Lycopersicon* includes both self-incompatible and self-compatible species, with the later varying in their degree of outcrossing. Tomato is self-pollinating, but can easily hybridize within the species or cross with wild relatives under appropriate conditions, thus permitting gene introgression from wild relatives.

Throughout the last century, tomato breeding was based on various standard methods, including the pedigree method of hybridization followed by selection and backcrossing of desired traits from one parent into another, which resulted in generation of improved tomato varieties and hybrids with high quality and yield. Tomato improvement has increased by the exploitation of exotic resources and the introgression of new valuable genes into the tomato genepool. Classical breeding has not only developed cultivars with monogenic and dominant resistance to certain plant pathogens or by the combination of F_1 hybrids, but has also enabled the acquisition of good, value-added agronomic traits, such as high fertility and fruit setting, earliness, uniformity, adaptation and firmness and long shelf life appropriate for shipping to distant markets.

The tomato flowers are normally perfect. There are four to eight flowers in each compound inflorescence. There is a light protective anther cone surrounding the stigma leading to self pollination. Emasculation is usually done one day prior to anthesis. At this stage, the sepals start getting separated and the anthers change from light to dark yellow colour. The stigma is fully receptive at this stage allowing for pollination even immediately after emasculation. Tomato is a typical day-neutral plant. It is mainly self pollinated but a certain percentage of cross pollination also occurs. If the pollen is shed as the style grows up through anther tube, self fertilization is mandatory. Self pollination also takes place when style is short and stigma is not extroverted beyond connivent anthers. A certain degree of cross pollination happens when stigma protrudes outside the level of extrovert anther.

The progeny of a single homozygous self fertilized plant forms the pureline. Such lines remain homozygous as long as self pollination takes place. The success of this breeding method depends on the existence of variability in the original population. This method is very much potential for creation of parental lines to produce F_1 hybrids.

Hybrid breeding is another important tool in improvement of tomato under protected condition. Hybrids offer opportunities for improvement in productivity, earliness, uniformity, quality and for rapid deployment of dominant genes for resistance to diseases and pests. For large scale production of F1 hybrids, there should be some naturally occurring mechanism for ensuring cross pollination between parental lines with minimum of self fertilization as genetic male sterility prevails in tomato. Shelf life of tomato fruit can be greatly improved by breeding for increased firmness, by utilizing "ripening inhibitor" genes or by a combination of both. Incorporation of improved firmness, a partially dominant character, resulted in cultivars with almost doubled shelf-life. As such ,fruits can be picked at later stages of ripening, their flavour and colour are either normal or improved in comparison with normal cultivars. The partially dominant gene 'rin' in rin/+ combinations was found to add about 20 to 30% to length of shelf-life. While this increase is of little value in normal soft genotypes, fruit of rin/+ in genetically firm backgrounds exhibit a shelf-life increased by 2-4 days. Flavour of these genotypes was normal and colour intensity only slightly reduced. The partially dominant mutant gene 'nor' has a stronger effect on shelf-life than 'rin'. Fruits of nor/+ genotypes picked at the early pink stage can be kept for about 3 weeks without cooling. The disadvantages of nor/+ types usually are the inferior flavour and decreased final colour ratings. It was found, however, that the general genetic background and certain mutant genes as well as growing conditions greatly affect these quality characters.

The combined effect of ripening mutant genes in rin/+ and nor/+ genotypes is a pronounced slowdown in fruit ripening. At 20^0 to 24^0C, such fruits may have a shelf-life of 4-5 weeks but their flavour is sometimes inferior and their colour orange to light red.

Varieties

Tomato is a high value vegetable crop for off season and main season production under polyhouse/greenhouse. The varieties available till date are many like Pusa Ruby, Pusa Early Dwarf, Hissar Lalima, Punjab Chhuhara, Hissar Arun, Arka Ashish, Arka Gaurav, Pant Bahar, Pant Tomato-3, etc. but none of them are the best suited for polyhouse cultivation.

The successful production of greenhouse vegetables depends extremely on the use of proper varieties. Although, seed costs may be somewhat higher for these varieties, this represents a relatively small investment when considering the total cost of production. The following is a list of currently recommended varieties of tomato for greenhouse cultivation:

- Tropic - 80 days, 226.79 g fruit, deep oblate shape, green shoulder, indeterminate vine, fairly large plant with good vigor. Resistant to *Verticillium* wilt, *Fusarium* wilt, *Stemphilium* Leaf mold, Tobacco mosaic Virus No. 1, *Altenaria* Stem Canker
- Dombello - 82 days, 198.44-226.79 g fruit, tomato hybrid, deep globe, greenback fruit color, large compact plant with good total production. Resistant to Tobacco mosaic, *Verticillium* wilt, *Fusarium* wilt race- 2, Nematode and silvering.
- Tuckcross 520 - 74 days, 198.44g, hybrid with deep oblate shaped fruit, green shoulders, indeterminate large plant. Resistant to *Fusarium* wilt and *Altenaria* stem canker.
- Cobra - resistant to *Fusarium* wilt races 1 & 2, *Verticillium* wilt, and tobacco mosaic virus—yield 3.72 kg/plant.
- Stresa - resistant to *Fusarium* wilt races 1 & 2, *Verticillium* wilt, tobacco mosaic virus, rootknot nematodes and *Cladosporium* leaf mold—yield 3.9 kg/plant.
- Capello - resistant to *Fusarium* wilt races 1 & 2, *Verticillium* wilt, tobacco mosaic virus and *Cladosporium* leaf mold—yield 3.81 kg/plant.
- Trust - resistant to *Fusarium* crown and root rot, *Fusarium* wilt races 1 & 2, *Verticillium* wilt, tobacco mosaic virus and *Cladosporium* leaf mold—yield 3.71 kg/plant.
- Caruso - resistant to *Fusarium* wilt races 1 & 2, *Verticillium* wilt, tobacco mosaic virus and *Cladosporium* leaf mold—yield 3.58 kg/plant.
- Belmondo - resistant to *Fusarium* wilt races 1 & 2, *Verticillium* wilt, tobacco mosaic virus and *Cladosporium* leaf mold—yield 3.58 kg/plant

The important genes favouring polyhouse cultivation were also exploited. Gene 'sp' responsible for indeterminate character was utilized for the development of these varieties. By adopting appropriate selection methods and hybrid breeding procedure, two varieties namely Pant Polyhouse Tomato-2 and Pant Polyhouse Hybrid Tomato -1 were released first time in India among the public sector. The morphological characters are as follows:

1. Pant Polyhouse Tomato-2

1. Tomato variety suitable for Polyhouse condition in India.
2. The variety is having 5-6 fruits per cluster.
3. The single fruit weight is 100-105 gm.
4. Its average yield is 1291 q/ha.
5. This cultivar has better storage quality because of its thick pericarp (0.9 – 1.0 cm)

2. Pant Polyhouse Hybrid Tomato -1

1. Tomato variety suitable for Polyhouse condition in India.
2. The variety is having 7-8 fruits per cluster.
3. The single fruit weight is 130-140 gm.
4. Its average yield is 1616 q/ha.
5. This cultivar has better keeping quality because of its thick pericarp (1.0-1.25 cm).

Grafting in Tomato

The most common genetic rootstock sources for tomato are tomato hybrids and interspecific hybrids *(S. lycopersicon×S. habrochaites* S. Knapp & D.M. Spooner), but eggplant rootstocks are also recommended for specific conditions, such as flooding or waterlogged soil (Black et al., 2003). It appears from currently available commercial sources that most rootstock cultivars originated from screening S. *habrochaites* lines and crossing selections with tomato to create hybrids to use as rootstocks. There are a few open pollinated tomato rootstocks available, which appear to result from direct selection in the intended environment.

While the potential genetic base for rootstocks of tomato would appear to be very large considering the vast array of closely related species that could be used, the actual genetic base appears to be limited. All currently available commercial rootstocks are limited to specific tomato genotypes with resistance to soil-borne disease and to *S. lycopersicon×S. habrochaites* hybrids. It would appear that the genetic potential of other *Solanum spp*. has not yet been fully exploited for rootstock development. Even the eggplant rootstocks which are sometimes recommended for a specific purpose are limited to a few genotypes. Reasons for not exploiting other species are unknown, but probably due to lack of a high level of resistance to soil-borne disease, which has previously been the driving force for rootstock development. The primary purpose for grafting tomatoes in Europe and the U.S. is to extend the harvest season in greenhouse production. In these situations, grafting increases vigor, but not at the expense of reduced fruit set which could lower yields. One advantage of greenhouse production is the ability (and even necessity) to prune the plants, which may reduce the negative yield effects of high vigor. Our observations have also been that most growers take advantage of the increased vigor of grafted greenhouse tomatoes by maintaining two main stems on grafted plants, thereby reducing the number of tomato plants in the greenhouse to half. This reduces the seed costs if the cost of the rootstock is less than the cost of the scion cultivar, although the cost of grafting will increase the cost per plant. In non-heated greenhouses where plants are grown in soil (e.g. southern Europe, Mexico) disease resistance is still the primary reason for grafting. These low-input greenhouses are often not heated and have limited cooling ability, which may further increase the necessity for disease resistance. The primary purpose for grafting tomatoes in open field production in Asia has been either for control of bacterial wilt for which there are a few resistant scion cultivars or for brown root rot, for which there are no resistant scion cultivars (Yamakawa, 1982). The particular environments for each of these diseases are quite different, as bacterial wilt prefers warm temperatures, which occurs in the open field in summer and brown root rot tends to occur in cooler conditions, such as found in winter greenhouse culture. This allowed specific rootstock selection for each environment. Bacterial wilt resistant rootstocks should have a certain level of heat tolerance while brown root rot resistant rootstocks need to be tolerant to cool temperatures. World Vegetable Centre, Taiwan reports screening multiple solanaceous varieties and wild accessions for resistance against bacterial wilt. They recommend using Hawaii 7996 as rootstock for tomato to control bacterial wilt except where flooding or saturated soils are expected, in which case they recommend the eggplant cultivar 'Surya' (Black *et al.,* 2003).

There is an increase in the use of grafting for heirloom tomatoes, especially in organic production (Rivard and Louws, 2008). In these situations, disease resistance is foremost whereas increased vigor is often not needed since too much vigor can negatively impact yields. One major limitation to grafting tomato is that Tomato Mosaic Virus (ToMV) resistance in the rootstock and scion must be compatible. Resistance to ToMV can be derived from the Tm-1 or Tm-2 alleles; the Tm-1 allele confers a tolerance or symptomless response in the host while the Tm-2 alleles confer a hypersensitive response (Hall, 1980). If a rootstock carries Tm-1 resistance and the scion carries Tm-2 resistance then the rootstock is capable of transmitting virus to the scion and inducing systemic necrosis (Yamakawa, 1982).

Beans

In beans, the potential for self fertilization is 99%. The male and female organs are mature enough for fertilization and once the fertilization process is accomplished, the flower opens. The dry and hot weather during fertilization period have a negative impact on fruit set. In case of beans, gene conferring pole type character rather than bush type can be exploited for polyhouse cultivation. A morphogronomic concept is defined as a result of interaction of several characters which determine the final plant architecture. These characters are: type of growth of the stem and branches i.e., determinate (in) or indeterminate (fin/fin+); number of nodes on the stem; length of internodes; climbing ability and branching pattern (position, number, length). Four prototypes are defined as follows:

1. Bush, determinate
2. Bush, indeterminate
3. Indeterminate prostrate with well developed branching but low or non existence of climbing ability.
4. Indeterminate with long guide and high climbing ability.

Breeding Behaviour

Generally, beans are self pollinated in nature but due to floral morphology as extrovert nature of pistil, it exhibits a sort of cross pollination.

Breeding objectives for beans under polyhouse condition

- High pod yield
- Nonstringy, long pods, flat or round in shape
- Early pod harvesting
- Pole type plant
- High number of green pods/plant
- High number of pod clusters/plant
- High number of primary branches/plant
- Free from interlocular space
- Photo-insensitivity
- Wider adaptability
- Resistance to diseases and insect pests

Some important genes of beans

Gene Symbol	Phenotype
Are	Resistance to anthracnose
Da	Straight pod
dw 1 dw 2	Duplicate gene causing dwarf plant
Ext	Interspecific gene for external stigma in *Phaseolus coccineus*
Fa	Basic gene for pod membrane
Fin (fin^+)	Indeterminate vs 'fin'
gas	Determinate plant growth
in	Causes both male and female sterility
itor	Determinate growth = (fin)
L	With *ram* triple branched inflorescenceLong vs short internodes
mo	Conditional resistance to bean virus
neu^+	Short day flowering response
ri	Confers resistance to bean virus 1
St	Stringy pods
T	Twining habit vs non-twining
te	Pod short (5-8 cm) and narrowC
To	ell wall fibre
tri	Produces three cotyledons
Uni	Unifoliate leaves

The varieties commercially grown and available in India on a large scale under bush type are Contender, Pusa Parvati, VL Boni 1, Arka Komal, Pant Anupama, Pant Bean 2 (UPF 626) and under Pole type are Kentucky Wonder and Canadian Wonder. Keeping in view of ideotype requirement for polyhouse cultivation, the pole type bean genotypes may be suited under protected environment.

Capsicum

In case of capsicum (*Capsicum annuum*), the yellow and red coloured fruits are preferred and have high market demand. Therefore, genes responsible for such characters can also be exploited under polyhouse cultivation. The varieties grown under polyhouse were Natasha, Swarna and Indra. The yield goes upto 896 to1000 q/ha when the crop of capsicum is taken under polyhouse cultivation. The other varieties available are California Wonder, Arka Mohini, Arka Gaurav, Arka Basant, Bharat, Kt-1, Pusa Deepti, Solan Hybrid-1, etc.

Breeding Behaviour

Both self and cross pollination occur in capsicum which extend upto 7-36% by bees and thrips. Pollination is also due to gravitational force. Cross pollination is also due to protogyny and heterostyly.

Breeding Objectives

- Earliness
- Indeterminate growth habit
- Desirable fruit shape and size
- Superior fruit quality
- Resistance to diseases and insect pests
- Resistance to biotic and abiotic stresses (heat, water stress, salinity, etc.)

Grafting in Capsicum

There are reports of pepper grafting, primarily to control *Phytophthora capsici* (Morra, 2004; Yamakawa, 1982) and nematodes (Kokalis-Burelle *et al.*, 2009; Oka *et al.,* 2004). However, pepper is not typically grafted commercially, with one exception apparently in winter greenhouse production in the Korean peninsula to control *P. capsici* (Kwon *et al.*, 2006). There are reports of successful grafts between pepper and several other genera of the Solanaceae family, indicating there is substantial potential for improved disease resistance in pepper by grafting.

Brinjal

Brinjal, *Solanum melongena* L. (Solanaceae), is often referred to as eggplant or aubergine in other regions of the world and is grown for its fleshy fruit. India is the second largest producer of brinjal in the world and contributes 26 percent of the global brinjal production. It is a versatile crop and has been grown in the country under different agro-climatic environments. It is consumed in every part of the country irrespective of the income of the people. In India, per capita consumption of vegetables is only about 190 g per day, which is 32 % less than the minimum dietary requirement of 280 g per day per person (ICMR – 2008). Brinjal is the most affordable as well as a nutritive vegetable. It is a rich source of calcium, phosphorous, potassium, fiber, folic acid, sodium and vitamins B and C.

Brinjal is usually a self pollinated crop but the extent of cross pollination is reported as high as 29% and hence, it is classified as often cross pollinated or facultative cross pollinator.

Breeding objectives:

- Earliness
- Large fruit (400-500 g) with dark purple colour having green calyx
- Indeterminate growth habit with robust plant stature.
- Low proportion of seed

- Soft flesh
- Lower solanine content
- Upright sturdy plant free from lodging
- Resistance to bacterial wilt, phompsis blight, little leaf, root knot nematodes, shoot and fruit borer, jassids and Epilachna beetle.

Qualitative genes in aubergine

Traits	Dominance
Hypocotyl	Purple > green
Stem	Purple> green>
Purple coloration of stem, petiole and vein	green
Stem colour	Purple > green
Stem and leaf	Purple stem> green leaf
Presence of spine	> absence of spine
Purple and spine	> green and spine
Plant height	Tall > dwarf
Fruit colour	Green (*Gm*)

Grafting in Eggplant

The primary purpose for grafting eggplant is to control soil-borne diseases, namely *Verticillium* wilt, bacterial wilt, *Fusarium* wilt, and root-knot nematodes (Goth *et al.*, 1991; Kalloo, 1993; Yamakawa, 1982). The first rootstocks used for grafting eggplant were selections from *S. integrifolium* (Yamakawa, 1982). *S. integrifolium* is highly resistant to Fusarium wilt and is more resistant to bacterial wilt than most eggplant cultivars and is highly compatible with eggplant and allows prolonged harvest (Tachibana, 1994). However, the level of bacterial wilt resistance is not enough to protect the scion when conditions favor the disease (Iwamoto *et al.*, 2007). Creating interspecific hybrids between *S. integrifolium* selections and eggplant genotypes with some resistance to bacterial wilt were successfully used to control bacterial wilt (Daunay *et al.*, 2001), but the level of resistance is still less than that found in some other *Solanum* spp. (Yamakawa, 1982). More recently, somatic fertile hybrids were created between *S. integrifolium* and *S. sanitwongsei*; both parents have resistance to bacterial wilt, although the disease resistance of the progeny was not reported (Iwamoto *et al.*, 2007). Tomato hybrids are popular eggplant rootstocks in Europe (Daunay *et al.*, 2001; Khah, 2005). The reason for this preference is probably due to the amount of genetic improvement made in tomato compared to eggplant rootstocks or the familiarity of growers with the tomato rootstocks. However, some researchers found that in disease-free soil, tomato rootstocks can have deleterious effects on yield and performance of eggplant (Honma, 1977). Morra (2004) reported that in Italy, where a large and diverse group of eggplant cultivars are grown, eggplant grafted onto tomato rootstocks produced variable results. *S. torvum* in particular is a promising rootstock for eggplant since genotypes in this species provide a high level of resistance to bacterial wilt, *Fusarium* wilt, *Verticillium* wilt and root-knot nematode. They also provide good vigor to the scion (Yamakawa, 1982). However, uniform seed germination is reported to be a problem that has limited its use (Ginoux and Laterrot, 1991). It is likely that new tomato rootstocks will be evaluated with eggplant cultivars. However, based on previous results, the best eggplant rootstocks are likely to come from more closely related *Solanum spp*. and interspecific hybrids among these species.

Problems and Prospects in India

Problems

Inspite of being advantageous, polyhouse cultivation still lags behind because of various problems conferred by vegetable growers. Problems requiring concerted efforts are:

- Lack of suitable varieties to be grown for polyhouse condition.
- Standardization of design of polyhouse and other structures for different temperature regions of the country.
- Cladding materials (UV grade LDPE etc.) are very costly. Intensive research is required to make cladding material affordable by average vegetable grower. Plastic films available in the country are not only costly but also have short life. The UV stabilized LDPE is not easily available.
- Lack of professional polyhouse manufacturers leading to use of improvised structures.
- Instruments for environment control in the polyhouse are not easily available.
- Vegetable production technologies including suitable varieties for different crops under different types of polyhouses for different temperature zones of the country are not worked out and documented.
- Lack of awareness among farmers pertaining to potential of polyhouse vegetable production.
- Import of polyhouses which are not suitable for Indian conditions.
- Due attention to these problems would lead to popularization of polyhouse vegetable production in temperate regions of the country.

Prospects

There are a number of opportunities in various temperate regions in India for commercial polyhouse vegetable production. In temperate areas, vegetable growers can increase their income by raising early crops in polyhouse structures mainly in low-cost polyhouses. Raising of vegetable nursery in polyhouse has many fold benefits like easy management, early nursery, protection from biotic and abiotic stresses, etc. The awareness for good nutrition and quality vegetables free from pesticides is rapidly increasing. The potential of polyhouse is emphasized. Polyhouse conditions provide many fold advantages over open field vegetable production. It conserves water and land, it is eco-friendly and does not require much sophistication. In 21st century, polyhouse vegetable production is likely to be a common commercial practice in the country not only because of its potential but because of its sheer necessity.

Polyhouse vegetable production is boon to cold desert region of the country and at high altitudes, where this technology has helped commercial cultivation and production of several vegetables during frozen winters, hitherto considered impossible. There are several avenues where polyhouse vegetable production could be of economic and academic advantage. Farmers having very small holdings can plan commercial vegetable production. Increased productivity of vegetables under polyhouse favours its early adoption in India.

To boost polyhouse vegetable production and breeding, strong research support is required. In vegetable production, hybrid seeds, transgenics, stress resistant varieties, micropropagated transplants and synthetic seeds are likely to replace conventional varieties. Breakthrough in electronics mainly for automation of farming practices like robotization, etc. would further facilitate polyhouse cultivation. Production of different vegetables throughout the year at a particular place would necessitate popularization or polyhouse cultivation. Barring flower industry, polyhouse farming is most suitable for particular strain of vegetables and to express its full potential would be the major research area in polyhouse cultivation of vegetable crops besides, structural engineering, poly photochemistry and plant nutrition.

Polyhouse vegetable production technology is highly potential for inaccessible areas and arctic deserts besides, boosting production in traditional growing areas. In future, organically grown vegetables would be in great demand in the country where polyhouse cultivation would be able to meet the demand of the consumer.

References

Ahn, S.J., Im, Y.J., Chung, G.C., Cho, B.H. and Suh, S.R. 1999. Physiological responses of grafted cucumber leaves and rootstock roots affected by low root temperature. *Sci. Hort.* **81**:397-408.

Arnó, J., 2005. Factors behind the increasing whitefly problems in greenhouse crops. *Newlsletter on Biological Control in Greenhouses Sting* **27**: 3-6.

Black, L.L., Wu, D.L., Wang, J.F., Kalb, T., Abbass, D. and Chen, J.H., 2003. Graftin tomatoes for production in the hot-wet season. In: International Cooperators' Guide. Asian Vegetable Research and Development Center.

Cohen, R., Elkind, Y., Burger, Y., Offenbach, R. and Nerson, H., 1996. Variation in th response of melon genotypes to sudden wilt. *Euphytica* **87:**91–99.

Cohen, R., Horev, C., Burger, Y., Shriber, S., Hershenhorn, J., Katan, J. and Edelstein, M. 2002. Horticultural and pathological aspects of Fusarium wilt management using grafted melons. *Hort. Science*, **37**.

Colla, G., Fanasca, S., Cardarelli, M., Rouphael, Y., Saccardo, F., Graifenberg, A. and Curadi, M., 2005. Evaluation of salt tolerance in rootstocks of Cucurbitaceae. *Acta Hortic.* **697**: 469–474.

Crosby, K.M., 2000a. Impact of Monosporascus cannonballus on root growth of divers melon varieties and their F1 progeny in the field. *J. Rio Grande Hortic. Soc.*, **52**.

Crosby, K.M., 2000b. Narrow-sense heritability estimates for root traits and *Monosporascus cannonballus* tolerance in melon (*Cucumis melo*) by parent–offspring regression. (In) Katzir, N., Paris, H.S. (Eds.), Proceedings of 7th EICARPIA Meeting on Cucurbit Genetics and Breeding. Acta Hortic., Leuven, Belgium.

Crosby, K.M., 2001. Screening *Cucumis melo* L. agrestis germplasm for resistance to *Monosporascus cannonballus. J. Rio Grande Hortic. Soc.* **53:**24–26.

Crosby, K.M., Miller, M.E. and Wolff, D.W., 2002. Screening plant introductions of *Cucumi melo* L. for resistance to *Monosporascus cannonballus.* In: Maynard, D.N. (Ed.) Cucurbitaceae. ASHS Press, Naples, FL, pp. 188–191.

Daunay, M.-C., Lester, R.N. and Ano, G., 2001. Eggplant. In: Charrier, A., Jacquot, M. Hamon, S. and Nicolas, D. (Eds.), Tropical Plant Breeding. CIRAD and Science Publishers Inc., pp. 199–222.

De Ponti O.M.B., 1975. Breeding Parthenocarpic pickling cucumbers (*Cucumis sativus* L.): necessity, genetical possibilities, environmental influences and selection criteria. *Euphytica* **25**: 29-40

Edelstein, M., Burger, Y., Horev, C., Porat, A., Meir, A., and Cohen, R., 2004. Assessing th effect of genetic and anatomic variation of Cucurbita rootstocks on vigor, survival and yield of grafted melons. *J. Hortic. Sci. Biotechnol.*, **79.**

El-Hafez, I.I.S and Baker, L.R. 1981a. Inheritance of Parthenocarpic yield in gynoecious pickling cucumber (Cucumis sativus L.) *Pissertion Abstract International,* **40:** 2923-2924

Giannakou, I.O. and Karpouzas, D.G., 2003. Alternatives to methyl bromide for root-knot nematode control. *Pest Manag. Sci.* **59:**883–892.

Ginoux, G. and Laterrot, H., 1991. Greffage de l'aubergine: reflexions sur le choix du porte-greffe. *PHM Revue Horticole* **321:**49–54.

Goth, R.W., Haynes, K.G. and Barksdale, T.H., 1991. Improvement of levels of bacterial wilt resistance in eggplant through breeding. *Plant Dis*. **75:**398–401.

Gyu, L.S., Ung, C.J., Young, K.K., Ho, C.J. and Beom, L.Y. 1997. Effect of rootstocks and grafting methods on the growth and fruit quality of tomato (Lycoperiscon esculentum Mill.). *RDA J. Hort. Sci.* **39**(2):15-20. (in Biological Abstracts).

Hall, T.J., 1980. Resistance at the TM-2 locus in the tomato to tomato mosaic virus. *Euphytica* **29:**189–197.

Herman, R. and Perl-Treves, R., 2007. Characterization and inheritance of a new source of resistance to *Fusarium oxysporum f. sp. melonis* Race 1.2 in *Cucumis melo. Plant Dis.* **91:**1180–1186.

Hochmuth, R.C. Greenhouse cucumber production. Florida Greenhouse Vegetable Production Handbook. **Vol. 3** (from internet site)

Kasrawi, M.A. 1988. Effect of silver nitrate on sex expression and pollen viability in parthenocarpic cucumber (*Cucumis sativus* L.). Dirast (Jorolan*).* **15** (II): 69-78

Lee, J.M. 1994. Cultivation of Grafted Vegetables I. Current status, grafting methods, and benefits. *Hort. Sci.* **29**(4):235-239.

Muigai, S. G. Muigai, M. J., Bassett, M. J., Schuster, D. J., and Scott, J.W., 2003. Greenhouse and field screening of wild *Lycopersicon* germplasm for resistance to the whitefly *Bemisia argentifolii*, *Phytoparasitica*, **31**(1): 27-38.

Oda, M. 1999. Grafting of vegetables to improve greenhouse production. Food & Fertilizer Technology Center, Extension Bulletin **480**:1-11.

Peterson, C.E. and Anhder L.D., 1960. Induction of staminate flowers on gynoecious cucumbers with giberellin A_3. *Science* **131:** 1673-1674

Pike, L.M. and Peterson, C.E. 1969. Inheritance of parthenocarpy in the cucumber (*Cucumis sativus* L.). Euphytica. **18:** 101-105.

Ponti, O.M. B. de 1976. Breeding parthenocarpic pickling cucumber *(Cucumis sativus* L.), necessity, genetical possibilities, environmental influences and selection criteria. *Euphytica.* **25**: 29.

Ponti, O.M. B. de. and Garresten, F. 1976. Inheritance of parthenocarpy in pickling cucumbers (*Cucumis sativus* L.) and linkage with other characters. *Euphytica*. **25**: 633-639.

Robinson, R.W and Catliffe, D.J 1971. Morphactin induced parthenocarpy in cucumber. *Science.* **171**: 1251-1255.

Singh, D.K and Ram, H.H. 2004. Isolation of parthenocarpic cucumber line. First Indian Horticulture Congress, on Improving Productivity, Quality, Post Harvest Management and Trade in Horticultural Crops Organized by Indian Society of Horticulture, held at IARI New Delhi, 6th- 9th November 2004. Book of Abstracts pp. 87.

Vuruskan, M.A. and Yanmaz, R. 1991. Effects of different grafting methods on the success of grafting and yield of eggplant/tomato graft combination. *Acta Hort.* **287**:405-409.

Zwinkles, L.A.M. 1987. The present situation and fruit development in breeding of parthenocarpic pickling cucumber. *Acta Horticulture*. **220**:129-130.

9

Recent Advances in Fertigation Techniques for Protected Cultivation of Vegetable Crops

A.S. Sidhu, M. Prabhakar, S.S. Hebbar, A.K. Nair and H.P. Bhagya

Abstract

Vegetables are health capsules, being the chief source of vitamins and minerals besides providing other dietary elements like carbohydrates, proteins and even fat and enzymes to some extent, antioxidants and fibre. Though India is the second largest vegetable producer (147 million tones in 2012) in the world, per capita supply and consumption of vegetables in India is very low (160 g day $^{-1}$). It is now realized that the problems of satisfying hunger and nutrition are much beyond the realm of food crops alone and therefore, to supplement and balance our diet, more vegetables are to be included. Vegetables not only provide nutritional security in rural areas but also financial security to small and marginal land holders growing vegetables in the hinterlands of urban markets. Commercialization of vegetable production is the scene of the day as they are money spinners in domestic markets and dollar earners in global markets. To compete in the domestic and global market and for year round supply of quality produce, protected or greenhouse cultivation of vegetables is imperative. Water and nutrients are the top most growth and production inputs in any crop cultivation and vegetables are no exception to this. In order to get the potential yields with matching quality of *vegetables grown under protected cultivation, the best of the management strategies to supply water and nutrients is needed. In this direction, drip fertigation is the best option for efficient supply of irrigation water and fertilizer nutrients for greenhouse grown vegetable crops. The studies carried in different parts of the world have clearly indicated that fertigation technology resulted in substantial yield increase with marked enhancement of water and nutrient productivity.*

Keywords: Protected cultivation, fertigation, drip irrigation, water soluble fertilizers, capsicum, tomato, cucumber, yield, quality

Introduction

Protected or greenhouse cultivation technology is an alternative to boost vegetable production. The yield advantage in terms of quantum jump in yield in a greenhouse-grown crop is indisputable. Increase in quantity, improvement in quality and regularity of supply make greenhouse cultivation the right approach to cater to special needs of the near and far markets. As agriculture have a major share in India's economy, the productivity should be substantially increased from the present levels, if India has to emerge as an economic power in the world. We need to constantly improve the productivity, profitability and sustainability of our major farming systems by bringing more and more vegetables under protected cultivation technology.

Though India and China developed protected cultivation technology almost at the same time, substantial progress took place in China with reports quoting nearly half of the vegetable production from protected cultivation whereas in India, it is beginning to happen with around 2500 ha of green houses used for production of vegetable crops. India's demand for water is likely to exceed the available source by the year 2020. This situation is due to rapidly increasing human population demanding more and more water for agriculture, industry and domestic use with a finite natural resource. Agriculture is the biggest consumer of water with about 90% of the renewable resource used for irrigation. Traditional irrigation methods are highly inefficient only about one-third of the applied water is actually transpired by the crops. Clearly, there is great scope for improved irrigation management. Intensification of agricultural production to meet growing market demand requires simultaneous application of irrigation water and fertilizers. Application of fertilizer in drip irrigation (fertigation) is an effective way to promote efficient use of these scarce and expensive resources.

Need to supply enough food for a growing population stimulated interest to increase irrigation efficiency. The first experiments leading to the development of drip irrigation date back to end of 19th century, but real progress was not achieved until the late 1950's and early 1960's (Keller and Bliesner, 1990). Rapid implementation of drip irrigation started in the 70's as a result of invention of cheap plastic pipes. Drip irrigation and similar water saving low volume high efficiency water application methods are grouped and termed as micro irrigation. Micro irrigation systems include emitters or drippers, row wetting devices, in-line drip laterals, micro-jets and micro sprinklers. The worldwide area irrigated by these systems in 2006, was about 6 million ha (Sne, 2006).

Adoption of drip irrigation methods with only partial soil wetting brought concomitant transition in restricting crop root system distribution mainly to the wetted zone. These limited root systems considerably modify classical fertilization management. The shift from broadcast fertilizer application to banded fertilization or to fertilizer added to the irrigation water was developed to meet the nutrient needs of drip irrigated crop. Chronologically, fertigation was an outcome of localized irrigation. Sustainable high yield with high yielding crops depends entirely on sustainable use of limited source of water and energy, specifically in developing countries with arid and semi-arid regions. This can only be attained with efficient use of water and fertilizers.

Fertigation is a tool to supply plant with its daily demand of water and nutrients as required by its specific growth stage throughout its development to achieve maximum efficiency of the fertilizer applied. By adopting this approach, readily soluble nutrients can be supplied directly to the root volume thereby maximizing nutrient efficiency and minimizing over fertilization and leakage to the underground water with possible damage to the environment. Fertigation is an attractive technology in modern irrigated agriculture which increases yield and fertilizer use efficiency. It is reported that with fertigation, N fertilizer use efficiency can be enhanced to 80–90 %. Through fertigation, water and nutrients are applied to the root zone of the crop, where they are mostly needed, normally resulting in a better water and fertilizer use efficiency than with conventional irrigation and fertilization methods.

There is widespread interest among farmers in India in fertigation. Nevertheless, information on the form and concentration of the nutrients required for different crops are presently inadequate. Moreover, low fertilizer recoveries due to extensive fertilization practiced during the last a few decades have created serious agricultural and environmental problems. High nitrate concentrations in groundwater and deterioration of some important quality parameters of agricultural products are the main concerns.

The chemical composition of soluble single-nutrient and compound fertilizers produced by the fertilizer industry are usually almost the same worldwide. On the other hand, application of these fertilizers is highly site-specific, depending on crop, variety, soil type, climatic conditions and water quality. Fertilizer demand in intensive vegetable production systems in protected cultivation is particularly variable, changing rapidly during the season and year and even within a day and night. Nutrient requirements of vegetable crops depend very much on the biological stage of growth, varying from seedling to harvest. Traditional surface irrigation systems result in a wet soil profile. Plant nutrients are distributed in this wet soil volume depending on their mobility, their adsorption and precipitation reactions with soil particles.

Movement of water in soil from a dripper point source progresses in both horizontal circular directions on the soil surface, as well as in vertical direction down the soil profile. This creates a wet soil volume with soil varying in moisture content over soil depth (Bresler, 1977). The interaction of soil particles with water is mainly physical in nature and involves adsorption and capillary forces which control the distribution of a unit volume of water in a unit volume of soil. Soil volume is subjected to mechanical compaction by field implements which affect pore size distribution and volumetric soil moisture content.

Advantages of Fertigation

Fertigation allows applying the nutrients exactly and uniformly only to the wetted root volume, where the active roots are concentrated. This remarkably increases the efficiency in the application of the fertilizer, which allows in reducing the amount of applied fertilizer. This not only reduces the production costs but also lessens the potential of groundwater pollution caused by fertilizer leaching. Fertigation allows adjustment of the amount and concentration of the applied nutrients to meet the actual nutritional requirement of the crop throughout the growing season. Under fertigation by increasing rates of N-fertilizer, the optimal crop yield was often reached at 50 % of the locally recommended N rate under conventional irrigation, thus, resulting in considerable saving of costly N fertilizer while at the same time reducing potential pollution hazard. Other advantages of the fertigation are: (1) saving of energy and labour, (2) flexibility of moment of application (nutrients can be applied to the soil when crop or soil conditions would otherwise prohibit entry into the field with conventional equipment), (3) convenient use of compound and ready-mix nutrient solutions containing small concentrations of micronutrients which are otherwise very difficult to apply accurately to the soil, and (4) supply of nutrients can be more carefully regulated and monitored. When fertigation is applied through drip irrigation system, crop foliage can be kept dry thus, avoiding leaf burn and delaying the development of plant pathogens. Drip and micro-irrigation have a characteristic not shared by any other irrigation method: fertigation is not optional, but is actually necessary. Fertigation provides the only good way to apply fertilizers physically to the crop root zone.

Fertigation in Greenhouse Grown Vegetables

The practice of supplying crops in the field with fertilizers via irrigation water is called fertigation. Fertigation- a modern agro-technique provides an excellent opportunity to maximize yield and minimize environmental pollution by increasing fertilizer use efficiency, minimizing fertilizer application and increasing return on the fertilizer invested. In fertigation, timing, amount and concentration of fertilizers applied are easily controlled.

Greenhouse studies (Goldberg *et al.*, 1971; Bar-Yosef and Sagiv,1982) in which the distribution of water and minerals were examined, showed that maintenance of nutrients within the sphere of plant roots with drip irrigation could be accomplished if proper rates and water application frequencies were used. Thus, drip irrigation method potentially provides better control of fertilizer nutrients that move within soil water resulting in reduced mineral leaching from the crop root zone. Plant analysis indicated that the treatments in which fertilizer N was applied through drip system, regardless of timing or growth stage, showed higher percentage of fertilizer N contents during most of the growing season than those of the furrow irrigated treatments. Fertilizers applied through trickle system can improve nutrient use efficiency (Miller *et al.*, 1981; Phene and Beale, 1976). Through fertigation, applied nutrients loss through leaching is around 10 per cent (Bresler, 1977; Stark *et al.*, 1983) compared to typical loss of about 30 to 50 per cent in the traditional system. It also increases flexibilty in scheduling of fertilizer application to fit crop needs (Miller *et al.* 1976; Phene and Beale, 1976; Phene *et al.*, 1979; Rolston *et al.*, 1981). Compared to traditional methods of fertilizer application, fertigation through drip resulted in 25 to 30 % yield improvements worldwide. It also improves the quality of vegetables harvested with the same amount of fertilizers (Bucks *et al.*, 1974; Miller *et al.*,1976). As the crop growth advances, we can change the dosage and ratios of fertilizers.

Requirements for Fertigation

In pressurized irrigation systems, pressure of the injected fertilizer solution has to be greater than that of the internal pressure in irrigation pipe line. The choice of fertigation equipment has to take into account cost, crop requirement and irrigation system capacity. Injection of fertilizer consumes energy to overcome the internal pressure of the irrigation network. Fertilizer injection equipment is classified into three principal groups, according to the means employed to obtain the higher pressure for the fertilizer solution: - injection by ventury device, differential pressure and positive pressure.

Venturi Injectors: Basically, these injectors work by means of a pressure difference between the irrigation line and the fertilizer stock tank. The system of injection is very low cost and works by creation of vacuum in the ventury, which is nothing but a conical constriction in the pipe. The vacuum is created by throttling of water, created by diverting a part of water from the main line. The fertilizer solution is sucked into the ventury through a suction pipe.

Fertilizer Injectors: Fertilizer injectors are piston or diaphragm pumps, driven by water pressure of the irrigation system and the injection rate is proportional to the flow of water in the system. A high degree of control over the fertilizer injection rate is possible and no serious head loss is noticed. This is the method of fertilizer dosing in most of the situations. These injectors are more expensive but are very accurate in proportioning fertilizer into irrigation lines regardless of water pressure. These injectors also have a much broader ratio with 1:100 and 1:200 being the most common. Thus, stock tanks for large application areas are of manageable size and these injectors have much larger flow rates. Injection by these proportioners is controlled either by a water pump or an electrical pump. Multihead installations permit feeding several fertilizers simultaneously without mixing. This is especially significant for fertilizers which are incompatible (forming precipitates, etc.) when mixed together in concentrated form. Dosatron features variable ratios (1:50 to 1:500) and plain water bypass. Amiad make or TMB make pumps are the other makes generally installed in the greenhouses. The most common power source for fertigation pumps are hydraulic energy, the device uses hydraulic pressure of irrigation water to inject nutrient solution. These pumps are suitable for fertigation in areas devoid of sources of electricity. Electric dosing pumps activate the fertilizer pump. These are common in glasshouses and in areas where electricity is available and reliable.

Fertilizer tank: In this system, a part of irrigation water is diverted from the mainline to flow through a tank containing fertilizer in fluid form or soluble solid form, before returning to main line. The pressure in the main line and the fertilizer tank is the same.

Other important equipments other than injectors are a filter to prevent dripper clogging by any solid particles from reaching the dripper and a back-flow preventing valve.

Fertilizers

Solubility of fertilizers in indigenous water source: Irrigation water contains various chemical constituents some of which may interact with dissolved fertilizers with undesired effects. The degree of acidity of fertilizer solution has to be considered in relation to its corrosiveness to irrigation system components. Reactions of soil particles with various chemical compounds delivered in the drip irrigation solutions, however, are very complex. They involve chemical interactions between the constituents of soil particles, which carry permanent electrical charges on their surfaces such as precipitation reactions with calcium carbonate (lime) in basic reactive soils and with aluminum and iron in acid soils.

Nitrogen

Nitrogen is the element consumed in the largest amount by plants. It is also soluble and mobile in the soil. It is subject to leaching. Plant quality and yield depend upon sufficient nitrogen in the soil during critical growth phases. When anhydrous ammonia or aqua ammonia fertilizers are injected into irrigation water, ammonia being quite water soluble. pH of the water rises and some of the ammonia escape and volatilize into the atmosphere. High pH of the irrigation water will also cause calcium and magnesium ions in the water to precipitate out of solution as lime in a very finely divided form (calcium and magnesium hydroxides and carbonates). The harder the water, the more lime precipitates which clog drip lines and filters, as well as valves, gates and sprinklers. Precipitate applied by the irrigation system in water tends to clog soil pores over time, reducing the soil's infiltration rate. Nitrogen (N) in fertilizer solutions is available mainly in three forms: Ammonium–N that has a positive electric charge (cation); Nitrate–N that has a negative electric charge (anion) and Urea–N that is a non-charged molecule.

These N compounds encounter a highly complex environment when they come in contact with the soil. The ammonium cation is adsorbed to the negatively charged clay particles and is slowly oxidized by soil bacteria to nitrate-N. The nitrate form of nitrogen does not react with soil exchange sites and is not held in soils. Nitrates move with other soluble salts to the wetted front. This is of particular interest since, NO_3 -N should always be applied with every irrigation and at the concentration needed by the fertigated crop to satisfy its requirement. Under irregular NO_3-N application, the fertigated crops might be under over-fertilization stress on the day of fertilizer application and under deficient stress due to leaching following the irrigation without fertilizer. Another part of nitrate moves with water and accumulates in very high concentration at the boundary between wet and dry soil zones. Urea, the non-charged molecule, is able to travel considerable distances in the soil with the moving water. Once in contact with soil enzyme urease, however, this molecule is rapidly converted to carbon dioxide (CO_2) and ammonia, which upon dissolving in water, results temporarily in local rise of soil pH. The ammonium form of N, derived from ammonium or urea fertilizers, is not nearly subjected to immediate leaching losses because temporarily, they may be fixed on exchange sites in the soil. Urea is a highly soluble, chargeless molecule, which easily moves with irrigation water and is distributed in the soil similarly to NO_3. At 25°C, it is hydrolysed by soil microbial enzymes into NH_4 within a few days.

Main consideration is given to nitrate N concentration in the soil volume occupied by active root system. Generally, distribution of fertilizer in relation to water in the soil profile is considered the key factor in determining nitrogen fertilizer efficiency. The advantages in terms of environment indicate that nitrogen fertilizers used in fertigation are used very efficiently by drip irrigation. Small amounts of fertilizers are supplied at regular intervals (daily or once a week) rather than giving in one big dose at planting or in 2 to 3 splits during the entire crop cycle. Nitrogen fertilization through trickle irrigation method is particularly attractive since N can be applied throughout the growing season at rates designed to meet the changing crop demand. By carefully regulating the amount of N applied with the trickle irrigation system, N losses due to leaching could be minimized (Phene and Beale,1976) and the efficiency of N fertilizer recovery could be relatively high (Phene *et al.*,1979; Miller *et al.*,1981 ; Stark *et al.*,1983). Adequate nitrogen can be applied to tomatoes using high frequency N fertilization without large denitrification losses (Stark *et al.*,1983), and early yields can be realised (Nicklow and Downes,1971). With trickle irrigated and plastic mulched tomatoes, blossom end rot incidence was less (Cook and Sanders,1991).

Phosphorus

Phosphate soluble fertilizers are prone to precipitation reactions with calcium (Ca) and magnesium (Mg) already in the irrigation line when the solution has a pH above 7 or when soluble iron (Fe) is present at low pH. Thus, even before phosphorus (P) emerges from the drip system; it has to be protected from precipitation, both inside the drip lines and in the fertilization tanks. Once in soil, the distance travelled by P is the smallest of all nutrients supplied by fertigation. Phosphorus fertigation has to take into account water quality, soil chemical composition and plant age.

Demand for high phosphate concentration for younger seedlings in most of the vegetable crops and limitation of phosphate amounts in irrigation system lead to pre-seeding/pre-planting incorporation of superphosphate and later, supply of water soluble phosphates through irrigation systems. Phosphorus is generally not recommended for application in trickle irrigation system because of its tendency to cause clogging and its limited movement in the soil. Use of acidic fertilizer sources such as urea phosphate and availability of completely water soluble solid fertilizers has made it possible to supply P through fertigation since, these sources should minimize formation of insoluble P salts and some reports indicate that this procedure stimulates plant growth and yield. These benefits are probably due to "band-like" application of P by drip irrigation, nutrients are not widely mixed, avoiding P fixation.

Contrary to N and K, phosphorus is readily fixed in most soils. Movement of applied P differs with the form of fertilizer, soil texture, soil pH and the pH of the fertilizer (Papadopoulos and Ristimaki, 1998). Under calcareous conditions, P is readily fixed. Phosphorus mobility in soil is very restricted due to its strong retention by soil oxides and clay minerals. Soil application of commonly available P fertilizers generally results in poor utilization efficiency principally because phosphate ions rapidly undergo precipitation and adsorption reactions in the soil, which remove them from the soil solution. Consequently, there is little or no movement of phosphate from point of contact with the soil. Therefore, there is inefficient utilization of applied P fertilizers. Continuous application of P through irrigation water was shown superior to applying P in adequate quantities as basic fertilization. Rauschkolb *et al.*, (1976), found that P movement increases 5 to 10 folds when applied through drip system, indicating that fertigation of P is particularly important.

Commercial standard P-fertilizers may also precipitate in the irrigation lines in reaction with ions, (such as Ca or Mg) in the irrigation water. Therefore, when choosing P fertilizer for irrigation, besides solubility, care must be taken to avoid P-Ca and P-Mg precipitation in the tubes and emitters. From this standpoint, acid P fertilizers (e.g., phosphoric acid, urea phosphate or mono-ammonium phosphate) are recommended. Calcium and Mg carbonate and orthophosphate precipitations may occur in the tubes and the drippers. In addition, high pH may reduce Zn, Fe and P availability to plants. The desired pH is below 7 and the range favoured by most cultivated crops is 5.5-6.5. pH of the irrigation water could be reduced or controlled by using Phosphoric acid or acid based fertilizers like urea phosphate and mono-ammonium phosphate. Use of acid fertilizers in drip systems may be beneficial in many ways other than the direct benefit from the added P, such as increased solubility of soil native P minerals, increased availability of other nutrients and micronutrients and prevention of chemical clogging of the fertigation system.

Potassium: Potassium (K) is the most stable form of all the major nutrients supplied by fertigation, always remaining in the same chemical form as a monovalent cation (K+). It is well established that better K nutrition tended to increase water content of the plant tissue. K sources such as potassium sulphate, potassium chloride and potassium nitrate are readily soluble in water, move freely in the soil and some of the potassium ions are exchanged in clay complex and are not readily leached away. Recent studies indicate that later supply of K, especially during fruit bearing stage, is likely to improve the crop quality and marketability of the produce.

Micronutrients: Micronutrients like iron, zinc, copper, manganese, can be applied as chelates or sulfate salts in trickle irrigation water. Their application requires careful and precise metering. Now, many of these are made available along with water soluble macro nutrients manufactured as specialty fertilizers by many firms.

Types of Fertilizer for Fertigation

1. Liquid fertilizers: These are the solutions which contain one or more plant nutrients in liquid form.
2. Solid water soluble fertilizer: These are 100% water soluble fertilizer, also referred as specialty fertilizers. These fertilizers generally contain two or more major nutrients as well as micro nutrients. Soluble fertilizer completely dissolves in water leaving no precipitate. Therefore, there is no chance of clogging of emitters due to fertigation.

Features of 100% Water Soluble Fertilizers

1. No chloride, no sodium and no salt build up.
2. Completely soluble in water.

3. Most of the fertilizers are acidic and hence, special acidification treatment for dripper clogging may not be necessary.
4. Blended with micronutrients

Fertilizes used in fertigation

1. Straight Fertilizers

Urea Phosphate (17:44:0), Mono Ammonium Phosphate (12:61:0), Mono potassium Phosphate (0:52:34), Potassium nitrate (13:0:46), Sulphate of potash (0:0:50 + 18% S),

Calcium nitrate, Magnesium nitrate and calcium magnesium nitrate.

2. NPK Formulations

20:10:10 + TE+ Mg, 19:19:19 +TE, 16:08:24 + TE + Mg, 13:40:13 + TE, 20:20:20, 6:12:36, 13:5:26 etc.

3. Chelated micronutrients

Fe EDTA (13%), Zn EDTA (15%), Fe EDDH-A (6%), Cheated combination of micronutrients, Kombi etc.

Fertilizer Solubility: Only 100% water soluble dry fertilizers should be used for fertigation systems. The fertilizer type used and amount required must be soluble enough to dissolve completely in the fertilizer tank water (Table 1). Conditioners which are used to prevent fertilizer pellets from absorbing moisture can create plugging problems. Most dry solid nitrogen fertilizers will absorb heat from the water when they are mixed since, cold water will not hold the same amount of soluble material as warm water. Consequently, it can be difficult to dissolve as much fertilizer as needed. Dry fertilizer products which significantly lower temperature of the water in which they are dissolved include urea, ammonium nitrate, calcium nitrate, and potassium nitrate.

Table 1. Solubility of fertilizers in water (Kg fertilizer/m^3)

Types of fertilizer	Solubility
Ammonium sulphate	71
Ammonium nitrate	119
Urea	110
Monoammonium phosphate (MAP)	23
Urea phosphate	96
Potassium sulphate	7
Potassium nitrate	32

Fertilizer Compatibility: If different fertilizer solutions are to be injected simultaneously into the irrigation system, you should perform a jar test beforehand to gauge compatibility. If the two solutions turn cloudy when mixed, there is a chance that injection of the chemicals will cause line or emitter plugging. Make sure you use the same dilution of fertilizer for the jar test that is to be used for fertigation. Compatibility can be affected by the order and way you mix things. Follow these basic mixing rules at all times.

1. Always fill the mixing container first with 50 to 75 % of the required water when mixing dry, soluble fertilizers.
2. Always add liquid fertilizer materials to water in the container before adding dry, soluble fertilizers. The additional fluid will provide some heat, in case certain nitrogen sources reduce temperature of the solution.
3. Always add the dry ingredients slowly with circulation or agitation to prevent the formation of large, insoluble or slowly soluble lumps.
4. Always put acid into water, not water into acid.
5. When chlorinating water with chlorine gas, always add chlorine to water, not water to chlorine.
6. Never mix an acid or acidified fertilizer with chlorine, whether chlorine is in the gas form or liquid form, like sodium hypochlorite. A toxic chlorine gas will form. Never store acids and chlorine together in the same room.
7. Do not attempt to mix either anhydrous ammonia or aqua ammonia directly with any kind of acid. The reaction is violent and immediate. Neither of these two materials should be used in drip irrigation systems.
8. Do not attempt to mix concentrated fertilizer solutions directly with other concentrated fertilizer solutions.
9. Do not mix a compound containing sulfate with another compound containing calcium. The results will be a mixture of insoluble gypsum.
10. Always check with the chemical supplier for information about insolubility and incompatibility.
11. Do not mix phosphorous containing fertilizers with another fertilizer containing calcium without first performing the jar test.
12. Extremely hard water (usually contains large amounts of calcium and magnesium) will form insoluble substances when mixed with phosphate, neutral polyphosphate or sulfate compounds.

Further, when preparing fertilizer solutions for fertigation, some fertilizers must not be mixed together. For example, the mixture of $(NH_4)_2SO_4$ and KCl in the tank, considerably reduce the solubility of the mixture due to K_2SO_4 formation.

Other forbidden mixtures are: Calcium nitrate with any phosphate or sulfate, Magnesium sulfate with di- or mono- ammonium phosphate, Phosphoric acid with iron, zinc, copper and manganese sulfate. Use of two fertilization tanks allows separating the fertilizers which interact and cause precipitation, placing in one tank the calcium, magnesium and micro-elements and in the other tank, phosphorus and the sulfate. The drip irrigation system needs to be completely charged or pressurized before fertigation begins. Farthest point from the pump should be at full pressure before injection of any fertilizer begins. If the system is not fully charged, air surges could occur and cause uneven application rates. The fertilizer should be injected ahead of the filters to ensure that any undissolved particles are filtered out before fertilizer enters the drip tape.

Hagin and Lowengart (1996) enumerated the main properties relating to suitability of fertilizer to the injection method as follows:

- Form: Soluble solid and liquid fertilizers are both suitable for fertigation depending on availability, profitability and convenience.
- Solubility: High and complete solubility is a prerequisite for fertilizers used in fertigation.
- Fertilizer solubility generally increases with temperature, depending on the fertilizer.
- Interaction between fertilizers in solution: When one type of fertilizer or more are prepared and mixed by the grower or are mixed in the irrigation line (but to a lesser extent), the compatibility between them must be checked.

Care to be taken while adopting fertigation

1. Fertilizer tank or injection pump may be used to inject the fertilizer
2. For precise placement of both water and fertilizer, it is necessary to use pressure compensating drippers or inline drippers instead of micro tubes.
3. Daily feeding is the most desirable, if not possible, feed alternate day.
4. Urea can be combined with WSF for meeting the N requirements.
5. Fertigation should be done during the end of irrigation period, run the drip system for 5-6 minutes after the completion of fertigation.
6. It is not advisable to use a very concentrated stock solution (generally not more than 10%).

Common N sources include ammonium sulphate, urea, ammonium nitrate, urea-ammonium nitrate, calcium nitrate, magnesium nitrate and potassium nitrate. Potassium can be supplied from potassium chloride, potassium sulfate, potassium thiosulfate or potassium nitrate. In case salinity is a problem, potassium chloride and potassium sulphate should be avoided and they should also be avoided if not clean and have extensive impurities. Potassium nitrate is the preferable form.

The choice of phosphorus products is more limited; phosphoric acid, urea phosphate or ammonium phosphate solutions are used most commonly. Mono-ammonium or mono potassium phosphates are also available. Liquid P fertilizers, except for good-grade phosphoric acid, may have impurities which complicate the already difficult task of eliminating chemical precipitation in the drip lines. However, with sufficient knowledge and attention to detail, good-grade phosphoric acid and ammonium phosphate solutions can be delivered successfully. The type of fertilizer for fertigation should be of high quality, with high solubility and purity, containing low salt levels and with an acceptable pH, and it must fit in the farm management program.

Water quality: Many water sources have high contents of calcium, magnesium and bicarbonates (hard waters), reaction of water is alkaline with pH values between 7.2 and 8.5. The interaction of these waters with fertilizers can cause diverse problems, such as formation of precipitates in the fertilization tank and clogging of drippers and filters. In waters with high calcium content and bicarbonates, use of sulphate fertilizers causes the precipitation of $CaSO_4$ obtruding drippers and filters. Use of urea induces precipitation of $CaCO_3$ because urea increases pH.

The main problem concerns phosphorus application: the presence of high concentrations of calcium and magnesium and high pH values lead to precipitation of calcium and magnesium phosphates. Recycled waters are particularly susceptible to precipitation due to its high bicarbonate and organic matter content. The resultant precipitates are deposited on pipe walls and in orifices of drippers and can completely plug the irrigation system. At the same time, P supply to the roots is impaired. When choosing P fertilizers for fertigation with high calcium and magnesium concentrations, acid P fertilizers (phosphoric acid or mono-ammonium phosphate) are recommended.

Clogging: This is especially critical for drip systems which must be kept free from suspended solids and microorganisms which plug the small orifices in the emitters. In case of clogging of drip system by bicarbonate precipitation, use of fertilizers with acid reaction partially corrects this problem. However, acid fertilizers cause corrosion of metallic components of the irrigation system. Therefore, periodic injection of acid in the fertigation system is recommended to dissolve the precipitates and to unclog the drippers. The following acids can be used phosphoric, nitric, sulfuric and hydrochloric acid. In Israel, HCl is widely used due to its low cost. Acid injection through the system will also remove bacteria, algae and slime. The irrigation and injection system should be carefully washed after the injection of acid.

Fertilizer dosing in fertigation: According to Sne (2006), to apply the same dose of fertilizers during the specific phenological stage of a plant, two different patterns of application can be made depending on the crop, soil type and farm management system:

- **Quantitative dosing:** A measured amount of fertilizer is injected into the irrigation system during each water application. Injection may be initiated and controlled automatically or manually.
- **Proportional dosing:** In this process, a constant predetermined ratio between the volume of irrigation water and volume of fertilizer solution is maintained, resulting in a constant nutrient concentration in the irrigation water.

Fertigation scheduling: While deciding the fertigation schedule following points need to be taken into consideration:

1. Total nutrient requirement of crops to produce target quality yields.
2. Part of the nutrient supplied in the form of organic manure.
3. Possible savings in the nutrient doses due to higher efficiency of added fertilizer through fertigation.
4. Relative requirement of nutrients at different growth stages of crop.

Requirement of nutrients varies from stage to stage of crop growth. It is essential to know the nutrient requirement of the crop based on growth stages. High level of nitrogen is required during vegetative growth stages to facilitate accumulation of dry matter in different plant parts. Its requirement is comparatively low during reproductive stage. Requirement of P is more during reproductive phase as it induces flowering, fruit set and seed development. Potash requirement is high during crop maturity stage.

In order to make correct planning of the nutrients supply to the crop according to its physiological stage, we must know the optimal daily nutrient consumption rate during the growing cycle that results in maximum yield and production quality. These functions are specific for each crop and climate. The optimal curve of consumption of nutrients defines the minimal application rate of a certain nutrient that is required to maintain a constant nutrient concentration in the soil solution.

Recommendation of a fertigation program

For each crop there are many fertilizer programs. Fertigation allows changing program during the growing season, adjusting it to suit fruit, flower, shoot and root development. A specific fertigation program is developed on the basis of leaf and soil analysis and tailored to suit the actual crop requirements. It is observed that recommended dose of each nutrient is different according to the physiological stage of the crop. The recommendations are different for tomato crop grown in open field or in greenhouse. Regarding soil type, the doses recommended for inert substrates are precise and expressed as concentration in the irrigation water (proportional dosification). On the other hand, in heavy clay soils, recommendations are expressed for quantitative dosification (in kg/ha) and recommended doses for phosphorus and potassium are not provided since, in this type of soil, these elements are adsorbed by clays and therefore, is very difficult to determine their concentration in the soil solution.

Recommendations for Tomato in greenhouse

For Substrate - soilless culture

Physiological stage	Concentration in the irrigation solution (dripper) (in ppm)				
	N*	P	K	Ca	Mg
Planting and establishment	120-150	40-50	180-220	100-120	40-50
Flowering	150-180	40-50	220-270	100-120	40-50
Ripening and harvest	180-200	40-50	270-300	100-120	50-80

* NH_4/NO_3 ratio=0.1-0.2

For Sandy soil

Physiological stage	Concentration in the irrigation solution (dripper) (in ppm)		
	N*	P	K
Planting and establishment	120-150	40-50	180-220
Flowering	150-180	40-50	220-270
Ripening and harvest	180-200	40-50	270-300

* NH_4/NO_3 ratio=0.3.

For Clay soil

Physiological stage	N	P	K (Kg/ha/day)
Flowering	2.0-2.5	*	0-2.5 *
Ripening and harvest	4.0-4.5	*	4-5.5 *

* = Depends on P and K levels in the soil

Fertigation Techniques in vegetable crops

Fertigation in Capsicum

Dose	:	200:200:200 kg NPK/ha
Basal application	:	50:50:50 kg NPK/ha
Fertigation start	:	21 days after transplanting
Interval of Fertigation	:	twice a week
Rate/m^2	:	2.2 g 19-19-19
Yield	:	127t/ha in 7 months

Tomato: Total: 250: 250: 250 kg of NPK / ha. Basal application is 50:50:50kg NPK/ha. Remaining 200:200:200kg/ha applied through fertigation starting from 21 days after transplanting for 6 month duration Polyhouse crop. Frequency of fertigation is once in three days. On an average, 4kg NPK/ha/day.

Capsicum: Total, 200: 200: 200 kg of NPK/ ha. Basal application: 50:50:50kg NPK/ha. Remaining 150:150:150kg/ha applied through fertigation starting from 21 days after transplanting for a 6 month duration Polyhouse crop. Frequency of fertigation is once in three days. On an average, 3kg NPK/ha/day.

English Cucumber: Total: 125: 100: 125 kg of NPK/ ha. Basal application: 25:25:25 kg NPK/ha. Remaining 100:75:100 kg NPK/ha through fertigation, to be given once in 3 days, starting from 10 days after transplanting ending at about 90 days after transplanting. On an average 4 kg each of N & K and 3 kg of P_2O_5/fertigation/ ha.

Muskmelon: Total: 125: 100: 125 kg of NPK/ ha. Basal application: 25:25:25kg NPK/ha. Remaining 100:75:100 kg NPK/ha through fertigation, to be given once in 3 days, starting from 15 days after sowing ending at about 90 days after transplanting.

Conclusion

Fertigation results in efficient use of water and fertilizers, produces higher yields, improves quality of production and protects environment. To ensure uniform distribution of water and fertilizers, the irrigation system must be properly designed and operated. The choice of suitable fertilizer is also very important and must be based on several factors like nutrient form, purity, solubility and cost. To implement a fertigation program, particularly under intensive greenhouse production, good knowledge is required of water and nutrient requirements over the growing season. Introduction of well tested and efficient fertigation techniques in greenhouse crop production systems will help to enhance the productivity of vegetable crops as well as saving precious water and fertilizer nutrients from being wasted in conventional agricultural systems.

References

Bar-Yosef, B. and Sagiv, B. 1982. Trickle irrigation and fertilization of iceberg lettuce. Israel – Ministry of Agriculture. *Special Publication*, **227**:220p

Bresler, E. 1977. Trickle-drip irrigation: Principles and application to soil-water management. *Advances in Agronomy* **29**:343-393.

Bucks, D.A., Erie, L.J. and French, O.F. 1974. Quantity and frequency of trickle and furrow irrigation for efficient cabbage production. *Agronomy Journal*, **66**: 53-57.

Cook, W.P. and Sanders, D.C. 1991. Nitrogen application frequency for drip irrigated tomatoes. *Horticultural Science*, **26** : 250-252.

Goldberg, D., B. Gornat and Bar-Yosef, B. 1971. Distribution of roots, water and mineralsa sa result of trickle irrigation. *Journal of American Society for Horticultural Science*, **96**: 645-648.

Hagin, J. and Lowengart, A. 1996. Fertigation for minimizing environmental pollution by fertilizers. Fertilzer. *Research*. **43**:5-7.

Keller, J. and Bliesner, R .D. 1990. Sprinkle and trickle irrigation. Van Nostrand Reinhold. New York.

Miller, RJ., D.E. Rolston, R.S. Rauschkolb and Wolfe, D.W. 1976. Drip application of nitrogen is efficient. *California Agriculture* **6**:16-18.

Miller, RJ., D.E. Rolston, R.S. Rauschkolb and Wolfe, D.W. 1981. Labelled nitrogen uptake by drip-irrigated tomatoes. *Agronomy Journal* **73**:265-270.

Nicklow, C.W. and. Downes, J.D. 1971. Influence of nitrogen potassium and plant population on the maturity of field seeded Tomatoes for once over harvest. *Journal of American Society for Horticultural Science*, **96**:46-49.

Papadopoulos, I. and Ristimaki, L. 1998. Nitrogen and phosphorus fertigation of tomato and eggplant. *In* Proc. of the HIS. Belgium, 2-9 Aug., 1998.

Phene, C.J. and Beale, D.W. 1976. High-frequency irrigation for water nutrient management in humid region. *Soil Society of America Journal*. **40:** 430-436.

Phene, C.J., J.L. Fouss and Sanders, D C .1979. Water-nutrient-herbicide management of potatoes with trickle irrigation. *American Potato Journal*, 56-59.

Rauschkolb, R.S., D.E. Rolston, R.J. Miller, A.B. Carlton and Burau, R.G . 1976. Phosphorus fertilization with drip irrigation, *Soil Science Society American Journal*, **40**:68-72.

Rolston, D.E., R.S. Rauschkolb, C.J. Phene, R.J. Miller, K.Urier, R.M. Carlson and Henderson, D.W. 1971.Applying nutrient and other chemicals to trickle irrigated crops. California Division of Agricultural Sciences Bullettin.1893.

Scaife, A. and Bar-Yosef, B.1995. Nutrient and fertilizer management in field grown vegetables. IPI-Bulletin 13. International Potash Institute, Basel, Switzerland.

Sne, M. 2006. Micro irrigation in arid and semi-arid regions. Guidelines for planning and design. Ed. by S.A. Kulkarni. ICID-CIID. International Commission on Irrigation and Drainage. New Delhi, India.

Stark, J.C., W.M. Jarre, J. Letey and Valoras, N.1983. Nitrogen use efficiency of trickle irrigated tomatoes receiving continuous injection of N. *Agronomy Journal*,**75**:672-676.

10

High Altitude Protected Vegetable Production

Brahma Singh

Abstract

High altitudes may be cold and humid or cold or arid ones making the requirements of protected cultivation entirely different in both the conditions. The present chapter is based on the environmental conditions existing in Leh region. Near sub-zero temperature with snow in higher altitudes makes vegetable production tough without use of techniques of protected cultivation. Protected cultivation has witnessed gradual developments from riverbed cultivation practices to use of several types of plastics (PP, PE, PVC etc.) and UV stabilized polythene sheets. Area under protected cultivation is spread worldwide mainly in countries like Holland, Spain, Isaral, China , Japan etc. Protected cultivation technology provides favorable climate for production of vegetable &/or their seedlings of superior quality even in the extreme adverse climate outside. Such structure exploits solar thermal energy to improve photosynthetic efficiency of the plants grown inside. This green house effect is very helpful in winters at high altitudes. Similarly, some modifications in these structures are needed to low the temperature during summers. A number of advantages with some limitations can be listed that are associated with the protected cultivation. At high altitudes, three methods viz. Polyhouse/Greenhouse /Net /shaded house, Low tunnels or row covers and plastic mulching can be used. Polyhouse is a framed structure with 200 micron UV stabilized low density cladding forming a greenhouse effect. Underground trench type polyhouse can also be made with polyethylene supported on wooden poles or GI pipes. Though a combination of above two known as polyench was found performing better in Ladhak area. Low tunnels or row cover use transparent plastic films or net over low hoops to cover rows of plants. Mulching is covering the soil around plants to conserve soil moisture, temperature, preventing weeds etc. using transparent and black plastics of 10-25micron thickness. Double colored plastic mulches are being tested for different crops and different seasons. In arid high altitudes, DRDO (through DIHAR, Leh) has worked extensively on protected cultivation and developed production technologies in the service of nation. Its contribution includes making possible the production of nursery, cucurbits and several other vegetables and their seeds in sub zero climatic conditions. Now, there is a need to develop more area specific, more appropriate, efficient and economic protection structures. Simultaneous development of tools, machinery and vegetable varieties suitable for use in protected conditions are among the major aspects of future work to be done in this field.

Keywords: protected cultivation, high altitude, low tunnel, plastic mulching

Introduction

India is the second largest producer of vegetables in the world, accounting for about 13 per cent of the world's vegetable production. In order to fulfill the demand supply gap, popularization of scientific and improved methods of vegetable cultivation and their seed production is of utmost importance with reference to specific agro-climatic conditions and unexplored niches in hilly regions of our country, which permit production of a wide variety of good quality vegetables. Ironically, the farmers in hilly areas are today no longer able to sustain themselves at the level of self-sufficiency. Reduction in the density of forest cover and overgrazing has led to soil erosion of unprecedented proportions. Land and water resources are becoming even scarce.

The only solution to overcome this situation is to popularize scientifically developed methods of vegetable cultivation practices for the improvement of agro-socio-economic conditions of the people. The necessary research and development programs must be carried out in the hilly region itself, this way the outcome of such work will be best suited to the agro-climatic condition of the region.

Greenhouse technology for production of vegetables has been found to be an appropriate intervention in the development of horticulture in these difficult terrains. Even though the present area under greenhouse in our country may be less, there is a large untapped potential in the hilly terrains, which would bring nutritional self-sufficiency besides enhancing the income of people and providing on-farm employment.

High Altitudes

In Indian Himalaya, high altitudes are of two types from their climate point of view. First one is cold and humid high altitudes which spread over mainly Uttarakhand, Sikkim, West Bengal and Arunachal Pradesh and other North East States. The second one is cold arid high altitudes mainly spread over Jammu & Kashmir Ladakh region, Himachal Pradesh-Lahual-Spiti and Kinnaur area, some parts of North Sikkim and Uttarakhand. Himachal Pradesh and Jammu and Kashmir have sizeable area under cold humid high altitude also. The climatic conditions in cold humid and cold arid high altitudes are different, necessitating different type of protected agriculture as is evident from the meteorological records of Leh given in Table 1.

Table 1. Meteorological data of Leh (Average of 25 Years)

	Mean temp		Extremes				Rain/snowfall (mm)				Max. wind	
Months	Max (°C)	Min (°C)	Max temp (°C)	Date/ year	Min temp	Date year	R.H (%) (at 9 am)	Mon-thly	No of Days	Heav-iest in 24 hrs	Date/ Year	Spead (knots)
Jan	3.1	-14.3	8.4	5/77	-30.2	25/76	66.3	14.2	6.6	18.0	31/75	54
Feb	2.2	-11.2	12.4	18/85	-30.0	18/75	58.2	10.5	6.4	19.4	28/88	50
March	7.5	8.8	18.4	30/77	-21.5	15/75	39.7	13.8	6.5	37.0	5/75	70
April	14.1	1.3	22.5	29/88	-7.8	6/78	45.4	3.2	5.0	27.8	15/75	83
May	18.2	4.5	27.6	17/90	-1.5	9/79	40.3	3.3	6.8	4.9	88/82	88
June	23.2	9.2	34.8	29/78	0.8	29/73	33.9	1.8	5.3	5.2	24/76	89
July	27.6	14.1	35.0	14/73	5.2	3/75	40.9	11.2	8.3	17.4	14/85	82
Aug	27.8	13.9	34.2	9/78	5.2	29/89	42.1	12.1	9.1	15.6	7/76	75
Sep	22.7	8.1	30.5	6/78	-7.8	29/82	41.4	6.8	5.3	46.6	26/88	75
Oct	15.8	-1.0	25.4	2/85	-7.0	23/90	56.8	1.7	1.5	21.2	13/87	63
Nov	9.6	-6.4	17.4	2/77	-17.5	29/78	51.1	3.9	1.1	21.1	4/81	60
Dec	3.7	-11.8	19.1	18/89	-20.5	31/74	50.1	9.0	4.4	15.0	26/77	62

Sub-zero temperatures result in snowfall in higher altitudes. It could result in dry cold or wet along with rainfall or snow. In Ladakh and Lahaul-Spiti, cold arid desert, permafrost occurs with frozen upper soil (mostly sandy). In these areas, ambient minimum temperatures are below or near freezing for almost five months. The relative humidity during this period is in the range of 45-60%. In Leh valley, average minimum temperature from November to April is sub-zero and can be as low as minus 16 °C. Wind velocity in the afternoon is very high resulting in dust storm or snow blizzards. This chapter is based mainly on the experience and experiments conducted at DIHAR, Leh.

Altitudes in Indian Himalayas range between 200 to more than 5000 meter above mean sea level. Winters in high altitudes are severe and prolonged, restricting vegetable production season from 7 to 2.5 months or less as given below. Higher the altitude lesser are the chances of vegetable production without using protected cultivation techniques and technologies. Protected cultivation technology is the boon in high altitudes for vegetable production Table 2.

Table 2. Vegetable production period at different altitudes

Altitude meters above mean sea level	Production Period	Production no. of Month
2670	April-October	7.0
3000	May-Mid October	5.5
3300	Mid May –Mid September	4.0
4000	Mid June –August	2.5

Historical Background and Present Scenario of Protected Cultivation of Vegetables

The protected cultivation technology is in use since several centuries. Riverbed cucurbits cultivation involving protection of young seedlings from cold winds in winter months by straw fence is an example in North India. In Europe, during winter months, underground horse or cow dung hotbeds covered with mica sheets were used for raising leafy vegetables. Commercial production of vegetables started in the18th century. In early 19th century, glasshouses in different designs came up in Europe and Asian countries, mainly in Netherlands and Japan, respectively. Subsequently, with the development of plastics, several designs of protected structures or greenhouses came up. Polyethylene polymer was first developed in 1930s and in 1950 it came in use in horticulture in form of plastic films and pipes for making structures and watering. Several types of plastics like polyvinyl chloride, polypropylene and polyesters were developed later keeping in view the requirements of protected cultivation such as transparent ultra violet (U V) stabilized polyethylene sheets as cladding material for greenhouses. A revolution in plastic technology helped in progress and popularity of protected cultivation.

Holland has a long tradition of protected cultivation under glass houses growing flowers and vegetables. More than 20 million hectares is under protected cultivation world over. In Europe, Spain is the leader covering 5100 ha mostly under low cost poly houses, about 150,000 ha under plastic mulches and over 20,000 ha under low tunnels and floating covers. Among the middles east countries, Israel is the largest exporters of cut flowers and vegetables grown under protected conditions coupled with micro irrigation and fertigation. Hi-tech greenhouses are being used for production of export quality cut flowers and vegetables. In Asia, China has the largest area under protected cultivation; over 2.5 million hectares under poly houses, 9.6 million hectares under plastic mulch and 0.9 million hectares under low tunnels and floating covers. More than half of vegetable produce come out of high tunnel poly houses. Perhaps due to this factor China is the largest producer of vegetables in the world today. Japan is next leading country producing fruits, vegetables and cut flowers under protected structures covering 40,000 ha. Countries like Kuwait, UAE, Korea, Canada and Russia have sizeable area under protected cultivation. In India, the area under protected cultivation is reported to be 2000 hectares including 500ha under net and shade houses. The area is on the increase due to the efforts made under National Horticulture Mission, Ministry of Agriculture, Government of India and other programmes supported by state governments and private sector.

Defence Research and Development Organization (DRDO) has a Field Research Laboratory (FRL) now renamed as Defence Institute of High Altitude Research (DIHAR) at Leh (Jammu and Kashmir) at 11500 ft above mean sea level which attempted protected cultivation in early 1960s for production of vegetables in frozen winters for the soldiers stationed there. This institute has demonstrated successful production of several leafy vegetables during frozen winter in Ladakh and multiplication of vegetable nurseries for open field production of several vegetables in summer which otherwise is not possible. The author had an opportunity to head this Institute during 1991 to 1995. Indo-American Hybrid Seeds (India), Bengaluru is the pioneer in India to make use of greenhouse technology in 1965 for commercial production of flower seeds, cut flowers and ornamental plants. Indian Petrochemical Corporation Ltd (IPCL) made significant contributions in protected cultivation by commercially producing U V stabilized low density polyethylene (LDPE) cladding and aluminum frames for poly houses. National Committee on use of Plastics (NCPA) in agriculture in 1981 under the Ministry of Chemicals and Petrochemicals further boosted protected cultivation. This was transferred to Ministry of Agriculture under Division of Horticulture in 1993 and was entrusted with the task of promoting use of plastics in agriculture projects like irrigation and protected cultivation. The NCPA operated through 22 Plasticulture Development Centers (PDC) spread over the country promoting protected cultivation (use of hail nets, shade nets, packaging of produce). China these days has large area under protected cultivation making it the largest producer of vegetables in the world.

Protected Cultivation of Vegetable Crops

Protected cultivation of vegetable crops refers to creation of favorable environmental conditions around the vegetable plants, offsetting or minimizing the detrimental effects of prevailing or expected to prevail abiotic and biotic factors. In the open field, vegetables face several abiotic and biotic stresses resulting in lower or poor or no yield. This condition is altered by using different protected structures for cultivation and /or nursery production. Vegetables which are transplanted using seedlings should be produced by raising seedlings under protected conditions. Protected conditions ensure production of healthy seedlings and use of healthy seedlings ensures better and higher production of vegetable crops both under protected conditions and open field conditions.

Principles Involved in Protected Cultivation

The basic principles of protected cultivation involved are greenhouse effects of heating cultivated space exploiting sunlight and ventilation for cooling and covered with plastic or glass or other suitable materials (polycarbonates, fiber reinforced polymer and others) which are transparent to incoming short wave radiation from sun impinging on outer surface but opaque (partially) to emergent long wave infrared radiation from soil, plant and structural surfaces, thereby tapping the heat, as a result enclosed surface maintains higher temperature than ambience. The tapped heat is gradually dissipated through conduction, convection/ventilation and radiation. By incorporating heat sinks, reduced ventilation and application of radiation shields either rate of cooling is slowed down or a supplementary heat source is placed inside. In summer months when the sun is too bright and ambience is too hot, shading, natural/forced ventilation, evaporative cooling prevent excessive temperature build up. Protected cultivation of vegetable crops refers to farming that creates favorable conditions around the plant, offsetting detrimental effects of prevailing biotic and abiotic factors. Structures and environment control measures are used to isolate the cultivable space from prevailing unfavorable environment and create plant growth conditions reasonably close to optimum. The fundamental principles involved in protected cultivation are green house effects for heating cultivated space using mainly sunlight and ventilation for cooling and air carbon dioxide regulation. The protected structure or greenhouse generally reflects back 43% of solar radiation incident upon it allowing transmittance of photo synthetically active solar radiation in the range of 400-700mm wave length which increases photosynthetic efficiency of the crops grown under the greenhouse. Moderately short to moderately long wave length (15-390mm), i.e. ultraviolet rays damaging the crops are absorbed by the claddings (glass, polyethylene film and others) of the structure. This facilitates better crop growth and yield under protected structures. The sunlight admitted inside the protected structure is absorbed by the crop, floor, and other objects inside the framework of the structure which in turn emit long wave thermal radiation in the infra red region (760-2500mm) and thus, temperature inside the structure rises. This phenomenon of tapping solar thermal energy in the structures is known as "green house effect". During low ambient temperature, this increase in temperature is most important function of protected structure for enhancing crop growth and yield. During summer, inside temperature rises higher than the optimum levels and necessitates lowering it below 35 degree Celsius through evaporative or appropriate ventilation cooling. The ventilation can be preferably natural or forced one or in between by putting solar fan(s). Besides, various methods of cooling the structure, as per requirement, are used. The economical and effective cooling adds to the production. At high altitudes captive power is hardly available in India. All manipulations in environment of structure are done with solar power or manually.

Advantages of Protected cultivation of vegetables

Protected cultivation of vegetables and flowers can reduce the amount of water and chemicals used in production of these crops under open fields. The comparative advantages are:

1. Productivity levels of vegetables are significantly higher than open field (4-10 times).

Productivity potentials of protected cultivation

Crop	Yield kg/sqm (q/ha)
Cucumber	25 (2500)
Tomato	14(1400)
Capsicum	12 (1200)

2. Under unfavorable climatic conditions, production of crops is possible that too with high productivity.
3. High input use efficiency is achieved.
4. Year round production of crops is possible to meet domestic and export demand.
5. Quality of produce is usually superior and offers better return.
6. Vertical growth of plants gives per unit increased income.
7. Off season production of crops gives higher return. In high altitudes production during non cropping season (frozen winter) is made possible. Likewise, production of coriander and palak under shade from May to August during summer in and around Delhi is very remunerative venture.
8. Production of early, healthy and transportable vegetable nurseries particularly of costly hybrid seeds with better germination percentage stand both for open and protected conditions is a proven advantage.
9. Safeguard against impact of abiotic and biotic stresses on plant growth and yield.
10. Use of modern technologies like hi-tech horticulture –hydroponics, aeroponics and nutrient film techniques is possible for higher returns.
11. It is a source of self employment generation.
12. Low temperature sensitive foliage plants are protected under shade net houses.

13. Suitable for urban and peri-urban vegetable farming
14. It makes vertical cultivation of vegetables possible using technologies like hydroponics, aeroponics and use of vertical beds for production.
15. Disease free seed production of costly vegetables becomes easy under protected structures.

Limitations

1. Manual or hand pollination in cross pollinated vegetables like cucurbits or development of their parthenocarpic hybrids/varieties.
2. Expensive, short life and non-availability of cladding materials.
3. Lack of appropriate tools and machinery.
4. Structure cost initially looks unaffordable. Farmers with zero risk affordability do not come forward to adopt it.
5. Inadequate support from planners and scientists- suitable varieties/hybrids and their production packages for protected production systems are either not available or very a few. Protected structures in use are not scientifically designed; hence potentials of structure are not fully exploited.

Protected Vegetable Production

The area under greenhouse cultivation, reported by the end of 20th century was about 110 ha. In India and world over 275,000 ha (Mishra, *et al.,* 2010). During the last decade, this area must have increased by 10 per cent if not more. In Europe, Spain is leading in protected agriculture with 51,000 ha mostly under low cost poly houses. In Asia, China has the largest area under protected cultivation, 2.5 million ha under poly house/greenhouse. Protected vegetable production is an important component of protected agriculture. Protected vegetable production is practiced throughout the world irrespective of altitude of the place since several hundred years. River bed production of early cucurbits prevalent in India since ages, is protected agriculture. It involves protection of production stages of vegetables mainly from adverse environmental conditions such as temperature, hail, scorching sun, heavy rains, snow etc. In fact the need to protect the crops against unfavorable environmental conditions led to the development of protected agriculture. This is now becoming important due to climate change. Greenhouse is the most practical method of achieving the objectives of protected agriculture, where natural environment is modified by using sound engineering principles to achieve optimum plant growth and yield. Besides, protected technology has potential to generate more produces per unit area with increased input use efficiency. There is a need to increase nutritionally rich vegetable production and productivity of seasonal and non-season crops in our country. Research results have shown that by adopting protected cultivation, productivity of vegetable crops can be increased by 4 to 10 times as compared to open environment. This aspect needs to be extensively exploited in India as has been done elsewhere in the world. To promote this, Indo-Israel protected vegetable production projects in the country are serving the purpose. NAIP program of ICAR is giving due importance to this aspects besides different public and private organizations. Areas having uncongenial environment for vegetable production can also be converted into potential vegetable production centers with the help of protected agriculture technologies and techniques as discussed in this article. Needless to emphasize that better quality produce is obtained under protected conditions.

Methods of Protected Vegetable Production in High Altitudes

The major protected cultivation methods at high altitudes of India in vogue are use of:

1. Poly houses/Greenhouse/net house/shade house
2. Low tunnels/row Covers
3. Plastic Mulching

Polyhouses / Greenhouses / Nethouses / Shadehouses

Poly house/greenhouse is a framed structure having 200 micron (800 gauges) UV stabilized transparent or translucent low density polyethylene or other claddings which create greenhouse effect making microclimate favorable for plant growth and development. The Structure is large enough to permit a person to work inside. The structure can be made in different shape and size using locally available materials or steel or aluminum or bricks or their combinations for its frame.

In Ladakh, poly houses are made above ground (poly house), underground (soil trench) and a combination of the two (polyench). Above ground poly houses are generally made of mud wall or unbaked brick wall on three sides. North side wall is made 7 feet high, east and west side walls are made with gradual slope to south having entrance on either side. Southern side is covered with polyethylene supported on locally available willow or poplar wood frames. Water for irrigation is stored inside but underground for convenience.

The underground trench type poly house is made with suitable dimensions, generally 5-10 x 3-4 x 1m with polyethylene cladding supported on wooden poles or GI pipes.

A combination of both-construction of poly house above trench, known as polyench is being found better than both in winter months for production of vegetables where soil and sun heat are harnessed for maintaining required higher temperature inside. Polyench can be single or double walled.

Polyhouses are constructed using GI pipe of 25-75 mm diameter with a wall thickness of 2mm. These structures are fastened by welding, nuts and bolts or clamped. Foundation for posts, size of hoops and perlins are worked out on engineering principles. Good cladding material (low density polyethylene, diffused or relatively translucent films, cross laminated, anti-fog, anti-drip, anti-sulphur types, fiber reinforced plastics, polycarbonates etc) is essential to ensure good life of greenhouse. Poly carbonate and FRP cladding green houses have also been found useful for covering large area. During winter months, solar heat is harnessed for production of leafy and other vegetables and vegetable nursery. The temperatures inside different protected structures during winter are higher than open field to the extent of supporting plant life.

Insect proof net and shading materials are used to keep insects at bay and to lower temperatures in summer if considered necessary. Net and shade houses are used for vegetable production as protected structures elsewhere in lower altitudes in the country.

Double Wlled Polyench

Ladakhi Polyhouse

Underground/Trench Polyhouse

FRP-Polyhouse

Underground Polyhouse/Soil Trench

The Ladakh region is located between 2700 to 4500 m altitude in the Western Indian Himalayas, close to the Tibetan/Chinese and Pakistan border. During the short summer period, Ladakhi communities concentrate on subsistence agriculture and collection of fuel wood in order to survive during the long and harsh winter. In the winter, snowfalls block the high passes, roads are closed and the people rely almost entirely on their own resources. All activities, even domestic, are very limited due to the cold. The market price of fresh products increases many fold and only the upper class modern families can afford the rare fruits or vegetables available on the market – transported by air from other areas.

Low Tunnels or Row Cover

Transparent plastic films or nets are stretched over low (1m or so) hoops made of steel wires, bamboo or willow twigs or cane or any other locally available suitable material to cover rows of plants in the field providing protection against unfavorable environment like low temperature, frost, wind, insect-pests etc. Different types of claddings are available in the market. Low tunnels with plastic mulch and drip irrigation are becoming popular for several vegetable crops production.

Plastic Mulching

Mulching is a practice of covering soil around plants which makes growing conditions more favorable by conserving soil moisture, maintaining higher soil temperature, preventing weeds and allowing soil micro flora to be favorably active. In other areas, organic mulches such as leaves, bark, peat, wooden chips, straw etc are used but in high altitudes particularly in arid high altitudes, plastic is used for mulching which has unimaginably and significantly contributed to vegetable production there. Plastic mulching is one of the widely used practices in protected agriculture particularly in vegetable production. As stated above, plastic mulches are mainly used for weed control, to increase soil temperature and to keep soil moist resulting in reduction in the use of irrigation water. These days plastic mulches are available in different colours and thickness keeping in view their multiple uses which are becoming popular among farmers. In plastic mulches, the most preferred colours are transparent (clear) and black, although wide variety of shades and colours are being used for specific purpose in the production of food crops. Currently worldwide, double colour plastic mulches viz., sliver + black, blue + black, yellow + black, white + black and red + black are being tested in production of different horticultural crops and more specifically vegetable crops in different seasons. After several investigations, following benefits of use of plastic mulches for commercial vegetables production have been observed:

1. Moisture retention by securing evaporation from soil surface.
2. Inhibits weeds growth- black, silver + black, yellow + black, blue + black,

white + black mulches greatly inhibit light penetration to the soil, due to this weed seeds do not germinate under these mulches. Clear or transparent mulch allows light penetration,weed seed germination and growth.

3. Reduces fertigation leaching by rain fall or runoff during rainy season.
4. Decreases soil compaction which provides better environment for seedling emergence and its root growth.
5. Protects fruit quality in vegetables by reducing rain-splashed soil deposits on fruit. In addition, mulches also reduce fruit rot caused by soil-inhibiting organisms, because it works as protected barrier between the fruits and the organisms.
6. Helpful in fumigation: mulches increase the effectiveness of soil fumigant chemicals.
7. Helps to control pests: Highly reflective mulches assist management of insects and diseases, specifically viruses that they may carry. Basically yellow + black is used if the whitefly is the vector for viruses; silver + black is used in case aphid is the vector and blue + black is used if thrips are the vector for virus.
8. Plastic mulches maintain a warm temperature during night which facilitates an early establishment of seedlings by strong root system or germination of seeds.
9. Soil water erosion is minimized.
10. Plastic mulches serve for longer period. They can be used for more than one season.

Looking at above and other uses and advantages, large scale use of different coloured plastic mulches is suggested in vegetable production. The use of 10-25 micron thick plastic mulches is suggested for different vegetable crops in different seasons and areas keeping in view the objective

Contribution of Protected Cultivation on Arid High Altitude Vegetable Production

Arid high altitudes of Ladakh and Lahaul and Spiti in early sixties used to grow root vegetables like radish, turnip, carrot, beet root; potato and mongol palak (beet leaf). After Chinese aggression (1962), induction of Indian defence forces in these areas necessitated local production of different vegetables. Defence Research and Development Organization (DRDO) through its laboratory, Field Research Laboratory, now Defence Institute of High Altitude Research, Leh did pioneering research. With the help of protected agriculture technologies it has been made possible to grow now all sorts of vegetables there during agriculture season (May to September or mid October). Perhaps, first glass house in high altitudes of the country was erected in Leh (11500 ft AMSL) in 1964.

First Glasshouse Constructed in 1964 at FRL (DIHAR), Leh

Since then DRDO has extensively worked on protected cultivation and developed appropriate production technologies in Himalayan high altitudes of India for the wellness of soldiers and society. During long frozen winters of Ladakh one could not have thought of vegetable production but for protected practices developed by DRDO, every farmer and others are producing different types of leafy vegetables almost round the year.

Some of the major contributions made by DRDO in developing protected vegetable production technologies are as follows:

1. Protected vegetable nursery production making cultivation of several vegetables possible

Early production of vegetable nursery under different protected structures during March and April (minimum atmospheric temperature is sub-zero) and transplanting them in May and June with and without plastic mulch extended agriculture period and made cultivation of cabbage, cauliflower, knoll-khol, broccoli, brussel's sprouts, tomato, brinjal, chilli, capsicum and onion possible. Use of plastic mulch enabled earlyness, quality and higher yield of these vegetables. In mulched crops, low pressure (gravity/slope) drip irrigation and fertigation are possible as experimented by DRDO. In this way most of the vegetables are being grown on large scale, making the area surplus in cabbage, something unbelievable. Early and late production of vegetables with the help of protected technology has also been standardized which extends period of locally produced vegetables-an important aspect there.

2. Making Cucurbits production possible in cold desert

Till early 1990s cucurbits cultivation in open in Ladakh was considered impossible. But growing seedlings in poly pouch under poly houses during April-May and transplanting them in open field with plastic mulch made it possible to grow almost all cucurbits in Leh. This has not only improved vegetable basket in the area but also added variety to food basket of local inhabitants and soldiers. Commercial production of cucurbits in cold desert of India is now possible through protected cultivation. Sarda melon imported in large quantity in the country can be produced in these areas with ease. Production of off season (August and September) muskmelon, watermelon, etc in open fields has also become possible. An early crop of cucurbits like squash, longmelon, etc. is also taken in poly houses.

3. Vegetable production in sub-zero atmosphere

As stated earlier, during winter these areas remain cut off with the main land due to heavy snow fall. Only air communication is on during winter months. Through air, transportation of bulky and perishable commodities like vegetables is not only expensive but very difficult. In Ladakh sector, Army alone spends several crores of rupees only on transportation of vegetables. Cost of transportation is more than the cost of vegetables. Hence, local production through protected cultivation is being successfully promoted there. This is being encouraged by harnessing solar energy, both thermal and photovoltaic, making heating of greenhouses possible. The geothermal energy sources available in the area are potential source of heating greenhouses. Remoteness of these sources is coming in the way of their exploitation.

Farmers in Ladakh used to laugh at the idea of producing vegetables there during long frozen winters but now every farmer is practicing it under one or the other protected structures. Ladakhi poly house is the most favored structure and supported by state Government/ Leh Council. Locally produced leafy vegetables under protected structures are seen in Leh market during frozen winters. Vegetables like mongol palak, mustard, lettuce, methi (fenugreek), coriander, leek, Chinese cabbage, mint, amaranth, karam sag, dum, kale, swischard, celery, parsley, radish, green onion, cabbage, knol-khol and orich/ustahak are produced in sub-zero atmosphere. This research is unique in the world.

4. Vegetable Seed Production

Seed production of biennial crops like temperate varieties of cole crops, root crops, and onion take two years or 18 months in these areas. In the first year, normal crop is grown and stored underground during long winters. In the second year, in summer they are planted for seed production. By the protected agriculture technology, now it has become possible to produce seeds of these varieties in half the time by raising early crop under protected structure and transplanting them in open fields for seed production. Pusa Himani radish, long day onions, Nantes carrot and others respond well to this technique. Production of seeds of temperate varieties of vegetables in India is a problem due to lack of consorted research and development efforts.

Future Prospects

To ensure nutritional security along with food security to the ever growing population of the country, it is essential to double the production of vegetable crops in the country. Major constraint is increased pressure on cultivable lands near metros where vegetables are generally grown. This is due to urbanization and industrialization which are also essential. Therefore, it is at most necessary to improve the productivity of vegetables by adopting protected cultivation in the country in general and at high altitudes in particular.

Protected cultivation of vegetables in high altitudes of Himalaya is practiced successfully indicating its potential to deal with conditions created by climate change scenario in the country. Protection against adverse climatic conditions for plant growth has become a universal necessity. Protection of plant growth and development against adverse physical (temperature, rain and wind) and biological (insects and diseases) factors through protected agriculture technologies is going to be common in near future because of climate change and advantages of protected cultivation. There is a need to develop area specific, most appropriate, efficient and affordable protected structures with cheaper and durable cladding materials. Emphasis would be shifted on development of suitable varieties and hybrids of vegetables for protected cultivation under organic and inorganic production protocols. Vegetable nurseries would be produced under protected structures both at individual farmer and commercial nurseries level. Tools and machinery for protected cultivation would be developed and become common. Vertical or multitier farming of vegetables would be developed to make use of protected space. High altitudes are likely to be harnessed for large scale vegetable production under protected structures. Human resource development on protected agriculture and Government support for its promotion would be taken up through State Agriculture Universities and Department of Horticulture. Plastic mulching coupled with drip irrigation in vegetable production is going to be a common practice because of their proven advantages.

Summary

High altitudes in India are reasonably populated with local tribes and troops. Vegetable production for them during winter months, when environment mainly temperature is unfavorable for their growth, has been discussed. Protected production technologies or green house technologies developed for these areas such as use of local poly house, both underground and above ground along with combination of both have been discussed. Production of leaf vegetables under subzero atmosphere, cucurbits and almost all vegetables in cold arid high altitudes of Ladakh using protected agriculture technology are mentioned in brief. Production of almost all vegetable crops during limited agriculture season from May to September in cold desert of Ladakh, considered remote possibly, has now become possible with the help of protected agriculture technologies. Future prospects of protected cultivation of vegetable crops in high altitudes and elsewhere have been reasonably highlighted.

References

Dhaulakhandi, A.B. and Singh, B. 1999. Winter performance of greenhouse attached passive solar heated hut at high altitude. *SESI, Journal* **9**(2):105-114.

Mishra, G. P., Singh, N. and Kumar,H. and Singh, S. B. 2010. Protected Cultivation for Food and Nutritional Security at Ladakh Defence Science Journal, Vol. 61, No. 2, March 2010, pp. 219-225.

NAAS 2010. "Protected Agriculture in North-West Himalayas". Policy Paper No. 47, National Academy of Agricultural Sciences, New Delhi. pp16.

Singh, B. 1995. Vegetable Production in Ladakh. Field Research Laboratory, Leh. India

Singh, B. and Dhaulakhandi, A. B. 1998. Application of solar greenhouse for vegetable production in cold desert in renewable energy. Energy Efficiency Policy and the Environment. Elsevier Science Ltd, UK, P2511-314

Singh, B., Dwivedi, S K. and Chaurasia, O.P. 2004. Improvement in production and productivity of horticultural crops in cold arid regions of India. Proceedings of the first Indian Horticulture Congress, 6-9 November, 2004, The Horticultural Society of India, New Delhi, India, viii+ 764p

Singh, B., Dwivedi, S. K. and Plajor, E. 2000. Studies on suitability of various structures for winter vegetable production at sub-zero temperatures. Acta Hort., **517**:309-14.

Singh, B., Dwivedi, S. K. and Sharma J. P. 2000. Greenhouse technology for winter vegetable cultivation in cold arid zones. In: Dynamics of cold arid Agriculture (Eds J. P. Sharma and A. A. Mir) Kalayani Publishers, Ludhiana. PP 279-293.

Singh, B., Dwivedi, S.K., Singh, N. and Paljor, E. 1999. Sustainable Horticulture practices for cold arid areas. In : The Himalayan Environment. eds. SK Dash & J Bahadur . New age International (P) Ltd, Publishers – New Delhi. pp 235–245.

Singh, B. and Dwivedi, S. K. 2002. Vegetable production potential in Ladakh. In: Vegetable growing in India. Eds. P. S. Arya and Sant Prakash. Kalyani Publishers, New Delhi. pp 87-93.

Singh, B., Dwivedi, SK. and Sharma, JP. 2000a. Greenhouse technology for winter vegetable cultivation in cold arid zones. In: dynamics of cold arid agriculture. Eds. J.P. sharma and A.A. Mir, kalyani publishers-Ludhiana, pp. 279-293.

Singh,B. 1999. Vegetable production in cold desert of India: a success story on solar greenhouses. Acta horticulture **534**: 205-12.

Singh, N. and Singh, B. 2003. Ladakh mein sabji utpadan (Vegetable Production in Ladakh. Field Research Laboratory, Leh.pp 139

Singh, B, Khokhar KS and Kumar M 2004. Use of plastic mulching for quality vegetables. *Intensive Agriculture*, **XXXXII** (1-2): 3-5

Singh, B, Kumar, M and Singh V P 2005. Cultivating sweet pepper in greenhouse is profitable. *Indian Horticulture*, **49**(4):15-17

Singh, B, Sirohi, N P S and S.P. Mehto 2005. Off-season cultivation of vegetables is remunerative under plastic low tunnels. *Indian Horticulture*, **49**(4): 20-21

Singh B, Kumar M and Sirohi N P S 2005. Growing parthenocarpic cucumber in greenhouse is ideal for peri-urban vegetable growers. *Indian Horticulture*, **49**(4): 22-23

Singh, B, Kumar, M and Yadav, H L 2005. Plug-tray nursery raising technology for vegetables. *Indian Horticulture*, **49**(4): 10-12

Singh, B, Kumar, M and Singh, V P 2005. Cultivating tomato in greenhouse is lucrative. *Indian Horticulture*, **49**(4): 13-14

Singh, B (2007) Summer squash cultivation under tunnels : A success story. *Indian Horticulture* **52**(4):28

Singh, B, Hasan, M and Tomar, B S 2009. Capsicum for farmers' prosperity. *Indian Horticulture.* **54** (5): 12-13

Singh, B, Singh, A K and Tomar, B S 2010. In Peri- Urban: Protected cultivation technology to bring prosperity. *Indian Horticulture.* **55**(4): 31-32

Singh, B (2011). Insect proof net house for cultivation for high value vegetables. *ICAR News.* **17**(3):6

Singh, B., Sabir N, Hasan M and Singh AK .2011. Greenhouse Cucumber: Production and Protection. Indian Agricultural Research Institute, New Delhi.TB-ICN.84/2011

Sabir, N., Singh, B, Hasan, M, Sumitha, R, Deka, S, Tanwar, R K., Ahuja, D B, Tomar, B S, Bambawale, O.M. and Khah, E.M. 2011. "Good Agricultural Practices for IPM in Greenhouses" A Bulletin. Editors IARI, New Delhi (in English). TB-ICN 23/2011

Singh, B, Sabir, N, Hasan, M and Singh, A K 2012. Greenhouse Cucumber: Production and Protection" A Bulletin. Editors IARI, New Delhi (in English). TB-ICN. 84/2011.

Singh B, 2012. Seed Production of summer squash in North Indian plains. ICAR News **18**(2), 14 -22.

11

Recent Advances in Raising of Vegetable Transplants under Protected Conditions

Chaurasia, S.N.S. and Naik, P. S.

Abstract

Production of healthy and vigorous seedlings is the foremost crucial step for raising a successful vegetable crop. Now a days seedlings are raised in trays under controlled atmospheric conditions called as transplants. As soil substitutes, artificial media made of perlite, vermiculite and peat moss are commercially available. A good rooting medium should be light weight, well drained, well aerated, and soilless with or without starter solution or added nutrients and with good water and nutrient holding capacity. The media should be sterilized before use. Selection of trays for cell shape, size and number depends on the type and nature of crop. Most of the vegetable transplants are now being produced in hardened plastic or polystyrene containers. Only good quality certified seeds should be used in media filled trays after priming, pelleting, coating or other seed treatments as per the crop requirement. Germination chambers as insulated rooms to maintain suitable conditions over large period of time may also be used. Generally vegetable crops need fertilizer application after expansion of first true leaf. Young seedlings are managed carefully to avoid any toxicity while taking care of their nutrient management. Water requirement also depends on growth stage and variation in weather conditions. Other seedling management practices include shading, thinning, weeding, disease/pest control, hardening off and subsequent transplanting. Containerized greenhouse plants can be grown using the conventional "rail" (or rack) system or the "float" system. Shipping factors such as mechanical injury, environmental conditions and length of storage affect plant vigour and establishments. IPM (integrated pest management) offers a practical way to effectively manage pests on vegetable bedding and transplants.

Keywords: protected condition, seedling, seed, germination

Introduction

The first step in successful vegetable production is to raise healthy vigorous seedlings. Young plants, whether propagated from seed or vegetatively require a lot of care particularly, during the early stages of growth. They have to be protected from adverse temperatures, heavy rains, drought, wind and a variety of pests and diseases. If small seeded vegetables are sown directly in the field, germination is often poor and the young plants grow very slowly and require a long time to mature. Also, the season may be too short for full development in the field. To overcome these problems, many vegetable crops are grown in nurseries before being transplanted in the field. Vegetable nursery is a place where plants are cared for during the early stages of growth, providing optimum conditions for germination and subsequent growth until they are strong enough to be planted out in their permanent place. A nursery can be as simple as a raised bed in an open field or sophisticated as a glass-house with micro-sprinklers and an automatic temperature control system. Although raising seedlings in a nursery has advantages, some vegetables do not transplant well, particularly root crops, and must be sown directly in the field for optimum results. It has to be noted, however, that transplanting seedlings interrupts their growth, which has the potential to reduce their vigor.

Advantages of growing seedlings in nursery beds

Although, many vegetable seeds can be sown directly in the field, experience has shown that raising seedlings in a nursery has a number of advantages as discussed below:

1. **Intensive care** - Seedlings receive better care and protection (from animals, weeds and pests) in the nursery. The average garden soil is not an ideal medium for raising seedlings especially from the point of view of soil tilth. At an early stage of development most vegetable crops require special attention which is not possible in the main field.
2. **Reduction of costs** - Fewer seeds are used for raising seedlings in the nursery than for sowing directly in the field because in the latter, seedlings have to be thinned to one, which is wasteful.

When expensive hybrid seeds are used, transplants therefore, become more economically attractive. Pesticides and labour are also reduced under nursery conditions as compared to planting directly in the field.

3. **Opportunity for selection** - Raising seedlings in a nursery affords the grower an opportunity to select well grown, vigorous, uniform and disease free seedlings.
4. **Extend a short growing season for late maturing crops** - Seedlings can be raised in a nursery under protected environment before conditions outside become suitable for growth and are transplanted in the field when conditions allow thus, reducing the amount of time spent in the field.
5. **Forced vegetable production for an early market** – Generally, prices of horticultural produce are attractive when production or supply is low. Vegetables can be grown 'out-of-season' in a nursery when conditions are not yet favourable. Such crops will thus, mature earlier after transplanting and hence, stand to fetch a higher price in the market.

Generally, in all parts of the country, raising of vegetable seedlings followed by growers in traditional open field condition method. During the process of seedlings raising by traditional method, 50 to 70% of the seedlings get damaged due to certain biotic and abiotic stresses. When the seeds are sown by broadcasting method in the open nursery beds (Seedlings become thin and are not suitable for transplanting). Continuous rain for several days, too low and too high temperature and humidity also play major/critical role in producing the seedlings. Healthy nursery raising is the most crucial step in vegetable production, which up to larger extent determine the productivity and profitability of vegetable cultivation. The major vegetable crops which are usually cultivated through transplanting method are: tomato, capsicum, cauliflower, cabbage, knol-khol, Chinese cabbage, Brussels sprouts, broccoli, etc.

A lot of advancement so far has been made in the field of seedling raising now a days. To reduce the cost of production and early establishment in the field, seedlings are now a days being raised in treys under controlled conditions round the year. These seedlings in treys are called by the name of 'transplants'. A "good" transplant is usually defined by consumer specifications. In general, a "good vegetable transplant" should be *stocky, green, and pest-free with a well-developed root system.* Once transplanted, the plant should tolerate environmental challenges and continue/ resume growth within no time to optimum yield. With this system, each transplant grows in an individual cell so there is less competition among plants and greater uniformity. Less labour is required for mixing and sterilizing soil, filling flats and pulling plants. Plug transplants establish better in the field because roots are not damaged in pulling.

Overly hardened or under-fertilized transplants may not establish quickly, which lead to delayed maturity and reduced yields. The "ideal" technique for growing transplants would be to raise the plant from start to finish by slow, steady, uninterrupted growth and with minimal stress. Since, ideal growing conditions rarely exist, plant growth may need to be controlled through manipulation of water, temperature and fertilizer.

Greenhouse grown containerized transplant is an increasingly popular way to establish vegetable crops. Transplants often result in better stands and earlier harvests; factors, which can increase profits to offset additional production costs associated with transplanting. However, if crops are lost due to frost or other factors, replanting further increases production costs.

Growers of greenhouse transplants generally fall into one of three categories: (1) producers of transplants solely for commercial sale, (2) growers of transplants for personal use and (3) growers of transplants for their own operations as well as for commercial sale. Time that must be devoted to the production of greenhouse transplants depends on the category in which a grower falls. Since, the third category is merely a hybrid/combination of the other two, categories one and two are discussed.

Recently, growers have made a transition to greenhouse-grown transplants using various types of containers, primarily plug trays. Plug transplant establishment is one of the key factors in producing high yielding tomato crops. Rapid early growth and complete plant survival are influenced by practices used by the greenhouse grower, and by the transplant management practices used by field tomato grower.

Greenhouse transplants have several advantages compared to field-grown transplants.

- They are produced more uniformly than field-grown plants.
- Favorable growing condition can easily be provided for raising seedlings successfully.
- Transplants often result in better stands.
- Their growth can be controlled more easily through fertility and water management and they can be held longer and harvested when needed.
- Because containerized plants are less crowded and healthier; stockier plants can be produced. Finally, container-grown greenhouse plants have a media-enclosed root ball, which retains moisture and root integrity at transplanting, reducing transplant shock.
- It is very easy and convenient to care of young tender seedlings growing in small areas against biotic and abiotic stresses.
- There is a higher germination rate and more high-quality plants can be produced in a greenhouse under controlled environmental conditions.

- Earlier harvest is attainable by using transplants than by direct seeding in the field. Generally, an earlier harvest is more valuable for the market.
- Costs associated with raising transplants can easily be recouped.
- Since, any variety of seed can be ordered for transplant production, a grower has a much wider selection of varieties.

Although vegetable bedding plants may only be in the greenhouse for a short period of time, it is important to produce a high quality pest-free transplant. Scheduling, plant nutrition, greenhouse management and pest management influence the quality of transplants. The basic differences between field grown seedlings and container grown transplants are as follows:

Traditional Method of Seedlings Raising (Grown under open field condition)	Recent Method of Seedlings Raising (Grown in plug trays under protected condition)
1. Use healthy and improved seed from a reliable seed source. If the seed is not treated, mix in 2-3 g thiram per kilogram of seed.	1. Use healthy and improved seed from a reliable seed source. If the seed is not treated, mix in 2-3 g thiram per kilogram of seed.
2. Disinfect the nursery area by solarization: Cover the beds with a plastic sheet for 24-48 hours. After removing the sheet, stir the soil with a disinfected rake.	2. Use clean trays. Disinfect trays with 1% solution of chlorine bleach. Wash the trays thoroughly with clean water before use.
3. Prepare seed beds about 3m x 1m in size and 20 cm high.	3. Use the correct cell size according to the crop. For example, cucurbit crops require larger cell size than other vegetable crops.
4. Prepare the nursery bed soil loose and friable and sow seed in lines about 5 cm apart and 1-2 cm deep. Adjust the planting depth according to the seed size.	4. Prepare growing media by mixing coco peat, perlite, and vermiculite in 3:1:1 ratio and fill the potting plugs. Sow one seed in each cell at a depth of 0.5 to 1.0 cm.
5. Cover the seed with a fine layer of well decomposed disease free compost or animal manure and mulch the nursery beds with paddy straw or dry leaves.	5. Adjust the pH of the mixture between 5.5-6.5 with lime to improve micronutrient availability. Avoid mixing soil into the media to minimize the occurrence of soil-borne diseases.
6. Water the nursery beds with a sprinkling can twice a day. After seedlings sprout, water once a day.	6. Fill the plugs tightly with the pre-moistened media.

7. Remove the mulch immediately after seedlings emerge from the soil.
8. Protect nursery beds from excessive rain and stagnant water by providing rain shelters. Protect seedlings from damping off and foliar diseases. Protect seedlings from insects and insect-transmitted diseases, especially viruses by covering nursery beds with 40-mesh nylon net.
9. Harden the seedlings by withholding water 4-5 days before transplanting. Hardening is facilitated by the application of potassium chloride (KCl) at 1 ml per 250-300 ml of water, or the application of a 10% sucrose solution 7-10 days before transplanting.
10. Drench the nursery beds with water one day before transplanting.

11. Transplant 4 to 6 weeks old seedlings, preferably in the evening.
12. Seedlings are grown in open field

7. Keep the trays in the mist chamber for sprouting the seeds.
8. Water the trays thoroughly to moisten the entire plug. This promotes root growth to the bottom of the plug. Avoid watering late in the afternoon to minimize occurrence of foliar diseases.
9. Harden the seedlings by withholding water 4-5 days before transplanting. If the seedlings are being raised in protected conditions, move them outside for several days to acclimate before transplanting.
10. Drench the plugs with water one day before transplanting.

11. Transplant 4 to 6weeks old seedlings, preferably in the evening.
13. Transplants are grown in trays

Location for Transplant Production

The transplant production facility should be located away from farming areas. The site should be levelled and well drained and have ready access to an abundant supply of quality water. The greenhouse should be located at sufficient distance from surrounding trees or buildings to prevent shadows during the day. Other important considerations concerning greenhouse location include (1) orientation (east supplies more overall light, north-south supplies more uniform light), (2) availability of labour, (3) good proximity to shipping routes, (4) easy access to utilities, (5) local zoning and tax laws, and (6) room for expansion.

Growing media

Media for raising seedlings

Generally, commercially prepared soilless media and used for containerized transplant production. The media must be free of insects, pathogens, nematodes and weed seeds. Most commercial mixes and components for commercial mixes meet these criteria.

Media have different properties that vary according to the contents. Be careful to choose a media best suited for the intended use. For instance, several media are available specifically for starting transplants from seed. Generally, they contain finer shredded peat particles than media used for bedding plants or potting soil. Check the media description that which media best suits your purpose. Select a medium that drains well and provides good aeration but still has moderate water-holding capacity.

Media are available with starter nutrients (charged) or without starter nutrients (non-charged). Although either type can be used, many growers prefer a non-charged media so they can add their own starter solutions and know for sure the fertility levels are correct. Also, growers using non-charged media know the exact amounts of nutrients, plants have received (provided they have precisely monitored fertilizer application/injection).

According to the way they are formulated, nutrient content of charged mixes varies. Some may contain only major nutrients while some may include a mixture of trace elements as well. Growers can get customized mixes by special order (contract) with a commercial formulator. Some charged mixes contain slow-release fertilizer that provides nutrients to plants after seed germination. Medium with slow-release nitrogen is generally not recommended since it may eliminate the option of withholding nitrogen to control plant growth. It is essential to know the concentrations of all the nutrients in a media for effective fertilizer management.

There are many sources and kinds of media. Desirable pre-packaged media are generally light-weight, well drained, well aerated, and hold moisture and nutrients well. Other ingredients such as washed granite sand and processed bark ash can also be found in some mixes. Perlite is a processed volcanic mineral that provides good drainage to the media while holding air in the rooting zone. Most of the water from perlite stays on the surface of the particles, making it readily available to plants. Vermiculite is a mica-type mineral that can hold large amounts of air, water and nutrients.

Media with finely ground perlite, vermiculite and peat moss are the best for starting vegetable transplants. Peat-lite mixtures, containing only peat moss and perlite, are not acceptable because the media are too coarse. Media without these components may not provide the needed nutrient and water-holding capacity to produce acceptable transplants. Additional ingredients are at the grower's discretion. Although pre-packaged media may be available in different sizes, it is usually shipped in bags or bales of either 3 or 4 cubic feet. The number of trays that can be filled by a bag/bale of media varies according to tray size and cell volume. Prices of media also vary according to contents, size of bags/bales and quantity of media purchased. We recommend transplant growers purchase commercially available media from a reputable company with an established quality control program.

It is possible for growers to mix their own soilless media. This requires additional time, labour and management, and any actual cost savings will likely be minimal. Except in special situations, commercially prepared media are preferred over home mixing. Do not use field soils for growing containerized plants. Field soils generally drain poorly and are often contaminated with diseases and weed seed. Even though this soil can be treated to help control pest problems, soil treatment is labour-intensive, costly and does nothing to improve soil drainage.

Selecting an appropriate media is especially important when growing transplants using the float system. Media containing clumps and large particles are not recommended for float production. In the float system, plants are irrigated by water drawn from the surface of the water reservoir through holes in the bottom of the tray into the media-filled cells. For this to occur, the media must form a continuous column from the bottom to the top of the cell. Because large particles prevent media from completely filling cells, they create air pockets. Cells containing air pockets do not take up water properly and stay too dry for good plant growth, such cells are called "dry cells." The media must have enough water-holding capacity to allow capillary action to draw up the water to the roots, but it must be porous enough to allow sufficient gas exchange.

Soil is the major medium for germinating seeds and growing seedlings, although it is not the best. There are artificial media made of perlite, vermiculite and peat moss, which are used as soil substitutes. For best results, a growth/rooting medium should posses the following qualities:

- sufficiently firm enough and dense to hold seeds in place during germination.
- sufficiently porous to let excess water drain away
- have a high water holding capacity.
- free from weed seeds, nematodes and other pathogens.
- high cation exchange capacity so that it can provide nutrients
- able to withstand sterilization treatment without being altered.
- not be toxic to plants.

Since, it may not be possible for one medium to have all these characteristics, different media are normally mixed together to obtain a 'near' ideal mixture. Materials found in the locality may also be used. Some media may be contaminated, therefore it may be necessary to sterilize them before they are used, to kill weed seeds, insects and numerous soil pathogens. This treatment includes the use of heat and/or chemicals. Steam sterilization and solarization are used in the heat treatment to kill all pathogens in the media. One of the advantages of this treatment is that the media can be used sooner after treatment and the treatment can be applied to the dry, wet or even cold media.

There is, however, a risk of the media structure or composition being modified by the heat. Chemical treatment which is also referred to as fumigation involves the use of insecticides which may be specific to certain pathogens or broad spectrum (controls a wide range of pests). These chemicals should be handled with caution as most of them are toxic to both humans and animals. Some of the commonly used chemicals are chloropicrin and vapam.

Growing media are designed to achieve high porosity and water retention while providing adequate aeration. Premixed media are common in the greenhouse industry. Suppliers offer a diversity of mixes that are available pre-packed (in bags, bales, super sacks) or in bulk. Recipes are specially formulated for propagation, specific crops or general greenhouse crops. Soilless media purchased in bags do not have to be pasteurized or fumigated before use. Preventative applications of biological fungicides or fungicides may be necessary with some vegetables that are prone to damping off. Growers can obtain commercially available mixes with different types of biological fungicides added to the mix.

Types of Growing Media

Growing media for vegetable bedding plants or field transplants in greenhouses contain a variety of soilless ingredients such as *peat moss, vermiculite, perlite, shredded coconut husks (coir), or composted materials plus starter nutrients and wetting agent.* Field soils are generally unsatisfactory for the production of plants in containers because soils do not provide the aeration, drainage and water holding capacity required. They also need to be pasteurized or fumigated to prevent the development of diseases and germination of weed seeds.

Compost-based mixes are also available commercially as a substitute for traditional soilless media, especially for organic production.

Peat: Peat consists of residues of a marsh swamp. Vegetative peat moss is available but should be broken into fine parts before used in mixtures or as media. It contains some organic nitrogen and is favourable for newly rooted cuttings or germinated seeds.

Sphagnum Moss: Sphagnum moss is dehydrated remains of acid hog plants and has three genera, growing naturally in damp humid forest lands. It is relatively sterile, light in weight and has a very high water holding capacity. Normally, it absorbs and holds water 20 times to its weight. Sphagnum is acidic in nature having a pH of about 3.5. It also contains a fungistatic substance which is useful to inhibit damping off. Moss is soaked in solution containing fungicide and is impregnated with nutrient solution before being used for propagation.

Vermiculite: This is a micaceous mineral which expands significantly when heated. Chemically, it is hydrated magnesium, aluminium, iron, silicate. When expanded, it is very light in weight. It is neutral in reaction and has good buffering properties. It is insoluble in water. Vermiculite is available in 4 Grades.

Perlite: This is grey white material having volcanic origin. It is neutral. It has no buffering reaction and it contains no mineral nutrients.

Leaf Mold: It is prepared by using fallen leaves of various tree species available locally, e.g. Oak, Silver oak, Maple, Azadirachta, Ficus, etc. It is prepared by stacking a few layers of leaves then covering them with a thin layer of soil and cow dung slurry. Some live culture of decomposing organisms is added to hasten the process of decomposition. The medium is ready for use after about 12 to 18 months of decomposition.

Saw Dust: It is a by-product or waste material from saw mills. The quantity and quality depend on the parent wood material. It is mixed during preparation of media.

Grain Husk: Several types of husks are available, paddy husk is one of the important wastages from rice mills. It is light in weight and cheaply available. It is suitable for mixing with other types of media.

Coco Peat: Coco peat, cow dung, etc. are also used as media. A mixture of a few media is always preferred and used in commercial nurseries. Many times soil is one of the main parts for mixtures. Media must be selected on the basis of availability, cost, ease in handling, etc. The media should be procured, stored and kept ready for use in nursery.

Media Sterilization

Steam treatment

Field soil is not recommended for vegetable transplants in greenhouses. However, if it is used, it must be treated to eliminate soil borne plant pathogens, insects, and weed seeds. Once the soil has been treated, care must be taken to avoid reinfestation. Treatment of soil with steam is preferred over chemicals because it is very effective and safe. All pathogens will be killed by steam and only a few of the hardiest weed seeds will survive. There are portable steam generators available. (Fig: Steam Generator)

The soil moisture content prior to steaming is important. Overly moist soil will take a long time to reach proper temperature. However, some moisture is necessary for effective killing of microorganisms as well as conductance of heat. Proper soil moisture for steaming is approximately the same for good planting conditions; soil squeezed in hand should crumble easily. If possible, the soil mass should be moistened evenly two to three weeks prior to the treatment. This will germinate difficult-to-control weed seeds, such as oxalis and clover, making them susceptible to heat. Prolonged steaming of soil at temperatures higher than 81°C can result in undesirable side effects such as overkill of beneficial soil microflora and accumulation of ammonium and manganese. Soil high in organic matter should be tested for ammonium after steaming. Several weeks may be necessary to allow for dissipation or conversion of ammonium. Incorporation of dolomitic lime and superphosphate, based on soil test, may reduce ammonium levels. High levels of ammonium can reduce and even stall plant growth.

Containers for Transplant Production

The Tray Cell

Cell size is generally a matter of economics. In crops such as onions or lettuce, where plants in number per acre are in tens of thousands, flats with small cells are preferred because more plants can be grown in less space. The size of cell depends upon the type and nature of crop. However, the yield in the field varies (Hall, 1989; Vavrina *et al.,* 1993). Despite these findings, larger cells are recommended for longer cycle vegetable transplants (>5 weeks) such as peppers (Weston, 1988) and tomatoes (Weston & Zandstra, 1986), while smaller cells may be the better choice for short crops (<4 weeks). In short crops, root growth may not completely fill a large cell, and damage may occur when "pulling," as soil falls away and exposes roots. Different cell shapes within the tray are available, including the inverted pyramid, the truncated cone, the bullet, and assorted beveled shapes.

Production of containerized vegetable transplants has changed dramatically in the past several years. Most container-grown vegetable transplants were produced in peat-based containers, but now, the vast majority are grown in hardened plastic or polystyrene (Styrofoam) containers. Generally, peat containers, clay pots, peat pellets, fiber blocks and plastic pots are not used for mass production. Therefore, it is limited to polystyrene and plastic types. Since, both plastic and polystyrene containers are a considerable financial investment, most are reused many times. Because of reuse, containers must be properly sanitized after each use or disease problems are likely to occur.

Most containers are sterilized using 10 percent chlorine bleach solution. Problems sometimes arise when chlorine rinses are used. If polystyrene containers are not aerated properly, chlorine can soak into crevices and cause toxicity in future plantings. Use proper rates and thoroughly rinse with water after washing to reduce the possibility of toxicity.

Unlike polystyrene containers, plastic containers do not have the small pores and crevices that may harbour disease organisms and soak up residual chlorine. Plastic containers have edges, however, that can be difficult to clean and may provide a suitable environment for pathogens. Plastic containers are usually made up of two parts: a reusable plastic tray or flat with inserts of various sizes and/or configurations. The inserts are typically used only once. Numerous types (shapes, sizes and configurations) of containers are available. Select the type that best fits your system. Plastic and polystyrene containers most often come in straight row arrangements. Polystyrene containers normally have inverted pyramid-shaped cells that taper toward the bottom. They may have cell sizes as small as 0.8 inch square or as large as 6 inches square. The number of cells in a container depends on the cell size. They may have from 12 to 338 cells per container. A polystyrene tray with a 1½-inch cell size has 128 cells, while a tray with a 2½-inch cell size has 72 cells.

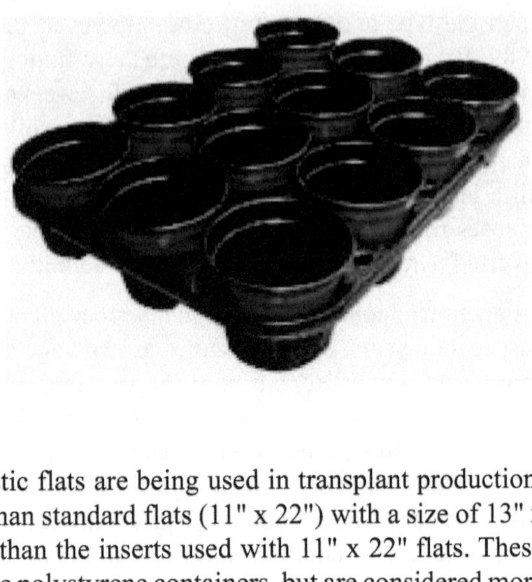

Cells in plastic trays are arranged similarly and hold the same volume of soil as the same size polystyrene cells. Since plastic cell walls are thinner, plastic trays typically have more cells per tray than similar polystyrene containers. For example, cell media volume of a 512-cell plastic container is roughly equal to the media volume of a similar sized 338-cell polystyrene container.

More recently, single piece plastic flats are being used in transplant production. These flats are typically larger than standard flats (11" x 22") with a size of 13" x 26". They have greater rigidity than the inserts used with 11" x 22" flats. These single piece flats are reusable like polystyrene containers, but are considered more durable and easier to clean. As with polystyrene and plastic inserts, a variety of sizes are available.

Different plant species require differing amounts of space, nutrients and water. Certain cell sizes are more suitable for some plant species than others. Larger cells hold a greater volume of media, which enable them to retain more water and nutrients. Therefore, transplants growing in larger cells require less frequent watering and fertilizing. This helps reduce the likelihood of moisture or nutrient stress. Also, larger-celled containers normally produce stockier and earlier plants. Because of larger root volume, these plugs experience less transplant shock in the field. Proper watering is more critical with larger sized cells however, especially in the early stages when over-watering can be a real problem resulting in a greater chance of root disease development.

Generally, smaller cells are used for plants such as cabbage, broccoli, cauliflower, collard, kale and lettuce. Trays with 1 to 1½-inch cells are well suited for producing transplants of these crops. These trays generally have from 200 to 338 cells per tray. Use larger cell sizes, 1½ to 2½ inches, for production of tomato, pepper, watermelon, muskmelon, cucumber and squash transplants. These trays generally have from 72 to 200 cells per tray.

Media Mixing

Proper mixing of the media is an important phenomenon in growing containerized transplants. With the advancement in nursery raising, now certain types of media mixers are available. This mixer not only mixes the different media in appropriate proportion but also screen out the oversize mixers.

Filling the Plug treys

Traditionally, the treys were filled manually, but with advancement so far made in raising containerized transplants, machines have been designed to fill the plug trays. These machines can fill about 400 trays per hour. This may be increased up to some extent. Fig below is showing the ways to fill the treys manually as well as automatically

Seeds and Seed Treatments

"Good seed doesn't cost, it pays!" This old adage applies doubly in vegetable transplant production. Poor germination or vigor can result in loss of plant size, uniformity within the tray, overall greenhouse productivity and revenue. Quality seed insures good germination, rapid emergence, and vigorous growth. Poor quality seed results in "skips" or excessive seeding to compensate for poor stand establishment, both of which diminish profitability. Purchase only high quality seed that is certified disease-free or treated to reduce the risk of contamination by disease-causing organisms.

Primed Seed

Priming (enhancing) good seed can improve seed quality and vigor (Perkins-Veazie & Cantliffe, 1984; Argerich & Bradford, 1989). Primed seed is biologically altered through addition of just enough water (and sometimes plant growth regulators) to allow the seed to undergo early stages of germination without radical emerging from the seed coat. In the "primed" state, seeds germinate more rapidly and emerge more uniformly over a greater range of temperatures, soil moisture conditions, and/or various stresses. This results in greater vigor, uniformity and rapid establishment.

Priming technology was developed to overcome germination and emergence problems in field plantings caused by adverse conditions such as cold soil, too much soil moisture, etc. However, enhanced-seed use in greenhouse can be very cost effective as well by alleviating the need to "play catch-up" within an order or flat as a result of uneven emergence (i.e., all plants are of the same physiological age when shipped and relatively uniform in size).

Pelleted and Coated Seed

Pelletization is a technology that uses a material such as clay to enlarge small or irregularly shaped seed and to make it spherical for ease in mechanical seeding. Pelletizing increases speed of seeding and singulation within the flat (one seed per cell); with the use of color coding it also can be used for variety identification. Pelletization was once thought to reduce seed oxygen availability in waterlogged soils (Durrant & Loads, 1986). Coated seed refers to a new technology that was designed to reduce dust and worker exposure to pesticides. These coatings are highly water permeable and are sprayed on the seed. Application of fungicides and fertilizers incorporated in the coating gives the grower additional benefits (Hummel, 1991).

Sowing the seeds in Plug Trays

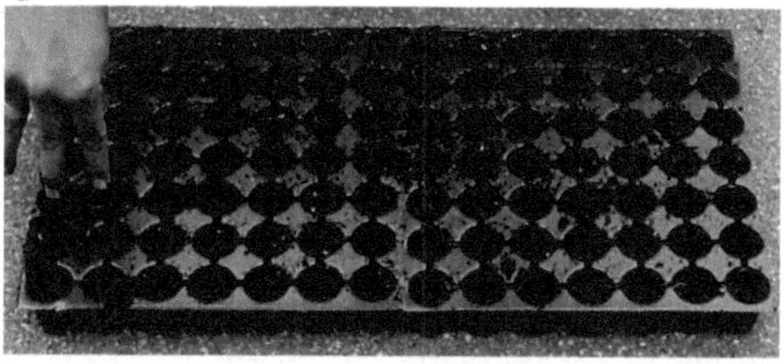

Earlier, the seeds in plug trays were sown manually and mostly by hand and took much time in filling. It was totally a time consuming process but now a days a lot of advancement has been made in this direction and seeds are now being sown in the plug trays directly by machines as shown in the figure.

Germination of Seed

Of the many factors which influence germination, water and temperature affect seed performance the most. Extremes in either of these factors can result in poor germination (Ellis & Roberts, 1981; Perry, 1982). Different seed lots sown in the same environment may not act alike. Wheeler and Ellis (1992), after studying onion seed quality, seedling emergence, and seed/seedbed interactions, reported that fluctuations in the seedbed environment were the most important factors in determining final seedling emergence.

Germination Chambers

Most vegetable crops benefit from use of a germination chamber. This is usually an insulated room in which temperature and relative humidity can be maintained at a precise level. The goal is to facilitate germination process in a confined area to minimize cost of heating a large greenhouse to germination temperature.

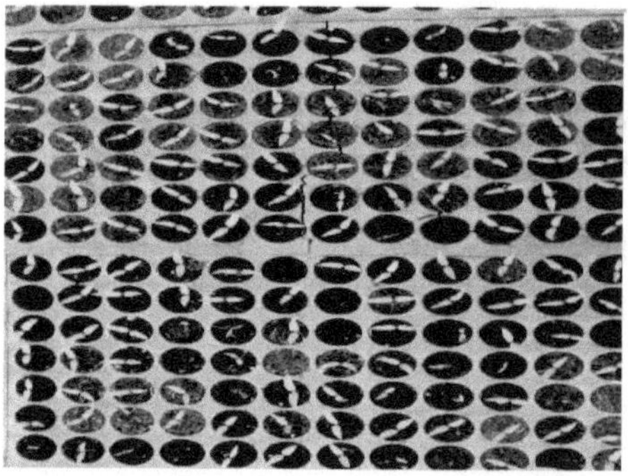

Air circulation is important to ensure uniform temperature and humidity throughout the chamber. There should be a thermostat to maintain the temperature regime. If temperature shoots up too high, variability between the seeds is accentuated, resulting in both uneven germination and transplant development. Germination time will vary between seed lots, so growers should check the trays regularly while they are in the germination chamber. The trays should be moved to the greenhouse after the seed coat has cracked and the shoot just starts to emerge. This will prevent excessive elongation. The time in the sweat chamber may only be 2 or 3 days.

Tempered Water: If possible, warm water should be used when watering plants during early growth-stages. Water should be heated to about 21⁰C. Heating the volume of water required to irrigate a large greenhouse may not be feasible.

For best results, a "controlled" temperature room should be used for germination. Such rooms are insulated to allow heating or cooling according to specific germination needs of the seed. This technique reduces fluctuations in temperature and moisture that can occur in an open house. Room design should allow fork-lift access for moving trays on pallets, if large-scale production is anticipated.

Table 1. Optimum temperature ranges for germination of seeds and growth of various vegetable transplants (temperatures are stated in ⁰C).

Crop	Germination (Temperature)	Approx. no. of Days to Emergence	Growing Temperature (Day)	Growing Temperature (Night)
Tomatoes	21-24	3-4	18-21	10-18
Peppers	26-28	4-6	18-21	12-18
Cole Crops	18-24	2-3	12-18	8-15
Vine Crops	24-30	2-3	21-24	12-18
Onions	18-24	3-4	16-18	8-15

Reference: Tips on Growing Bedding Plants, Ohio Florists Association

Starting seeds too soon, will result in overgrown plants of poor quality. Following are the guidelines for growing vegetable bedding plants. This will vary according to different growing conditions and should serve only as a guide.

Germination Tips

Warm temperatures and uniform moisture are needed to ensure successful germination and get the plants off to an even start. Many germination chamber systems are commercially available including custom built germination units. Many growers use bottom heat or root zone heating to provide warm, even temperature. Rubber tubing or mats with hot water are placed on the bench top under the plants. A weed mat barrier is placed on the top of the bench to help spread the heat with skirts on the side to help contain the heat. In all systems, it is important to remove flats from the germination chamber as soon as radicles break through the seed coat to avoid seedling stretching. Experience and experimentation with your total seeding system is the key to uniformity and success.

Cole Crops (Cabbage, Broccoli, Brussels sprouts, Cauliflower)

To prevent premature seeding or bolting, avoid exposing transplants to temperature below 10°C for long periods (a week or more). Cold temperature causes development of premature heads or "buttoning" in cauliflower and broccoli. Any stress or check in growth results in a "wire stem" and plants will not become well established in the field or garden, resulting in reduced yields and performance.

Eggplant

Eggplant seed can be directly sown into 50 cell trays to shorten the time needed to produce transplants by approximately one week. Germinate seed in flats at 21° to 24°C. Eggplants are susceptible to chilling injury and should not be grown below 8°C. Any stress or check in growth will result in tough woody stems and transplants that will have a tough time getting started later in the field or garden.

Tomatoes

Tomato seeds germinate are best at 21°C. As soon as there is any evidence of germination, they should be removed from mist and bottom heat. The ideal root-zone temperature is 26 to 29°C during the first four weeks of growth and 20 to 26°C during the fifth and sixth weeks. Optimal growing-on day temperatures are 18 to 24°C with minimum night temperature of 16°C. Exposure of tomato plants to temperature below 16°C will likely result in rough fruit (catfacing) on the first few clusters. Transplant young seedling into 2" to 3" containers when they have two true leaves and grow on until planted in the field. For earliest production, some growers finish their transplants into 6" or larger containers.

Peppers

Seeds germinate at 28° to 32°C. Note that germination is very slow at lower temperature. Seedlings develop well at 24°C daytime and 18°C night temperatures. Seeds may be directly sown into 72-cell trays for early production. Peppers are prone to damping off diseases, especially if the media is compacted. They are also susceptible to transplant shock.

Vine Crops

Cucurbits do not transplant well, and are the best to sown in the final container. After germination, excess plants can be thinned.

Seed Sowing-Depth in Containers

In most commercial facilities, flats are automatically filled with soilless mix, "dibbled" or pressed to a depth of about 1/4 inch, seeded, then covered with soilless mix or vermiculite. Proper planting depth is more important for seeds with minimal food storage reserves (cabbage family) and those that need light for germination (lettuce, celery). Variable planting depth can result in problems such as uneven stands and "leggy" plants.

Types of Nursery Facilities

There are three main facilities normally used for raising seedlings in a nursery. Choice of a particular one will depend on the available resources and prevailing environmental conditions.

*Greenhouses - environment fully controlled

*Nethouse - environment partially modified

*Open field - where climatic conditions are normally favourable for the crops grown.

Fertilization in Containerized Plants

The amount of fertilizer needed at any given time depends on the kind of vegetable transplants being grown, the frequency of fertilizer application, the stage of growth, the growing environment and the amount (if any) of fertilizer in the media. As a general rule, the more often plants are fertigated, lower the required nutrient concentration for any given fertigation. In general, vegetable crops will not need fertilization until after the expansion of first true leaf. The inherent fertility of peat (a component of the soilless mix) can carry the crop further than one might anticipate. Therefore if a nutrient charge is added to the medium prior to germination, the salt index level of this charge should be very low. High salt levels can cause complications in germination and early growth. Container-grown plants are usually fertilized with soluble fertilizer, which has been dissolved in a stock. This stock solution is metered into the irrigation lines and mixed with water during application. Application of fertilizer and water through the irrigation system is called fertigation. If a charged media (contains fertilizer) is used, fertigation is usually delayed one or two weeks. Otherwise, begin fertigation when the first true leaf begins to develop. During fertigation, apply nutrient solution until it runs out to the bottoms of the flats. This ensures that all plants are adequately fertilized (assuming the nutrient concentration is correct). In addition, this helps prevent salt injury by leaching residual fertilizer salts from the media.

The goal in vegetable transplant production is to produce a sturdy, compact plant which when transplanted, will establish and grow quickly and produce an optimum yield. A good rule of thumb is to begin fertilization only after the first true leaf has expanded to 1/2 to 3/4 inches in length.

Excessive fertilization, especially early in transplant growth, will result in "leggy" plants that may defoliate and establish slowly in the field. Minimal fertilization will result in stunted plants with retarded growth in the field. Proper fertilization directly impacts field performance.

No fertilizer formula or prescribed application is available that covers all transplant crops or all environmental conditions. Growers generally develop their own blends and application schedules. Some growers like to have phosphorus, lime (Ca), some starter nitrogen (N), and micronutrients provided in the soilless mix. Others prefer to dispense all nutrients through liquid fertilization to provide for "greater control" of plant growth. Good vegetable transplant crops can be produced using either method. The method selected depends on specific growing needs. For example, leaching of fertilizer is necessary to slow excessive growth, liquid fertilizers are more easily washed out of a mix than are granular forms.

Slow release N sources are not generally used for vegetable transplants since, plant height is difficult to control with constant N. Usually, a single application of a full-complement micronutrient package provides all the elements required for seedling growth.

Fertilizers used for liquid applications should be 100% soluble in water. Rate of application depends on nozzle size and trolley speed in automatic watering systems. Frequency of fertilizer application varies from grower to grower. Some systems deliver 250-400 ppm N once weekly or 125ppm N twice weekly; however, 30ppm N of a totally water-soluble fertilizer applied at every watering should meet most crop nutrient demands (Vavrina & Hochmuth, 1994). Generally, summer-and fall-grown transplants should receive more frequent, lower-rate feedings (e.g., 30 ppm), while winter- and spring-grown transplants can receive a fewer, higher-rate feedings. Amounts of fertilizer needed per 100 gallons of water to give 50 and 100 ppm N solutions are given in Table 2.

Fertilization, watering, temperature and sunlight each are important components in controlling plant height. The grower's ability to be flexible in fertilizer application schedules will usually result in plants of greater acceptability for commercial use.

Choosing Fertilizers

Factors to be considered when choosing fertilizers include the ratio of ammonium to nitrate-N, trace element starter charge, content of calcium and magnesium, and potential acidity or basicity. Commonly used fertilizers include 15-0-15 (Dark Weather Feed), 15-15-15, 15-16-17 and 20-10-20 or Cal-Mag 15-5-15.

Peat-Lite Specials (15-16-17, 20-10-20)- These fertilizers are among the most popular for routine fertilization of vegetable bedding plants. Both are high (>50%) nitrate fertilizers. However, these fertilizers also have elevated trace element levels which may raise iron (Fe) and manganese (Mn) to toxic levels at low pH. Both are acid-forming fertilizers, but 20-10-20 has the greater potential acidity.

15-15-15 Geranium Special- "Triple 15" is a good alternative to the Peat-Lite Specials for crops sensitive to trace element toxicities. Trace element levels supplied by this fertilizer are lower than the Peat-Lite Specials. Otherwise, at the same rate of nitrogen (N), plant response will be very similar to 15-16-17. This is an acid-forming fertilizer also; the potential acidity is slightly greater than 15-16-17.

20-20-20 General Purpose- Growers who use this fertilizer on soilless media risk ammonium toxicity because nitrogen in this fertilizer is 75% ammonium and urea. Some growers who use media containing soil do not appear to have problems. If 20-20-20 is used, the growing medium should be tested frequently for ammonium. 20-20-20 supplies trace elements and has the greatest potential acidity of fertilizers. Tomato, eggplant and pepper plants are especially sensitive to ammonium. High ammonium levels, especially in soilless mixes, can reduce plant growth and cause yellowing of the foliage.

Low Phosphorus (P) Fertilizers (20-0-20, 20-1-20, 15-0-15) These fertilizers can be tried as an alternative to chemical growth regulators for vegetable bedding plants. This technique of growth control is sometimes called "phosphorus starvation." It is generally believed that more P than necessary is being applied to greenhouse crops. Too much P may cause plants to stretch and P is a ground water pollutant. Unfortunately, in terms of height control, these fertilizers may be of no benefit if they are applied to a growth medium containing superphosphate or a high starter charge of P. Also, there is a risk of P deficiency if the fertilizers are used continuously with low P growth media. Low P fertilizers are quite different in many ways. 15-0-15 and 20-0-20 supply Calcium (Ca). 15-0-15 is a basic (raises pH) fertilizer containing about 95% nitrate and 20-0-20 is a neutral fertilizer and is 50% nitrate. 20-1-20 is an acidic fertilizer and it does not supply Ca, but it is about 70% nitrate.

Calcium nitrate and potassium nitrate (15-0-15). Use of this fertilizer combination greatly reduces the chance of trace element toxicities. Some growers alternate its use with Peat-Lite Specials on a 2 to 3 week basis to supply Ca and to counter the acidic effect of the Peat-Lites. However, both superphosphate and a trace element fertilizer must be incorporated in the growing medium if this combination is to be used as the sole fertilizer.

Guidelines for Rates and Frequency of Fertilizer

Small, slow-growing plants should receive lower rates or less frequent application until they are well-established. Care should be taken not to over-fertilize vegetable bedding plants to avoid overgrown plants. Young seedlings are especially vulnerable to injury from high soluble salts.

While plants are in the plug or seedling stage, use a complete water soluble fertilizer at the rate of 50 to 100 ppm N every time plants are watered and use clear water (no fertilizer) every third watering. Use the lower rate (50 ppm) early and the higher rate (100 ppm) later if the seedlings are to be held in the flat or tray three or more weeks before transplanting. Shortly after transplanting, as plants approach rapid growth, increase the rate to 200 ppm N at every watering or 300 ppm N once every 7 days, watering with clear water 2 or 3 times in-between each fertilization.

Fertilizer Solution Volume: The volume of fertilizer solution applied has a dramatic effect on growth of the vegetable bedding plants. As the volume of water-soluble fertilizer increases, quantity of nutrients delivered to the plant also increases resulting in an increase in height, stem diameter and plant weight. Doubling the volume applied also doubles the amount of each nutrient potentially available to the plant.

Nutritional Problems

Vegetable bedding plants and transplants are subject to the same nutrient disorders as other plants. Early in production, serious problems are: high soluble salts, trace element toxicities, and ammonium toxicity. Late in production, particularly in cell packs, plants may develop nitrogen deficiency symptoms as the earliest indication of insufficient fertility.

Soluble Salts- Injury to vegetable bedding plants from excess salts seems to be the most common problem shortly after transplanting. Seedlings are much less tolerant than established, rapidly growing plants. Some soilless mixes may contain enough "starter charge" to cause excess salts problems in the first a few weeks after transplanting, particularly when a water-soluble fertilizer is also applied. Excessive drying, poor drainage, and uneven watering are factors which can aggravate this problem. Check roots of plants often and conduct regular soil tests to identify and prevent problems. It is difficult to diagnose a soluble salt problem by symptoms alone. Often nutrient deficiencies and root diseases cause the same symptoms. Therefore, a greenhouse (not field) soil media test is advisable.

Trace element toxicities- Iron (Fe) and/or manganese (Mn) can be accumulated to toxic levels by tomato plants. Symptoms appear as numerous small dark spots and mottling of the foliage. The potential sources of excess Fe and Mn are: trace element fertilizers in the mix, water-soluble fertilizers with elevated trace elements level, and sometimes irrigation water. Low growth medium pH aggravates the problem by increasing Fe and Mn availability. Toxicity can be avoided by keeping the pH in the range of 5.8 to 6.0 for susceptible crops and by the use of fertilizers with lower trace element levels.

Ammonium toxicity- This is less common today because most growers use water-soluble fertilizers that supply about 50/50 ammonium and nitrate to fertilize plants in soilless media. Tomato, eggplant, and pepper are more sensitive to ammonium nitrogen, but many other vegetable bedding plants can be harmed if ammonium become excessive. Too much ammonium during early spring (February or March) in low light and cool media conditions can be toxic to plants. Plant growth may be reduced with yellowing of the foliage.

Organic Nutrient Management

Quality of the planting mix is important to insure proper plant health in organic production. Conventional growing media that contains synthetic ingredients (wetting agent, chemical fertilizer) cannot be used in organic production of field transplants, and vegetable bedding plants. However, acceptable growing media can be composed from a wide variety of approved materials. These organic blends may be purchased off-the-shelf, custom-blended by manufacturers, or produced on-the-farm.

Most commercial potting mixes contain synthetic fertilizers and wetting agents that do not meet organic standards. One alternative is to arrange a special order from a commercial supplier who agrees to exclude starter fertilizers and wetting agents and then, plan to add your own. Purchasing a commercially prepared organic mix is the easiest way to get started and most growers choose this option to reduce the risk of soil-borne diseases. Common components such as peatmoss, perlite, vermiculite, and coconut coir are generally acceptable for organic certification, but check with your organic certifier. Compost, being the most renewable, is a preferred material for many organic growers.

Since, compost can vary from batch to batch, there can be variation in the performance of mixes, even when using the same recipe. If making or buying compost for your potting mix, try to have it ready at least 6 months before you need it. This allows time for the compost to mature so that nutrients are stabilized, phytotoxic compounds have degraded and disease suppressive beneficial microbes have a chance to increase. It is very important to test the compost before you use it to determine its pH, available nutrients and soluble salt levels. Organic potting mixes may contain from 20 to 50% compost by volume depending upon the crop, container size and growing conditions. 100% compost is generally not advisable.

Watering Transplants

Amount and frequency of watering will vary depending on cell type, growing media, greenhouse ventilation and weather conditions. It is important to water thoroughly, and moisten the entire plug, which will promote root growth to the bottom of the plug. If the plug is not watered thoroughly, root growth will be confined to the top of the plug. Allow the plug to dry down before watering, but do not let the plant wilt severely, as this will damage the roots.

Plug transplants should be watered thoroughly in the morning, but should not be watered late in the afternoon. If the plants remain wet overnight, disease problems increase. If an overhead watering boom is used, it is advisable to remove and rearrange the nozzles occasionally to avoid the "streaking" that results from variations in output from different nozzles.

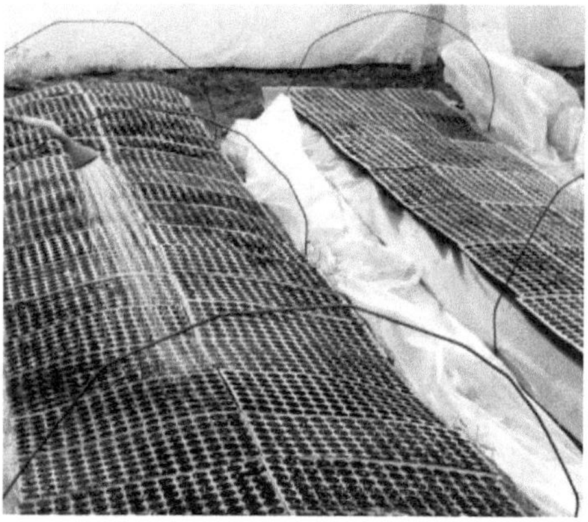

Knowing when to water is generally determined by experience. An accomplished grower can determine when to water by picking up a tray and judging the amount of moisture by hand weighing. Water requirements change as plants grow and with varying weather conditions. Seed should be kept moist but not soaking wet; as saturated plugs generally have less oxygen (oxygen is very important during germination). Generally, for established transplants, water should be applied only when surface of the medium is dry to touch. Excessive watering leads to succulent plants with restricted root growth.

A standard technique in both the bedding plant and vegetable transplant industries is to water until "run through" is achieved. This means to apply enough water so that each cell of the flat has sufficient water to allow some of it to drain out of the bottom. This technique attempts to saturate each cell, which in turn provides for more uniform growth within the flat (i.e., growth differences due to an inequality of water between neighboring cells are reduced). Overhead irrigation should be done in the morning, if possible, to allow the foliage to dry before night. Prolonged periods of wet foliage encourage disease. Chlorination (1-2 ppm chlorine) is suggested for ebb and flow systems to reduce bacterial and fungal contamination.

Grafting

Grafting of vegetable transplants is a common practice to control soilborne diseases and nematodes, for both field and greenhouse grown crops. In addition, grafted vegetables can produce higher yields and have improved tolerance to environmental stresses, soil salinity and low soil temperatures. Grafting of chile peppers (*Capsicum annuum* L.) is a recent practice where *C. annuum* scions are grafted onto *C. annuum* rootstocks which have soilborne disease and nematode resistance. Intergeneric grafting of herbaceous crops is an old art, and became standard practice in Japan and Korea in the late 1920s for introducing disease resistance to watermelons.

Watermelons were grafted to bottle gourd (*Lagenaria siceraria Standl.*), and melon (*Cucumis melo L.)* onto wax gourd *Bemin casa hispida* rootstocks for disease resistance. Because grafting can give increased disease and pest resistance and vigor to some herbaceous crops, it is useful in low input sustainable horticulture. Furthermore, because soil sterilization can never be complete, grafting has become an essential technique for continuous cropping of fruit-bearing vegetables grown in greenhouses. Despite widespread use of grafting, information about scion and rootstock compatibility for some horticultural crops is still lacking. As a rootstock for Solanaceous crops, tomato cultivars can provide a broad range of disease resistance to soil borne diseases and nematodes. Several tomato cultivars, e.g., Celebrity, have a wide range of resistance to soil borne diseases, as well as southern root-knot nematode. If tomato could be used as a rootstock for Capsicum, it would add another avenue in the production strategy for providing soil borne disease and nematode resistance. The tube graft in which a silicon tube is used to hold the scion and rootstock together during healing is especially popular for tomato grafting. The process is fast and large number of grafted seedlings can be managed easily throughout the healing process. Tube grafting makes it possible to graft small plants grown in plug trays. This is an advantage because smaller the plants, more are the plants that can be fitted into healing chambers or acclimation rooms.

Seedling management

1. **Shading-** Shading should be done to protect the young seedlings from high heat intensity in sunny areas and also from from heavy rain. Shade can be provided by polythene nets or even grass. The shade should be removed some days before transplanting to allow the seedlings to acclimatize to field conditions.

2. **Thinning-** This is a way of regulating plant density in rows and in holes. During thinning, weak, diseased plants are pulled out to allow healthy seedlings to grow well. It is normally done when seedlings have formed a few true leaves.

3. **Insect pest and disease control** - This is a continuous process from seedling emergence to transplanting. It is normally done by physical means but chemicals can also be used if the need arise.

4. **Weeding-** This is done by physical means when weeds emerge.

5. **Hardening-off-** Transplants must be 'hardened-off' so that they can withstand the transition from a relatively sheltered and protected environment to a sometimes harsh open situation. Generally, hardening is imposed from about 1 to 2 weeks prior to transplanting seedlings, by gradually exposing them to higher (or lower) temperature and the higher light intensity prevailing in the field. It should, however, not involve any treatment that may

reduce the rate of photosynthesis, such as nutrient stress. Care should be taken not to over-harden plants, as this may delay maturity and in some instances even reduce crop yields.

6. **Transplanting-** This refers to the operation of lifting seedlings from the seedbeds or containers and transferring them to the field where they will grow and mature. The main aim during transplanting should be to interrupt growth as little as possible, and if the operation is not carried out properly it can severely check growth or in extreme cases, cause death of the transplants. Most vegetable seedlings are ready to be moved 4-8 weeks after sowing. The seedbed should be given a thorough soaking about 6-12 hours before the plants are moved to ensure that they are fully turgid and that the roots retain plenty of soil when the plants are lifted. The main field should also be irrigated at the same time so that the planting holes can be opened up easily and the plants easily 'firmed'. The best time to carry out transplanting is in the late afternoon or early evening, as this allows the plants some time to get partially reestablish before having to face heat stress during the day. A cool cloudy weather is ideal for transplanting. It is always wise to raise about 30% more seedlings than are actually required so that the weak ones can be discarded and any casualties replaced. The adaptability of vegetables to transplanting varies widely between crops. Transplanting success depends on how rapidly a plant is able to regenerate those parts of the root system that are lost/damaged during transplanting.

Transplant Age

The "ideal age" for finished vegetable transplants has eluded researchers for decades. For example, good commercial yields of tomatoes have resulted from transplants ranging in age from 2 to 13 weeks (Vavrina & Orzolek, 1993). Studies conducted with broccoli, pepper and watermelon have shown similar results (Lamont, 1992; Weston, 1988; Vavrina et al., 1993). Perhaps the key to productivity in vegetable transplants resides not in age, but in the quality of the transplant. Providing plants with adequate resources to maintain appropriate growth in a limited space may be all that is necessary.

Transplant Height

Growers have no legal chemical tool available for controlling plant height. Certain plant growth regulating chemistries can accomplish height control (Vavrina & Armbrester, 1990). Plant height can be controlled to a certain extent by judicious use of water and fertilizer. Only with experience can one learn when to withhold water and fertilizer for height control, as either method can greatly reduce the quality of plant.

Transplant Root Pruning

When roots in overhead irrigation get sufficiently long and grow out of the tray through drainage hole, they prune naturally upon exposure to the air. Pruning occurs because air actually dehydrates or dries out the roots. Branch roots will then begin growing in the soilless mix of the plug. In a continuous float system, however, root growth goes unchecked (Leskovar *et al.,* 1994). Sometimes, even if water source is removed daily, humidity of air space beneath the trays may be high enough to support root growth. This mat of roots extending below the tray interferes with shipping and handling, and if mechanically removed, it leaves numerous sites for disease entry. Upon "pulling" the plants, further damage may occur, resulting in greater die-back after transplanting Therefore, a continuous float system for vegetables is not advised. Additionally, the flotation troughs should be sufficiently deep (>8 inches) to allow for good air flow to dissipate the humidity after drainage.

Finishing and Hardening Transplants

Hardening-off transplants is important, especially if they are to be planted under stressful, early-season conditions. Vegetable transplants should be hardened-off by:

1. Reducing temperature in the greenhouse through ventilation. Do not reduce temperature below 10°C on crops that are sensitive to chilling injury. Air movement also helps in the hardening process.

2. Reducing watering to let the plants wilt slightly. Do not let plants wilt excessively. Do not harden-off transplants by withholding fertilizer as this can result in nutrient deficiencies and can delay field establishment. Holding plants outside for several days allows plants to become acclimated to the field conditions while they are still in the trays. Plants should be held in an area that is exposed to full sunlight, but is protected from drying winds. The plants should be checked regularly and watered as required. If risk of frost is a forecast, the plants should be moved inside.

Transition from the greenhouse to the field involves changes in light, temperature and wind. Vegetable transplants benefit by a gradual "hardening" off period before they are transplanted into the field. Gradual exposure to outdoor growing conditions and reduced watering at the end of growing period with some protection from wind and temperature but full exposure to light, can increase the survival rate of transplants in the field. Three to six days are adequate to acclimate transplants.

Care must be taken to not "over harden" young transplants. Cool-season crops exposed to very low temperatures can result in bolting (in cabbage) or buttoning (in broccoli or cauliflower). Warm-season crops generally are hardened at temperatures higher than those of cool-season crops. Cold temperatures can set back warm-season crops and can induce disorders such as catfacing in tomatoes.

Most transplants will be sufficiently hardened for field setting if strict measures are applied to maintain plant height. However, additional hardening can be accomplished by increasing the light intensity, reducing the irrigation inputs, or lowering the greenhouse temperature by 5-10°F. Experienced growers may manipulate or reduce fertilizer inputs to achieve hardening. This technique is not advised for the novice grower for reasons of quality loss. Growers should slowly harden plants by letting them acclimate to "new" conditions, beginning about one week before anticipated sales. Over-hardening can result in slow field establishment and reduced yield.

Method of shifting Container-Grown Plants in Greenhouses

This system is the most intensive and produces the earliest plants. During the past decade, growers have shifted from field to greenhouse production. This trend is expected to continue. Consequently, value of the field-grown transplants will continue decreasing and value of the greenhouse-grown containerized transplants will continue increasing.

Containerized greenhouse plants can be grown using the conventional "rail" (sometimes called "rack") system or the "float" system.

Rail System

In the rail system, rows of rails, usually aluminum T-rails, are precisely spaced to support the trays (flats) on each end. In this system, often referred to as the "Speedling System," trays have drainage holes in their bottoms and plants are watered and fertilized from above, usually by an overhead boom.

Growers also use various other types of benches which support trays at various heights. There are even moving benches available that use space more efficiently since, only one isle is needed that can be created anywhere in the greenhouse as the tables are rolled out of the way. Some operations put flats directly on the ground. Usually this is on a concrete floor that will often have in-floor heating.

Float System

With the float system, growers fill polystyrene (Styrofoam) trays with media and then seed them. Trays are then irrigated and placed in a warm environment. After seeds germinate, trays are placed on a reservoir of water where they "float," either continuously or intermittently, until they are ready to go to the field. Soluble fertilizer is dissolved in the water and, as plants float, they take up water and nutrients (fertigation) from the reservoir. The float system requires less labor and management for irrigation and fertilization. Another advantage is that foliage does not get wet during irrigation. This can help reduce the incidence and severity of foliar diseases. One of the two water management systems may be used for float production of transplants. They are the *intermittent float system* (sometimes called *ebb and flow*) and *continuous float system.* The intermittent float system costs more to build and operate. Water is pumped into the reservoir to float and subsequently, irrigate the plants for a brief period. Then water is pumped or drained back into a holding tank. In the continuous float system, water remains in the reservoir entire time.

The float system appears to have potential for vegetable transplant production, but it has not been fully vetted for this purpose. Several kinds of vegetable transplants, including eggplant, tomato, pepper, cabbage, broccoli, cauliflower, pumpkin and squash have been produced using the float system. In some cases, such as seedless watermelon production, this type of system is not suitable since these seeds need to be germinated in moist but not wet media. There is evidence that this system produces transplants more quickly than the conventional method, but the plants do not do as well upon transplanting. At this time, we do not recommend this system other than for experimental purposes. Whether one uses the float system or the rail system, the major advantage of producing transplants in greenhouses is that temperatures can be controlled and moderated for earlier, less stressed transplants.

Drainage from the trey

Hole in the bottom of the tray should be large enough to allow good drainage when watering from overhead, and provide good capillarity (upward movement of water through the spaces within the soilless mix) when using ebb and flow irrigation. Physical properties of the soilless mixes are extremely important when considering drainage. Over the 4-6 weeks time frame of transplant production, some soilless mixes may compact. Compaction may lead to water logging of the medium, which reduces oxygen supply reaching the roots. Lack of oxygen can lead to physiological problems such as root hair loss and spindly growth. These conditions emphasize the need for good drainage within the plug.

Shifting of transplants from Greenhouse to field

Shipping factors such as mechanical injury, environmental conditions and length of storage can affect plant vigor and establishment. Mechanical injury most often occurs if the plants are pulled from the trays and packed in boxes. Leskovar and Cantliffe (1991) reported that tomato transplants shipped in the tray yielded more extra-large fruit than did transplants pulled from the tray and packed in boxes. Exposure to ethylene can also damage vegetable transplants. Ethylene is a plant hormone that when present at high levels can cause leaf drop, root decay, senescence, or loss of chlorophyll. Sources of ethylene gas include the exhaust of propane-powered forklifts and heaters and senescing (overripe) fruits and plants.

Transplanting in the Greenhouse

The practice of germinating seedlings in flats and transplanting them to containerized trays is a time tradition. In general, this process is the best accomplished when the first true leaf is 1/4 to 1/2 inch long. "Prick" the plants from the flat using a small tool (pot stake, small knife, dibble, or similar tool) to support the hypocotyl section of the stem and adjacent root system. Transfer the seedling to the new container (with a well-drained medium) and set slightly deeper than originally grown to support the stem. Shade may be required to protect newly "spotted" (transferred) seedlings from wilting. Water lightly.

The plug transfer method involves transferring a small plug to a larger plug (Marr & Jirak, 1990). Less damage occurs with such a technique. Either transplant procedure is labor intensive and may not be warranted in large-scale operations, since quality transplants can be grown in containerized trays from seed to maturity.

Cleaning of Trays before Re-use

Transplant trays should be cleaned and sterilized before re-use to avoid buildup of soil borne diseases such as damping-off. Adequate cleaning can be achieved by washing trays with soap and water, followed by a chemical dip in either a quaternary ammonium (2-5%) or sodium hypochlorite (1-2%) solution for at least 20 minutes, rinsing in clean water, and allowing trays to dry thoroughly before re-use. Steam sterilization at 160-180°F for 40-60 minutes is the most effective method of combating tray borne diseases however, purchasing sterilization equipment may be costly, and excessive temperatures and run times in the steam process can "melt" styrofoam trays. Additionally, steam sterilization causes the styrofoam bead structure to separate, accelerating age cracks and resulting in fissures that young roots can penetrate. These roots often tear off during plug removal. Cleaning Styrofoam trays with the steam sterilization process will result in a tray life of 4-5 uses, while nonsteam cleaning methods enable approximately 20 uses. (Fig: Pro tray washing machine).

Pest Management in Green house

There are only a limited number of insecticides and fungicides labelled for greenhouse-grown vegetable bedding plants. Integrated pest management (IPM) offers a practical way to effectively manage pests on vegetable bedding plants and transplants. Through the use of sound cultural practices, monitoring techniques, accurate problem identification, and timely implementation and evaluation of appropriate management strategies, growers can improve their production while minimizing their reliance on routine pesticide application. IPM utilizes many different management options; genetic, cultural, physical, mechanical, biological and chemical. Routine crop inspection alerts growers to developing pest and cultural problems while they are still minor and can be easily managed. Early detection and intervention are the foundation of an IPM program.

Specific Insect Pests and Mites

Common insect pests on vegetable bedding plants include aphids, fungus gnats, shore flies, whiteflies, thrips and two-spotted spider mites.

Aphids

Several species of aphids can occur on vegetable transplants, but the most common are green peach, melon and foxglove. Aphids are small, 1/16-inch in length, round, soft-bodied insects that vary in color from light green to pink or black. The green peach aphid is yellowish-green in summer; pink or yellowish in fall and spring. Winged forms are brown with a large dusky blotch on the abdomen. Melon aphids are greenish-yellow to very dark green with black mottling and short dark cornicles or "tailpipes" (tubular structures on the posterior part of the abdomen). In greenhouses, aphids are usually females that produce live young called nymphs.

Each female can produce 50 or more nymphs. Nymphs mature to adulthood and begin reproducing in as little as 7 to 10 days. Adults are usually wingless, but some will produce wings when population reach outbreak levels. Honeydew can cover leaves and provide a food source for a superficial black fungus known as sooty mold. Aphids are present on weeds and winged aphids may enter the greenhouse through vents. Aphids can also transmit certain viruses.

Whiteflies

The sweetpotato whitefly *(Bemisia argentifolii)* and greenhouse whitefly *(Trialeurodes vaporariorum)* may infest vegetable bedding plants. However, greenhouse whitefly is the most common species. Both adult and immature whiteflies have piercing and sucking mouthparts, are able to remove fluids and produce honeydew which results in sooty mold fungus. Winged adult whiteflies are 1/16-inch in length, and found on the undersides of the youngest, most tender leaves. Females may lay from 150 to 300 eggs, which hatch into first-instar nymphs in about a week. The crawlers move for a short distance before settling down to feed. After three molts, a pupal stage is formed from which adults emerge in about six days. Whiteflies complete their egg to adult cycle in 21 to 36 days depending upon greenhouse temperatures.

Thrips

The most injurious species is the western flower thrips (WFT). They often do considerable damage before they are discovered, because thrips are small, multiply rapidly and feed in plant buds in which they can remain undetected. WFT is also vector to spoviruses. Feeding marks from the rasping mouthparts of thrips appear as white streaks on the leaves. Infested new growth may curl under and leaves are often deformed. Adult WFT are about 1/16-inch long, with narrow bodies and fringed wings. Females are reddish brown and males are light tan to yellow. The wingless immature larval stages are light yellow. Female thrips insert eggs (several hundreds per female) into plant tissue. The tiny yellowish larvae molt twice and feed on plant fluids as they mature. Larvae drop off the plant into the soil and pass through two stages, after which adults emerge. The egg to adult lifecycle can be completed in 7 to 13 days depending upon greenhouse temperature. During warmer temperatures, development is more rapid than at cooler temperatures.

Spider Mites

Two-spotted spider mites can be found on vegetable bedding plants. Adult females are approximately 1/50-inch long, and slightly orange in color. All mobile stages are able to pierce plant tissue with their mouthparts and remove plant fluids. Most spider mites are found on the underside of leaves. Feeding injury often gives the top leaf surfaces a mottled or speckled, dull appearance. Leaves then turn yellow and drop. Large populations produce visible webbing that can completely cover the leaves. Eggs are laid singly, up to 100 per female, during her 3 to 4-week life span. Eggs hatch into larvae in as few as 3 days. Following a brief larval stage, several nymphal stages occur before adults appear. Egg to adult cycle can be completed in 7-14 days depending upon temperature. Hot and dry conditions favor spider mite development.

Disease Management

Diseases of vegetable bedding readings include *Botrytis* blight, damping-off, *Alternaria* blight, late blight, powdery mildew, downy mildew, bacterial diseases such as bacterial leaf spot, bacterial canker, black rot, and viral diseases such as Cucumber Mosaic Virus (CMV), Tobacco Mosaic Virus (TMV), and Tospoviruses. Effective control of diseases requires accurate identification. Failure of disease control is often because the cause cannot be accurately identified. Symptoms caused by poor cultural practices can also mimic disease symptoms. Fungicides cannot correct problems caused by high soluble salts, poor aeration or a nutrient imbalance. An integrated approach to disease management involves the use of resistant cultivars, sanitation, sound cultural practices and proper use of the correct pesticide.

Seed treatments are useful for many vegetable crops to prevent root diseases, as well as certain diseases carried on or within the seed. There are two general types of seed treatment: eradicative and protective. Eradicative seed treatments use hot water or chlorine to kill disease-causing agents on or within the seed. They are useful in controlling certain seed-borne bacterial diseases such as bacterial leaf spot on pepper and tomato and bacterial canker on tomato. Protective seed treatments use fungicides on the seed surface to protect the seed against decay and soil-borne organisms such as damping off caused by *Pythium, Phytophthora* and *Rhizoctonia*. For more information regarding seed treatments, contact your seed sales representative, Extension vegetable specialist or plant pathologist.

Specific Diseases

Botrytis blight

Botrytis can cause leaf blight, stem cankers, damping off and root rot. Plants may be attacked at any stage, but the new tender growth, freshly injured tissues and dead tissues are the most susceptible. *Botrytis* blight produces characteristic gray fuzzy appearing spores on the surface of infected tissues.

Air currents and splashing water can easily disseminate the spores. In general, germination of spores and infection is dependent on a film of moisture for 8 to 12 hours, relative humidity of 93% or greater and temperatures between 13° and 19°C. After infection, colonization of plant tissues can occur at temperatures up to 21°C.

Botrytis diseases can only be managed by a combination of methods including manipulation of environmental conditions (temperature, humidity and duration of leaf wetness), sound cultural practices and use of fungicides. Fungicides alone cannot control *Botrytis* and this pathogen has a long history of fungicide resistance development. Control weeds and remove plant debris before and during production. Dispose off diseased plants and debris in a plastic trash bag. Keep the bag closed to help prevent spreading spores to uninfected plants as the bag is removed from the greenhouse. Cover trash cans to prevent the airborne spread of spores from diseased plant tissue. Reduce humidity and leaf wetness duration to prevent spore germination.

Late Blight

Late blight is caused by the water mold. The fungus typically overwinters in potato cull piles or in soil where plant tissue has not completely frozen and is not considered a problem for locally grown tomato seedlings. Common symptoms on tomatoes are sunken, dark green or brown lesions on leaves and brown lesions on stems, sometimes white fungal growth develops under moist conditions. Leaf lesions begin as irregularly shaped olive green to brown spots and quickly grow larger spots that are consistently small are most likely *Septoria* leaf spot. Confirm Late Blight by submission of a sample to a diagnostic laboratory. Many of the same fungicides used for Botrytis blight will help protect tomato seedlings from late blight.

Damping-off of Seedlings

Damping-off is a common disease of germinating seeds and young seedlings. Several fungi are capable of causing damping-off including *Rhizoctonia, Alternaria, Sclerotinia* and the water molds, *Phytophthora* and *Pythium.* Soil-borne fungi generally do not produce air-borne spores but are easily transported from contaminated soil to pathogen-free soil by infected tools, hose ends, water-splash and hands. Young seedlings are the most susceptible to damping-off. However, later in the crop cycle, same pathogens may cause root and stem rot. Symptoms of damping-off include seedlings failing to emerge or wilting, often with a stem lesion that appears water-soaked or dark, necrotic and sunken at the soil line. Pathogens usually spread radially from a central point of origin so plants often die in a circular pattern. Vegetable seeds that are germinated in poorly drained, cool soils are especially susceptible. Young plants that do emerge are weak and often wilt at or below the soil line. Cabbage, cauliflower, tomato and pepper seedlings may be girdled by brown or black sunken cankers. Damping-off must be prevented because it is difficult to stop once symptoms occur. There are several strategies to prevent damping-off. Use only certified disease-free seed from reputable seed companies. Disinfect all flats, cold frames, pots and tools. Avoid overwatering, excessive fertilizer, overcrowding, poor air circulation, careless handling, and planting too deeply.

Downy Mildew

Infected leaves develop a diffuse yellowing that is easily confused with nutrient deficiency. The fuzzy, dark growth makes leaf undersides appear dirty. Management of environmental conditions such as temperature, humidity and duration of leaf wetness, sound cultural practices and fungicides will help prevent disease development. It is vital to reduce humidity and leaf wetness duration to prevent spore germination. Provide good air circulation and reduce humidity within the canopy. Proper planting dates, fertility, watering and height management will prevent overgrown plants, thereby reducing humidity within the canopy.

Powdery Mildew

Powdery mildew may occasionally occur on vegetable transplants including tomato, eggplant and other solanaceous crops, as well as cucurbit crops. Faint, white mycelium may develop on leaves and stems, with yellow margins. Most growers are all too familiar with powdery mildew when it develops on curcurbits in the field. Powdery mildew that affects certain cultivars of ornamental verbena can also infect curcurbit seedlings including squash, cucumbers and pumpkins. Greenhouse growers who produce curcurbit transplants as well as verbenas should be especially careful to separate these two crops. It is possible that this powdery mildew could affect the cucurbit transplants that may not have otherwise become infected until the fruit was beginning to form in the field.

Black rot

Black rot, caused by the bacterium *Xanthomonas campestris* pv. *campestris* occurs where cruciferous plants are grown. All Brassicas can be severely affected. The bacterium enters the leaves by colonizing the hydathodes (water pores) and moves from the leaf margins inward. Lesions may also begin at wounds. Diseased tissue is often V-shaped, flaccid, tan to yellow and with blackened veins. The blackened veins are diagnostic and are the best seen by holding the leaf up to the light. These bacteria can be introduced on infected seeds, infected transplants purchased from another operation or in the field on crop residues. These bacteria can also survive on weeds in the same family as the host crop. Management of these bacterial diseases is similar and includes the following strategies:

Buy certified disease-free seed from a reputable source. Use hot water-treated seed. Ideally, the seed should be custom-treated by the seed company. Seed companies may treat the seed upon request. There is a risk that germination percentages will be reduced if the seed crop is grown under stressful environmental conditions. Promptly remove infected plants and adjacent plants to prevent further infection and avoid unnecessary handling of plant material. Avoid overhead irrigation, splashing or periods of extended leaf wetness.

Viral Diseases

A few viral diseases of vegetable bedding plants include cucumber mosaic virus (CMV), tobacco mosaic virus (TMV) and the tospoviruses, INSV and TSWV. There is no control for plants infected with a virus. It is important to have the virus disease accurately identified. Serological techniques such as ELISA (enzyme-linked immunosorbent assay) are now available to accurately identify a wide range of viruses.

References

Argerich, C.A. and K.J. Bradford. 1989. The effects of priming and aging on seed vigor in tomato. *J. Exp. Bot.* **40**:599-607.

Aldrich, R. A. and J. W. Bartok. 1994. Greenhouse Engineering, NRAES-33. 199 pp. Available from :www.nraes.org,

Biernbaum, John. 2006. Greenhouse Organic Transplant Production. Michigan State University http://www.msuorganicfarm.com/Transplants.pdf .

Durrant, M.J. and A.H. Loads. 1986. The effect of pellet structure on the germination and emergence of sugar-beet seed. *Seed Sci. and Technol.* **14**:343-53.

Ellis, R.H., and E.H. Roberts. 1981. The quantification of aging and survival in orthodox seeds. *Seed Sci. and Tech.* **9**:373-409.

Gaston, Michelle (Ed.). 1999. Tips on Growing Bedding Plants (4th ed.). Ohio Florists' Association (OFA), Columbus, OH, 43215. 164 p.

Hall, M.R. 1989. Cell size of seedling containers influences early vine growth and yield of transplanted watermelon. *HortSci.* **24**:771-73.

Hummel, N.W., Jr. 1991. Coated seed. *Grounds Maint.* **26**(8):20, 22, 26.

Lamont, W.J. 1992. Transplant age has little effect on broccoli head weight and diameter. *HortSci.* **27**:848.

Leskovar, D.I., and D.J. Cantliffe. 1991. Tomato transplant morphology affected by handling and storage. *HortSci.* **26**:1377-79.

Leskovar, D.I., D.J. Cantliffe, and P.J. Stoffella. 1994. Transplant production systems influence growth and yield of fresh market tomatoes. *J. Amer. Soc. Hort. Sci.* **119**:662-68.

Marr, C.W., and M. Jirak. 1990. Holding tomato transplants in plug trays. *Hort Sci.* **25:**173-76.

McGovern, R.M., C.S. Vavrina, and L.A. McKay. 1993. Effect of transplant tray type and tomato cultivar on the incidence of Fusarium crown and root rot in tomato transplants. Proceedings of the *Fla. St. Hort. Soc.* **106:**173-75.

Perkins-Veazie, P., and D.J. Cantliffe. 1984. Need for high quality seed for effective priming to overcome thermodormancy in lettuce. *J. Amer. Soc. Hort. Sci.* **109**:368-72.

Perry, D.A. 1982. The influence of seed vigor on vegetable seedling establishment. Sci. Hort. **33**:67-75.

Pundt L. and T. Smith. 2007 Growing Vegetable Bedding Plants.

Schnelle R. and J. Barrett. 2009. Sumagic and Tomato Transplants. Greenhouse Product News **9**(11).

Vavrina, C.S., and K. Armbrester. 1990. Performance of plant growth regulators on tomato transplants and subsequent field production. Proc. Nat. Symp. Stand Estab. Hort. Crops, April, 1990. pp. 167-72.

Vavrina, C.S., and G. Hochmuth. 1994. Nitrogen regimes for fresh market tomato transplants in Florida. *HortSci.* **29**:480.

Vavrina, C.S., S. Olson, and J.A. Cornell. 1993. Watermelon transplant age: Influence on fruit yield. *HortSci.* **28**:789-90.

Vavrina, C.S., and M. Orzolek. 1993. Tomato transplant age: A review. *Hort. Technology,* **3**(3):313-16.

Vavrina, C.S., C. Villagomez, and K. Armbrester. 1992. The effect of Alcosorb RD, a hydrophilic gel, hydrated with fertilizer solutions on bell pepper transplants. SWFREC Res. Rpt. IMM92-04.

Weston, L.A. 1988. Effect of flat cell size, transplant age, and production site on growth and yield of pepper transplants. *HortSci*, **23**:709-11.

Weston, L.A., and B.H. Zandstra. 1986. Effect of root container size and location of production on growth and yield of tomato transplants. *J. Amer. Soc. Hort. Sci.* **111**:498-501.

Widders, I.E. 1989. Pretransplant treatments of N and P influence growth and elemental accumulation in tomato seedlings. *J. Amer. Soc. Hort. Sci.* **114**:416.

12

High Value Vegetable Production in Greenhouse

M. Prabhakar, S.S. Hebbar, A.K. Nair and J.K. Chinnu

Abstract

Greenhouse vegetable cultivation is gaining popularity for reasons such as high productivity of quality produce round the year with enhanced input use efficiency. As agriculture has a major share in our economy and if India has to emerge as an economic power in the world, the productivity should substantially increase from the present levels. We need a new and effective technology which can constantly improve the productivity, profitability and sustainability of our major farming systems. One such technology is the green house vegetable production technology. At Indian Institute of Horticultural Research, Bangaluru, research on greenhouse vegetable production is in progress since the year 2000, resulting in the development of many technologies which have made possible to realize high yield levels in tomato, capsicum and cucurbits that are comparable to international standards of greenhouse productivity. Tomato and bell pepper yields have touched 300 tonnes and 200tonnes/ha/year, respectively and cucurbits 80 to 100 tonnes/ha/crop of 3 months. Vegetable cultivation in cost effective greenhouse is a technical reality in India, which is being adopted by resourceful farmers in and around big towns and cities as a peri-urban horticultural system. Such production systems have not only extended the growing season of vegetables and made their availability round the year but also opened the possibilities *of growing new vegetables such as coloured capsicums, cherry tomatoes, seedless watermelons, icebox melons, sweet cucumbers, coloured cabbage, etc., besides tomato, cucumber, eggplant and climbing beans. Quality seedling production is one more important greenhouse activity in the present agricultural scenario as more and more F_1 hybrids are being adopted by the farmers.*

Keyword: Protected Cultivation, Tomato, Capsicum, Cucurbits

Introduction

Hybrid seed production of selected high value vegetables such as solanaceous, cole crops and cucurbits under protected environment is also a major step to increase vegetable production and farm income in India. Greenhouse vegetable production, however, is a highly intensive enterprise requiring substantial labor and capital inputs. Because of this, potential growers should have a holistic vision of all the factors necessary for a successful enterprise.

A greenhouse is generally covered with a transparent material like polythene or glass. Depending upon the cladding material and its transparency, major fraction of sunlight is absorbed by vegetable crops and other objects. These objects in greenhouse emit long wave thermal radiations for which cladding material has lower transparency, with the result that solar energy is trapped inside and raises the temperature inside the greenhouse. This is popularly known as greenhouse effect. This rise in temperature in greenhouse is responsible for growing vegetables in cold climates. During summer months, air temperature in greenhouse is to be brought down by providing cooling device. In commercial greenhouses, besides temperature, controlled humidity, carbon dioxide, photoperiod, soil temperature, plant nutrients etc. facilitate round the year production of desired vegetable crops. Controlled climatic and soil conditions provide an opportunity to vegetable crops to express their yield potentials.

In the world vegetable scenario, India's share is 14.35 per cent of total world production of vegetables and occupies second position next to China with a production of over 146 million tones (2010-11) from an area of 8.4 million hectares with a productivity of 17.2 tonnes/ha (Anon., 2011). Vegetables being a rich and good source of vitamins, minerals and food fiber, occupy an important place in the food basket of Indian consumers from nutritional security point of view. With rising per capita income, especially of those dwelling in urban areas, there is rapid growth in consumption of vegetables which calls for supply of good quality produce round the year. The growing population and their improving economic status are causing a rapid growth in demand for vegetables in India. Protected cultivation of vegetables makes availability of vegetables round the year. Protected vegetable cultivation is one of the potential sources of enhancing income and a reliable means for the reduction of poverty and malnutrition and thereby a step towards augmenting livelihood of farming families in India (Singh *et al*, 2007). Protected cultivation also enables vegetable growers to realize greater returns/unit of land and offer other benefits, like early harvest, longer harvest duration, reduced leaching of fertilizers and eco-friendly management of pests, weeds and diseases (Manish Kumar *et al.* 2007). Through protected cultivation, using different structures such as low tunnels, polythene house, shade net house, net house, fiberglass or polycarbonate or glasshouse, etc. round the year supply of vegetables can be made possible.

Greenhouses – World Scenario

Greenhouse vegetable cultivation is spread all over the world, covering about 5 to 10 % of vegetable cultivated area in most of the countries in Europe, Asia and USA. The major objective is for very high productivity of quality produce during off seasons with minimum usage of pesticides. Increases are expected to continue for at least the next a few years, according to trade sources, which report that many greenhouse growers expect to increase their growing area anywhere from 20 to 50 percent because of increasing demand.

There are more than 50 countries now in the world where cultivation of crops is undertaken on a commercial scale under cover. United States of America has a total area of about 4000 ha under greenhouses mostly used for floriculture with a turnover of more than 2.8 billion US $ per annum and the area under greenhouses is expected to go up considerably, if the cost of transportation of vegetables from neighbouring countries continues to rise. The area under greenhouses in Spain has been estimated to be around 25,000 ha and Italy 18,500 ha, used mostly for growing vegetable crops like watermelon, capsicum, strawberries, beans, cucumbers and tomatoes. In Spain, simple tunnel type greenhouses are generally used without any elaborate environmental control equipments mostly using UV stabilized polyethylene film as cladding material. In Canada, the greenhouse industry caters both to the flower and off-season vegetable markets (Papadopoulos and Gosselin, 2007). The main vegetable crops grown in Canadian greenhouses are tomato, cucumbers and capsicum. Hydroponically grown greenhouse vegetables in Canada find greater preference with the consumers and could be priced as much as twice the regular greenhouse produce.

The Netherlands is the traditional exporter of greenhouse grown flowers and vegetables all over the world. With about 89,600 ha under cover, the Dutch greenhouse industry is probably the most advanced in the world. Dutch greenhouse industry however, relies heavily on glass framed greenhouses, in order to cope up with very cloudy conditions prevalent all the year round. A very strong research and development component has kept the Dutch industry in the forefront. The development of greenhouses in Gulf countries is primarily due to the extremity in the prevailing climatic conditions. Israel is the largest exporter of cut flowers and has wide range of crops under greenhouses (15,000 ha) and Turkey has an area of 20,000 ha under cover for cultivation of cut flowers and vegetables, mainly tomato (Yildiz Dasgan *et al*, 2004). In Saudi Arabia, cucumbers and tomatoes are the most important crops, contributing more than 94% of the total production. The most common cooling method employed in these areas is evaporative cooling. Egypt has about 1000 ha greenhouses consisting mainly of plastic covered tunnel type structures. Arrangements for natural ventilation are made for regulation of temperature and humidity conditions. The main crops grown in these greenhouses are tomatoes, cucumbers, peppers, melons and nursery plant materials. In Asia, China and Japan are the largest users of greenhouses. The development of greenhouse technology in China has been faster than in any other country in the world. With a modest beginning in late seventies, the area under greenhouses in China has increased to 48,000 ha in recent years. Out of this 11,000 ha is under fruits like grapes, cherry, Japanese persimon, fig, loquot, lemon and mango. The majority of greenhouses use local materials for the frame and flexible plastic films for glazing. Most of the greenhouses in China are reported to be unheated and use straw mats to improve the heat retention characteristics. Japan has more than 40,000 ha under greenhouse cultivation of which nearly 7500 ha are devoted to only fruit orchards. Greenhouses in Japan are used to grow wide range of vegetables and flowers with a considerable share of vegetable demand being met from greenhouse production. Even a country like South Korea has more than 21,000 ha under greenhouses for production of flowers and fruits. Thus, greenhouses permit crop production in areas where winters are severe and extremely cold as in Canada and USSR, and also permit production even in areas where summers are extremely intolerable as in Israel, UAE, and Kuwait. Greenhouses in Philippines make it possible to grow crops in spite of excessive rains and also in moderate climates of several other countries. Thus, in essence greenhouse cultivation is being practiced and possible in all types of climatic conditions.

Status in India

If India has to emerge as an economic power in the world, our agricultural productivity should equal those countries, which are currently rated as economic power of the world. We need a new and effective technology which can improve continuously the productivity, profitability and sustainability of our major farming systems. One such technology is the green house technology. Although it is centuries old, In India, research on greenhouse vegetable production has been going on for the past few years which has brought in remarkable change in agricultural production techniques which has made possible to realize yield levels in tomato and capsicum which are comparable to any international levels of greenhouse productivity. Presently in the greenhouse, main vegetables cultivated are bell pepper, tomato, cucumber, eggplant, climbing beans, melons and some exotic vegetables. The major regions where green house vegetable production is prevalent on commercial scale are around big cities such as Bengaluru, Pune, Ahmadabad, Chennai, New Delhi , Hyderabad and in Punjab, Haryana, Himachal Pradesh, Uttarakhand, Gujarat, etc.

In general, tomato yields are much higher in greenhouse (375 tonnes/ha/year) than open field condition (100 tonnes/ha/year). Maximum greenhouse yields of tomato are estimated at 500-600 tonnes/ha/year elsewhere in the world. Yields are increasing worldwide because of availability of better cultivars and more intensive use of technology. The trials conducted at IIHR, Bangaluru also showed that the yield of 300 tonnes and 200 tonnes per hectare per year can be obtained in tomato and bell pepper respectively.

Vegetable cultivation in cost effective low and medium-cost greenhouse is a technical reality in India and slowly it is being adopted by resourceful farmers in and around big towns and cities as a peri-urban horticultural system. This system can be adopted for producing safe vegetables on terraces and roofs in urban and peri urban vegetable farming (Prabhakar *et al*, 2010). Such production system has not only extended the growing season of vegetables and made their availability year round the year but also throwing open the possibilities of growing new vegetables such as cherry tomatoes, colored capsicums, etc. Quality seedling production, which is slowly catching up, is one more profitable venture and necessity in the present agricultural scenario as more and more F_1 hybrids are adopted by the farmers. Hybrid seed production of selected high value vegetables such as capsicums and cucurbits under protected environment is also a major step to increase vegetable production and farm income in India. Greenhouse vegetable production, however, is a highly intensive enterprise requiring substantial labor and capital inputs. Because of this, potential growers should carefully consider, with a holistic vision, all the factors necessary for a successful enterprise.

Structure and Maintenance

When considering greenhouse designs, three major factors should be considered: load limitation, light penetration and cost. Bracing along sides of the greenhouse and roof should be sufficient to withstand wind, particularly in the spring. Bracing along the roof also should be sufficient to withstand crop load if tomato or cucumber vines are to be supported by twine attached to the bracing. A concrete footing is preferred for a permanent greenhouse. A wide door at one end of the greenhouse will ensure easy access for equipment. Most greenhouse crops grow best in light whose wavelengths range from 400 to 700 nanometers. This range of wavelengths is called *photosynthetically active radiation* (PAR). Most greenhouse coverings will accommodate these short waves of visible light. Polyethylene and fiberglass tend to scatter light, while acrylic and polycarbonate tend to allow radiation to pass through directly. Scattered or diffused light tends to benefit plants by reducing excess light on upper leaves and increasing reflected light to lower leaves. Plastic glazed greenhouses have several advantages over glass greenhouses, the main advantage being cost. Plastic also is adapted to various greenhouse designs, generally resistant to breakage, lightweight and relatively easy to apply.

The common structure that is being used in India is naturally ventilated poly cum net house or simple net house. In some cases, only shade nets are used to protect the crop from hot sun and winds during summer. Very simple and reasonably cheaper, stone pillars or wooden or bamboo poles with clear polyethylene or insect proof nylon nets as cladding material are also being used for vegetable production. Normally, poles of 10 to 12 feet height are suitable for most of the vegetables. For net house cultivation, UV stabilized nylon nets of 40–50 mesh size are used. For longer life and prevention of sagging, good support with criss-cross wiring of GI wire is essential. Greenhouse structures require constant maintenance and repair. Many of the selected greenhouse covers must be replaced on a regular basis. Cooling and watering systems must be maintained and routinely serviced. In addition, contingency plans and backup systems must be in place in case any of these major systems should break down. Even a one-day loss of cooling during summer or water during a critical period can result in drastic reduction in crop yields.

Along with the essential skills, capital and labor to build, maintain and grow a crop, producers must develop markets willing to pay the relatively high prices necessary to make the enterprise economically viable. Greenhouse-grown vegetables cannot compete with comparable field-grown crops based on price; therefore, greenhouse-grown vegetables often are marketed to buyers based on superior quality and off-season availability.

Vegetable Seedling Production Using pro-trays and Protected Structures

Nowadays, the entire vegetable production system has been receiving attention because of the necessity to improve yield and produce quality without causing damage to the environment. To ensure high productivity with good quality produce, raising and supply of quality vegetable seedlings through use of good seeds and appropriate cultural practices are very essential. This has become a necessity in view of higher cost of hybrid seeds. One of the most important changes taking place is the way vegetable seedlings are being produced. In the past, all seedlings used for transplanting were produced by the growers themselves as open field raised bed nurseries. At present, vegetable transplants production all over the world is done by specialized companies or private individual entrepreneurs. In India, many young entrepreneurs and progressive farmers are coming forward to take up quality seedling production and supply to the vegetable farmers based on the requirement or orders. Raising of seedlings under protected structures of polyethylene or 40-mesh nylon or shade net covering on nursery beds facilitates year round production of good quality uniform and healthy seedlings. Use of water soluble specialty fertilizers, bio-fertilizers and bio-pesticides at nursery stage has good potential for producing healthy seedlings. Use of micro irrigation and fertigation helps in maintaining uniformity in growth of seedling by maintaining good moisture and nutrient regimes in the growing media. Since, the hybrid seeds are expensive, this method reduces cost by minimizing seed waste.

Use of protected structure for vegetable seedlings as a commercial venture is a recent phenomenon. It is gradually attaining status as a commercial venture in India to produce pest free, healthy and uniform seedlings of tomato, capsicum, cauliflower, cabbage, chillies, brinjal, melons and gourds (Singh *et al*,2005 and Prabhakar and Hebbar, 2006). In commercial nurseries, most commonly used pro-trays are 98 celled and the most common growing media used is sterilized cocopeat, which helps in preventing nursery diseases especially, seedling damping-off. Since, it is low in mineral nutrients, the media need to be supplemented with nutrient solution.

Design of the protected structures depends on the availability of materials and the investment capability of the farmer. Growing plants under a cover protects the crop from damaging rains and winds, scorching radiation, temperature and pests and diseases. The most commonly used method to reduce radiation and temperature is by shade net covering and may be with fogging or misting facilities. These are ideal for the production of vegetable seedlings as they are comparatively more economical and feasible. Use of low cost structure involving granite pillars or wooden poles, shade net and 40 mesh nylon net to prevent some vectors which transmit viral diseases are in practice (Prabhakar *et al*,2012). Naturally ventilated polyhouse is better than net house, as the damage due to rain fall battering can be avoided during rainy periods. Hardening of vegetable seedlings before transplanting in the main field is very essential for reducing transplant shock and also to have better crop stand. Plants should be gradually hardened, or toughened for at least 3 to 5 days before taking it to the main field, depending on the crop. Hardening can be done by controlling irrigation or by removing shade net or directly exposing to open field condition. At present price, the structural cost works out to be Rs 150 to 160 per square meter with a production cost of about 15 to 20 paisa per seedling excluding the cost of seed. Half an acre of protected structure can accommodate about 8000 seedling trays and about 8 lakh seedlings can be raised per batch.

Greenhouse Vegetable Production

The central idea behind greenhouse vegetables is to provide safe food of high quality. The objective is to produce vegetables throughout the year with significantly less water and pesticides than is commonly observed in open field grown vegetables. Year-round production is the key to maintain the greenhouse's profitability. However, this does not necessarily mean that growers should be producing the same crop year-round. (Summer tomatoes bring more money than winter ones do). Another option would be to raise a crop other than fresh vegetables, such as vegetable seedlings or use for production of F_1 hybrid seeds.

Site Selection

Several factors should be considered in selecting a suitable site. Wherever practical, the site should be level or nearly so. Sloped sites are not completely out of the question, but they do raise construction costs considerably because the site must be leveled or terraced to accommodate the new structure. The site should be near a power source and have a ready supply of usable water. Typically, water from a well is better than surface water because there is less likelihood of disease and algae contamination. Availability of labor is very important. Although, much of the work in crop production is repetitive and tedious, one should not expect suitable labour performance at minimum wage levels. Labour that is conscientious and skilled, or at least trainable, is less expensive and more reliable in the long run. Access to good roads and proximity to expected markets should also be considered.

Selection of Cultivars

Cultivars suitable for open field condition are usually not suitable for polyhouse cultivation, as relatively long duration and high-yielding hybrids are ideal. Selection of cultivars and hybrids depend on plant type and their growth habit. Tomato and capsicum cultivars and hybrids should be indeterminate type. The plants are grown upright as a single stem rather than a bush. Cucumber cultivars should be unique. They should have only female flowering habit, with dark green parthenocarpic (seedless) fruits free of bitterness.

Grafting

Grafting is an emerging technique and is becoming popular throughout the world, especially for the green house cultivation of vegetable crops, mainly Solanaceous and Cucurbitaceous ones, from the point of resistance against the soil borne pathogens, including the root knot nematode (*Meloidogyne* spp) *(*Ricárdez *et al.*, 2008). The use of rootstocks resistant to soil-borne fungal pathogens used together with chemical means of nematode control provided promising results for their further use in integrated strategies as alternatives to methyl bromide (Giannakou and Karpouzas, 2003). *A*part from obtaining better yield and quality, it is especially successful in cucurbits like melons for resistance against diseases. Grafting involves union of suitable varieties of greenhouse crops used as scions over the resistant root stocks usually from the same family. The grafted seedlings are conditioned in the dark and cool shade nets are prepared specially for the purpose before planting in seedling trays and finally they are transplanted in the greenhouses.

Salinity is one of the major abiotic stresses that reduces plant growth and crop productivity in many vegetable production areas of the world. Grafting represents an interesting tool to avoid or reduce yield losses caused by salinity stress in high-yielding genotypes belonging to *Solanaceae* and *Cucurbitaceae* families. Grafting of vegetable seedlings is a unique horticultural technology to overcome soil-borne diseases and pests and to add extra vigour to the plants under various environmental stress conditions. The technique was practiced for many years in East Asia, and more recently it has become a common pest management technique worldwide. In 1990s, nearly 60% of open fields and greenhouses for producing muskmelon, watermelon, cucumber, tomato, and eggplant were reportedly planted with grafted seedlings in Japan and 81% in Korea.

There are three primary techniques used for grafting: tongue approach grafting, cleft grafting and tube grafting. Cleft grafting and tube grafting are quite similar in which the shoot of the fruit producing variety is completely cut off from its own roots and attached to the severed stem of the rootstock plant (The name 'Tube Grafting' originated because when the technique was first developed, a tube was used to attach the shoot to the root. Clips are now used to make this graft). Tube grafting is quicker and less complicated to do than cleft grafting because it only requires a single straight cut on both the root and shoot portions of the graft. Also, because a fewer intricate cuts are involved, this technique can be used on very small seedlings. With both cleft and tube grafting, the newly grafted plants must be protected from drying out until the graft union gets healed. This usually involves covering the plants with a plastic cover to reduce light and retain moisture and then misting the grafted plants periodically during the day. Tomato grafts heal quickly and the plants begin to be acclimated back into the greenhouse after 4-5 days and usually plants are fully set and in the greenhouse 7-8 days after the initial grafting.

Growing methods

Different methods can be used to produce vegetables in a greenhouse. Mostly, soil is the common growing medium but some times from the point of better maintenance of water regime, root zone aeration, good control of soil borne pathogens, etc. have made to adopt other growing media as preferred alternatives. Systems utilizing solid substrates other than soil are more properly termed 'soilless culture'. Environmentally friendly substrates are the main tendency in soilless culture. The inorganic substrates, rockwool and perlite generate high yield and quality, which is difficult to overcome by the organic substrates. Potential materials require a specific management to improve qualities of the final product (Inden and Torres, 2004).

Soil Culture

In soil culture, also known as ground culture, vegetables are raised on level ground as well as raised beds. Soil culture is more popular with growers than other methods. One of the reasons to prefer soil culture to other media is that soil-building practices are an already familiar concept based on decades of research and experience. Production techniques used in ground culture are similar to intensive market gardening methods used in the field. Organic fertilizers and soil amendments are extensively used to feed the soil. A major concern in ground culture systems is soil sterilization or pasteurization for control of soil-borne pathogens. Methods of disinfection chemical application such as 4% formalin solution drenching, solarization, steam, electrical heat and biological control. To maintain good physical condition of the soil, substantial organic manure application @ 50 tonnes per hectare is good. Even mixing of other materials like digested saw dust, rice husk or coir pith @ 5 to 10 % volume of top 9 inch soil layer gives good results.

Bag Culture

In bag culture, plants are grown in a soil-less medium contained in lay-flat or upright polyethylene bags (10-15 liters in size) and fed with liquid fertilizers through drip irrigation lines. Media can be peat/vermiculite, sawdust, cocopeat or coirpith, rice hulls, peanut hulls or mixes of any of those. Bag culture has become the preferred method of greenhouse vegetable production as it is easy to establish and manage successfully wherever soil borne diseases make growing of vegetables a difficult task. Bags are has most often placed in double rows in the greenhouse. Bag culture is well suited to the production of upright and vining crops like tomatoes, cucumbers and peppers. Bag culture has become the preferred method of greenhouse vegetable production in many parts of the U.S., because it is easy to establish and manage successfully. One problem with bag culture is the potential for excess fertilizer solution to leach out of the drainage holes in the bags and into the greenhouse soil.

Hydroponic Systems

The various methods of hydroponic production have become the de facto standard for producing greenhouse vegetables. These methods are very clean, with no organic material present. In addition, they give the grower complete control over the crop's nutritional needs. This allows the grower to maximize growth and fruit production. The disadvantage of this method is that it requires calculating and measuring exact amounts of fertilizers which usually means handling several different fertilizer compounds. Mistakes in calculations or quantities used are more likely to show up as deficiencies or toxicities in the crop than in other methods employed. A constant feed of nutrient solution as shown in Table 1, has to be followed for hydroponic system.

Plants can be grown in various media in these systems; all are essentially nutrient free. In one system called the nutrient film technique (NFT), plants are placed in a polyethylene tube that has slits cut in the plastic for the roots to be inserted. Nutrient solution is pumped through this tube, bathing the roots. The solution re-circulates, and nutrients are added as depletion occurs. Other systems use rockwool as the supporting medium. More recently, polyethylene bags of perlite have been used as the supporting medium. Commonly available organic media which are comparatively resistant for degradation such as cocopeat can also be used as media in hydroponics. Regardless of the production system, most will require some type of automation to supply water. This is particularly important in any of the hydroponic systems.

Whatever may be the growing medium, the cropped area should be mulched to control weeds, retain soil moisture and prevent soil compaction. Plants can be set in black plastic mulch or organic mulches can be used if it is sterilized. Use only quality seedlings raised in pro-trays and grown under cover in order to ensure that they are free from any pests and diseases.

Fertigation

After realizing the importance of micro-irrigation system, the next most essential input to boost the productivity and quality is the fertilizer. Fertigaton generally means application of solid or liquid fertilizers via pressurized irrigation system; thus, forming nutrient containing irrigation water as per the crop need with desired concentration.

Soil fertility should be determined and managed with soil testing. As with any field soil, the soil pH should be adjusted to 6.0 to 6.5. Soluble salts buildup can be particularly severe with the soil system. To minimize this problem, use fertilizers that do not contribute excess soluble salts. This would include calcium nitrate, potassium nitrate, triple superphosphate, di-ammonium phosphate, potassium sulfate and sulphate of potash-magnesia.

Water and fertilizers are normally applied in a targeted fashion through drip irrigation system. Liquid fertilizer applied through irrigation lines–a technique known as fertigation is a common practice in greenhouse production. Nitrogen and phosphorus are the two primary nutrients injected through this system. Fertigation can provide supplemental doses of these nutrients, especially at critical periods during the growing season such as fruit filling. However, fertigation is not a substitute for complete soil-based fertility program. The studies conducted on fertigation requirement of greenhouse vegetable crops have clearly indicated that substantial yield increases with nitrogenous water soluble fertilizers as compared to response in open field grown vegetables (Mahajan and Singh, 2006).

It is found better to use water soluble solid fertilizers for fertigation but from economic point of view, traditional fertilizers such as urea and Muriate of potash are also used as sources of nitrogen and potash. Of course, if phosphorous has to be supplied through irrigation water, one has to use water soluble fertilizer such as Mono-Ammonium Phosphates or Urea Phosphates. Single super phosphate can not be used as it results in clogging of drip systems as it is water insoluble. For greenhouse crop, constant feeding is commonly followed containing all the required elements as per the crop growth requirement (Table 1).

Table 1. Final nutrient solution concentration at various stages for tomato, pepper and cucumber (ppm)

	N	P	K	Ca	Mg	S	Fe	B	Cu	Mn	Zn	Mo
Tomatoes*												
Transplant to first cluster	65	50	120	80	40	56	2.8	0.7	0.2	0.8	0.2	0.05
First cluster to second cluster	75	50	120	92	40	56	2.8	0.7	0.2	0.8	0.2	0.05
Second cluster to third cluster	95	50	150	104	40	56	2.8	0.7	0.2	0.8	0.2	0.05
Third cluster to fifth cluster	114	50	150	127	48	66	2.8	0.7	0.2	0.8	0.2	0.05
Fifth cluster to termination	148	50	208	127	48	66	2.8	0.7	0.2	0.8	0.2	0.05
Peppers†												
Seedling to termination	225	40	350	175	40	56	2.8	0.7	0.2	0.8	0.2	0.05
Cucumbers†												
Seedling to first fruit	133	62	150	130	50	70	205	0.44	0.05	0.62	0.09	0.03
First fruit to termination	240	62	150	260	50	70	2.5	0.44	0.05	0.62	0.09	0.03

* Hochmuth, G. (1991). † Fertility levels are based on Canadian recommendations. Lower levels of fertilizer may be required, particularly during early stages of production.

Types of fertilizer for fertigation

1. Liquid fertilizers: These are the solutions which contain one or more plant nutrients in liquid form.
2. Solid water soluble fertilizer: These are 100 percent water soluble fertilizer, also referred as specialty fertilizers. These fertilizers generally contain two or more major nutrients as well as micro nutrients. Soluble fertilizer completely dissolves in water leaving no precipitate. Therefore, there is no chance of clogging of emitters due to fertigation.

Fertigation scheduling: While deciding the fertigation schedule, following points need to be taken into consideration.

1. Total nutrient requirement of crops to produce target quality yields.
2. Part of the nutrient supplied in the form of organic manure.
3. Possible savings in the nutrient doses due to the higher efficiency of added fertilizer through fertigation.
4. Relative requirement of nutrient at different growth stages of crop.

It is known that requirement of nutrients varies from stage to stage of crop growth. Therefore, it is essential to know the nutrient requirement of the crop based on the growth stages. High level of nitrogen is required during the vegetative growth stages to facilitate accumulation of dry matter in different plant parts. Its requirement is comparatively low during reproductive stage. Requirement of P is more during reproductive phase as it induces flowering, fruit set and seed development. Potash requirement is high during crop maturity stage.

Fertilizers used in fertigation

1. Straight Fertilizers

Urea Phosphate (17:44:0), Mono Ammonium Phosphate (12:61:0), Mono potassium Phosphate (0:52:34), Potassium Nitrate (13:0:46), Sulphate of Potash (0:0:50 + 18% S), Calcium Nitrate, Magnesium Nitrate and Calcium Magnesium Nitrate.

2. NPK Formulations

20:10:10 + TE+ Mg, 19:19:19 +TE, 16:08:24 + TE + Mg, 13:40:13 + TE, 20:20:20, 6:12:36, 13:5:26 etc.

3. Chelated micronutrients

Fe EDTA (13%), Zn EDTA (15%), Fe EDDH-A (6%), Chelated combination of micronutrients, Kombi, etc.

Features of 100% water soluble fertilizers:

1. No chloride, no sodium and no salt build up.
2. Completely soluble in water.
3. Most of the fertilizers are acidic and hence special acidification treatment for dripper clogging may not be necessary.
4. Blended with micronutrients.

Special care to be taken while adopting fertigation:

1. Fertilizer tank or injection pump may be used to inject the fertilizer.
2. For precise placement of both water and fertilizer, it is necessary to use pressure compensating drippers or inline drippers instead of micro tubes.
3. Daily feeding is the most desirable, if not possible, feed alternate day.
4. Urea can be combined with WSF for meeting the N requirement.
5. Fertigation should be done during the end of irrigation period, run the drip system for 5-6 minutes after completion of fertigation.
6. It is not advisable to use a very concentrated stock solution (generally not more than 10%).

Organic Production: Organic greenhouse production is also attractive because many greenhouses are already virtually pesticide free. However, developing fertility practices acceptable to certifying agencies, while at the same time maintaining current yields is a major obstacle. Table 2 summarizes restrictions on construction, substrates, containers, fertilization, and pest control for organically certified production.

Pest Management in Greenhouse

Modern greenhouse design has tried to minimize or eliminate insect and disease problems so that plants can be grown pesticide free under many conditions. Some of the innovations have included using ultra-fine screening over all openings to prevent insects, including thrips, from entering the house. Building a vestibule at all entrances is also helpful in preventing insects from entering the greenhouse. Preventing insects from entering the greenhouse is the best way of controlling insects problem. Plants should, however, be inspected daily for insect infestations. If pesticides are to be used, care must be exercised in selecting materials

Biological control methods for pests in commercial greenhouses is becoming more and more popular and is the most preferred way. Increasing concern over the exposure of greenhouse workers to pesticides, improved monitoring of pesticide residues in food and stricter regulations governing the use of pesticides on food

Table 2. Summary of general restrictions on greenhouse production for organic certification.

Construction and Operation

General	Allowed	Restricted	Prohibited
Transition period may be required if taking over existing conventional greenhouse. Whole house must be certified, but can have uncertified houses on same farm if product distinguishable	Proper ventilation required, but emergency use of portable heaters allowed	Pressure treated lumber only allowed on foundations and end walls (must be isolated from growing media and plants)	No asbestos, urea, or formaldehyde; No biodegradable plastic

Substrates & Containers

Allowed	Restricted	Prohibited
Perlite , Rockwool;Peat compost preferred;Coir; Containers that can be recycled	Fiber containers with no prohibited material	No synthetic fertilizers, fungicides, wetting agents, sterilants in potting mix

Fertilization

Allowed	Restricted	Prohibited
Compost (N, K, Micros);Colloidal soft rock and hard rock phosphate (P); Guano (P), Wood ash (K); Rock dusts [granite, feldspar, greensand](K); Natural potassium sulfate (K); Limestone, gypsum, calcium oxide, (Ca);Dolomitic limestone, Kierserite (Mg); Kelp meal, liquid or powdered seaweed extract, rock powders	Sodium Nitrate (N); Fish emulsion (N, P); Bone Meal (N, P); Calcium Chloride (Ca); Epsom salts (Mg); Chelate, acid treated sulfate, or oxide materials	Synthetic Sources; Muriate of potash; Quick lime or hydrated lime

(Contd.)

Pest Control

General	**Allowed**	**Restricted**	**Prohibited**
Resistant Crop Varieties	Insect Traps; Insect Barriers; Hand, Mechanical, orHydraulic Removal; Biological Controls (Predatory Insects);	Oil Sprays; Insecticidal Soaps; *Bacillus thuringiensis* (BT); Copper; Sulfur; - Approved Non- synthetic Fungicides	Synthetic Pesticides,Fungicides, and Nematicides

crops also contributed to renewed interest in a non-toxic alternative to chemical Pest control. The recent return to biological methods is also prompted by the development of pesticide-resistant pest populations, high cost of pesticides, the difficulty in observing pesticide withdrawal times while maintaining harvest schedules and the reduction in yield due to the phytotoxic effects of chemicals and of course most important, the environment concern. The availability of beneficial organisms is an important aspect of biological control in such a large country as India for which SAUs or ICAR Institutes or Private Pest Control Research Laboratories can be contacted.

Normally, as an IPM package neem cake @ 5 kg/ square meter of growing area and *Trichoderma*, *Pseudomonas*, *Paecilomyces* enriched organic manure @ 1 kg for every one tonne of FYM added, is recommended.

Steps involved in Pest and Disease management:

- Using clean and healthy seed and virus free seedlings raised in plastic flat trays under cover.
- Removal of all debris of the previous crops.
- Sterilizing all surfaces likely to be contaminated.
- Growing resistant varieties.
- Practice of crop rotation using diverse crop species.

Production practices

Bell Peppers: Peppers require comparatively long growing period to reach transplant size and normally, transplants of 35 to 40 days old are used. Paired row method of planting is adopted with 60 cm x 30 cm spacing and between paired rows, a spacing of 90 cm is maintained, accommodating about 44000 plants per ha. Compared to other vegetables, it grows well with higher humidity conditions in the range of 60 – 70 per cent. Capsicum plants are pruned to retain four stems. Pruning is done at weekly intervals, starting from 15 to 20 days after transplanting (DAT). Capsicum plants show dichotomous branching. The tip of the plant splits into two at 5th or 6th node, 15 to 20 days after transplanting. These two branches again split into two giving rise to four branches by 25 to 30 DAT (Hebbar *et al*, 2011). Only these four main branches are maintained till the end of the crop. At every node the tip splits in to two giving rise to one strong branch (the branch with thick stem and larger leaf) and one weak branch (the branch with thin stem and smaller leaf). The weak branch is removed, retaining the strong branch. This operation needs to be done once in a week. From 4th month onwards, the pruning operation will be done once in 10 days. All the shoots appearing at the base of the plants are removed as they are not productive. One to two fruit will set per node. Plants are often vibrated or trellis wires tapped to ensure proper pollination and fruit set.

Tomatoes: Tomatoes selected for greenhouse production are indeterminate types. Indeterminate tomatoes continue to grow and set fruit throughout their life cycle. Varieties selected for greenhouse production fall into two broad categories: large beefsteak or slicing types and cluster or hand types. The larger types are harvested singly. The cluster or hand types are harvested in clusters of four to seven fruits and are sold with the cluster stem still attached. Seedlings for transplanting are ready around 24 to 26 days after seeding. Make sure that the medium the plants will be grown in is thoroughly moistened with water or dilute nutrient solution. If growth bags of perlite are used for production, they should be moistened as described above, and the irrigation emitters should be placed in each bag before slits are cut to allow excess solution to drain away. Paired row method of planting is adopted with 60 cm x 45 cm spacing and between paired rows a spacing of 90 cm is maintained accommodating about 30,000 plants per hectare. Indeterminate tomato varieties are trained to a single stem and supported by twine. The plants must be pruned and the fruit thinned throughout the crop cycle. As the plants grow up the string, lateral branches are removed from leaf axils while the main stem is allowed to grow continuously. Fruits must be thinned as they form in the flower clusters. The variety will determine how many fruits should be left at each flower cluster, normally, 4-5 fruits per cluster are retained. The flowers must be vibrated or shaken each day that flowers are produced to ensure pollination. The best time to do this necessary operation is between 10 a.m. and 3 p.m. During overcast or rainy days, pollen will not be dry enough to adequately pollinate the crop; therefore, plants must be shaken during sunny dry days.

Cucumbers: Cucumber varieties grown in greenhouses are known as European or English or baby cucumbers. These special cucumbers are parthenocarpic, that is, they produce fruit without the need for pollination. In fact, care should be taken to avoid pollination by non-parthenocarpic cucumbers because this will result in bitter and misshapen fruit. These varieties are gynaceous and multi pistillate. Cucumbers are a warm season crop; therefore, greenhouse temperatures should not be allowed to drop below 18°C. Seed germination can occur with media temperatures above 16°C, but optimal germination occurs at 30°C to 35°C. Typically, cucumbers will be germinated in flats of rock wool or foam blocks, usually with a single seed per well to ensure plants can be transplanted with minimal root damage. Plants should be transplanted to final spacing in the greenhouse when they are large enough to handle without damage. They will usually be ready for transplanting in two to three weeks under optimal conditions. Cucumbers are vining crops, so they are grown with a trellis system. As the plant is trained, all fruits below 24 to 30 inches are removed to maximize foliar growth. Above this height, fruits are allowed to set. Several different trellis systems have been developed to produce greenhouse cucumbers. These systems, using either one or two wires strung approximately 8 feet above the crop, are referred to as single and V-cordon systems. Both single and double rows of plants can be used with the single cordon system. In either case, the plants are trained up a string to a suspended wire where the plant is draped over the wire. At this time, the plant may be pinched and two laterals allowed growing down to the ground, or the primary stem may be allowed to continue growing to the ground. In either case, all other laterals are removed as they appear and at least one fruit is allowed to develop at each node. In the V-cordon system, two wires are suspended so that the plants can be alternately strung to one wire or the other. This results in a V-shaped system of plants growing to the respective wires. In all other ways, they are treated the same as the single wire system in terms of pruning laterals and draping over the wire. In addition, the in-row spacing will differ between the two systems. In the single cordon system, plants are spaced 12 to 18 inches apart. If a single wire with a double row is used, the plants are spaced 60 cm apart with 60 cm between the double rows (Hebbar *et al,* 2012). With the V-cordon system, the plants are spaced 30 cm apart with plants alternatively strung to one wire or the other. Between-row spacing for the double-row single-wire system is 120 cm to 150 cm; for single-row single-wire system, 90 cm to 120 cm. With the V-cordon system, between-row spacing is 120 cm. Fruit matures continuously as the plants grow. Flower opening to harvestable fruit will take between 10 and 14 days. Therefore, harvesting is a continual process, usually done three times a week. Harvested fruit must be protected from desiccation and bruising because the skin is rather tender and prone to water loss. Greenhouse cucumbers are individually shrink-wrapped to restrict desiccation and extend shelf life.

Muskmelon: It is a vining plant in the cucurbit family. This warm season crop is sensitive to cold temperatures. In North Indian plains, early summer harvests can be achieved by following transplanting of musk melon crop in low tunnel cultivation. To grow in peak winter, it requires heated greenhouses. Muskmelon requires warm climate to grow well and produce high quality fruits. Experiment conducted at IIHR, Bangaluru during winter months using 10 hybrids showed that growing of muskmelon in naturally ventilated polyhouse increased the yield by two folds compared to open field cultivation. General varieties meant for open field cultivation were tried and that required hand pollination for the production of fruits inside the polyhouse. Muskmelons are vining crops, so they are grown with a trellis system. Two main stems are maintained and made to grow to a height of about 8 feet by supporting the plants with plastic or gunny twine. Each branch was allowed to set two fruits and generally, 4 fruits per plant were maintained. Two rows are planted on 100 cm wide beds. Beds are spaced 50 cm apart. An inter plant spacing of 45cm is maintained accommodating a plant population of 30,000/ha. A fertilizer dose of 120-100-120 kg NPK/ha is followed. A yield of about 75 t/ha can be realized from 100 days duration crop in mild winter of *southern tropics.*

Harvesting and Handling: Harvesting of tomato, pepper, melons and cucumber will be a continual process throughout the growing cycle. Tomatoes and peppers are harvested with stems attached and are packed that way to avoid bruising and damage. Care should be taken so that tomatoes are not stored or transported with other vegetables such as peppers, lettuce or cucumbers. Tomatoes produce ethylene during the ripening process which can damage other vegetables. Harvesting should be done during cooler period of the day and the vegetables should be protected from heat. Harvested material should be pre-cooled as soon as possible if this is part of the postharvest procedure. Table 3 shows the best pre-cool method and storage conditions for optimal storage life.

Table 3. Optimal storage conditions for greenhouse-grown vegetables*

Vegetables	Temperature (°C)	Relative humidity	Storage life (days)
Cucumber	10-13	95%	10-14
Pepper	08-13	90-95%	14-21
Tomato	10-13	90-95%	07-21

*Hochmuth, G. (1991).

Economics of Greenhouse Vegetable Production: Greenhouses are expensive to build and operate. A commercial greenhouse of one acre with shade nets and foggers, fertigation and ventilation systems will cost between Rs 30 to 35 lakhs to erect and equip. On the other hand, one acre sized low cost naturally ventilated simple greenhouses–like stone or wooden pole and plastic or nylon net houses can be constructed for Rs 6 to 8 lakhs. On an average, five labourers per day per acre are required for proper crop care and upkeep. In order to keep the maintenance cost as minimum as possible, areas with minimum extremities of weather conditions such as Bangaluru, Pune, Nasik and surrounding places, are ideal for year round production of vegetables. As the cost of cooling is quite high, regions with hot summer conditions will add to the production costs, making the commercial vegetable production an economically viable proposition provided the market prices for the produce fetches higher premium price.

Conclusion

Although, greenhouse production is an intensive undertaking, it can be very satisfying and rewarding. One advantage of greenhouse vegetable production is the relatively small amount of area required compared with open field culture to achieve the similar target yield and the produce will be of superior quality. In addition, the return on investment can be good if the requisite markets can be found.

References

Anonymous. 2011. Indian Horticulture Database. Natonal Horticulture Board, Gurgaon, India.

Giannakou, I.O. and Karpouzas, D.G. 2003. Evaluation of chemical and integrated strategies as alternatives to methyl bromide for the control of root-knot nematodes in Greece. *Pest Management Science*, **59**:883-892

Hebbar, S.S., Balakrishna, B., Prabhakar, M., Srinivas,V., Nair, A.K., Ravikumar, G.S., Ganeshan, G., Sharma, Debi, Sudhakar Rao, D.V, Doijode, S.D., Hegde, M.R. and Rao, M.S.2011. Protected Cultivation of Capsicum. Technical Bulletin No. 22 (Revised Edition). Indian Institute of Horticultural Research, Hesaraghatta, Bangalore, Karnatka.

Hebbar, S.S., Prabhakar, M, Nair, A.K. and Srinivasamurthy, D. 2012. Protected cultivation of cucumber hybrids in naturally ventilated Polyhouse during winter. National Seminar on protected cultivation of vegetables and flowers-A value chain approach at G.B.Pant University of Agriculture and Technology, Pantnagar, 11-12 January, 2012 pp: 46.

Hochmuth, G.1991. Florida Greenhouse Vegetable Production Handbook Vol. 3. Florida Cooperative Extension Service Publication SP-48

Inden, H and Torres, A. 2004. Comparision of four substrates on growth and quality of tomato. *Acta Horticulturae*, **644**: 205-9.

Manish Kumar, Kohli, U.K., Gupta, S.K. and Amit Vikram. 2007. Effect of growing media, irrigation regime, fertigation and mulching on productivity of tomato in naturally-ventilated polyhouses in hills. *Indian Journal of Agricultural Sciences* **77**(5): 302-304.

Mahajan,G and Singh, K.G. 2006. Response of Greenhouse tomato to irrigation and fertiagation fertigation. *Agricultural Water Management*, **84** (1-2):202-206.

Papadopoulos, A.P. and Gosselin, A.2007. Greenhouse Vegetable Production in Canada. *Chronica Horticulturae*, **47**(3): 23.

Prabhakar, M. and Hebbar, S.S. 2006. Enhancing seed use efficiency through quality seedling production in vegetable crops. Proceedings of National Symposium on Improving Input Use Efficiency in Horticulture held during August 9-11, 2006 in Bangalore. pp: 5-7.

Prabhakar, M., Hebbar, S.S. and Nair, A.K. 2010. Quality vegetable seedling production. In: A Hand Book: Production technology of vegetables. Eds: M.Prabhakar, S.S.Hebbar and A.K.Nair. Indian Institute of Horticultural Research, Bangalore pp 138-143.

Prabhakar,M.,Hebbar,S.S. and Vasugi,C.2010.Future and prospects of organic vegetable gardening on terraces. In Souvenir on "Green your terrace-the organic way".National Seminar on Organic Terrace gardening during 9-10 September 2010. Bangalore.pp:16-20.

Ricárdez, M., Rodríguez, N., Díaz, M. and Camacho, F. 2008. Influence of Rootstock, Cultivar and Environment on Tomato Yield under Greenhouse. *Acta Horticulturae* No.: **797**.

Singh, B., Kumar, M., Singh,V. and Mehto, S.P. 2005. Plug-tray nursery raising of cucurbits in soilless media: a sustainable technology for off-season crop production under northern plains of India. *Acta Horticulturae*: **742**.

Singh, B., Kumar, M. and Sirohi, N.P.S.2007.Protected Cultivation of Cucurbits under Low-cost Protected Structures: A Sustainable Technology for Peri-urban Areas of Northern India. *Acta Horticulturae*: **731**.

Yildiz Dasgan H., Oguz Özdogan A., Osman Kaftanoglu and Kazim Abak.2004. Effectiveness of Bumblebee Pollination in Anti-Frost Heated Tomato Greenhouses in the Mediterranean Basin. *Turkish Journal of Agriculture and Forestry Sciences*, **28**(2):73-82.

13

Protected Cultivation of Vegetables in Tropics

Pradeepkumar, T. and Ajay Bhardwaj

Protected cultivation is a specialized agricultural system in which a certain control of the soil–climate ecosystem is exercised modifying its conditions (soil, temperature, solar radiation, wind, humidity and air composition). Plants are cultivated by means of these techniques modifying their natural environment to prolong the harvesting period, alter the conventional cropping cycles, increase yields, improve product quality, stabilize production and provide products when open field cultivation is limited (Wittwer and Castilla, 1995). It is the most contemporary approach to produce mainly, horticultural crops qualitatively and quantitatively and has spread extensively the world over in the last few decades. Protected cultivation also known as controlled environment agriculture (CEA) is highly productive, conservative of water and land and also protective of the environment (Jensen, 2002). Protected cultivation is now becoming popular in Kerala experiencing humid tropical climate with the adoption of locally suitable green house technologies.

Protected cultivation has rapidly expanded in many regions across the globe as early as 1960, when plastics were introduced/used in agriculture and subsequently during 1970's, when the rise in oil prices resulted in increase of energy costs. About 115 countries in the world are into greenhouse vegetable production commercially (Sabir and Singh, 2013). The world scenario shows the area under protected cultivation to be nearly 6,23,302 ha. while total estimated world greenhouse vegetable production area is 4,02,981 ha. Of the total world greenhouse vegetable area, soilless/hydroponic culture systems account for 95,000 ha (Hickman, 2011).

The advent of protected cultivation technology in India materialized during the early nineties, post globalization. In India, area under protected cultivation is presently around 25,000 ha while the greenhouse vegetable cultivation area is about 2,000 ha. Faced with constraints of land holdings, rapid urbanization, declining crop production, declining biodiversity and ever increasing population, demand for food, especially vegetables has increased manifold and protected cultivation has offered a new dimension to produce more in a limited area. Today Dutch protected cultivation is one of the most intensive farming systems in the world with high levels of output by using the latest technologies (Goncharova, 2004).

Use of plastics has transformed the scenario of horticultural crops production, which has lead to another revolution after green revolution (Wittwer, 1993). The major revolution in the adoption of protected cultivation technology is the development of inexpensive and light weight plastics as green house cladding material and at present, it has virtually replaced the traditionally used cover of glass. At that time, the choice of glass covering material was wise and opportune due to the fact that it is indestructible (except violent means), conserves heat, possesses high transmittance and sustains snow load. After the discovery of plastic films and its extensive use as a covering material in agriculture for greenhouses, low/high tunnels, row covers and portable low tunnels *etc increased.* In contrast to glasshouses, plastic greenhouses are adopted in all the continents, especially Asia and USA.

During last decade, there is significant increase (5-10 times) in area under plastic greenhouses in different countries. The first use of polyethylene as a greenhouse cover was in 1948, when Professor Emery Myers Emmert at University of Kentucky used this material instead of glass. In India, green house technology started in 1980 and initially it was used for research only. The National Committee on Use of Plastics in Agriculture (NCPA) recommended specific trials of greenhouse technology in various regions of the country in 1982. India has varied agro-ecological situations in different regions, so the cost effective low and medium-cost greenhouses, low tunnels and other related facilities have to be developed accordingly (Pathania, 2010). Greenhouses are built in Ladakh and other cold dessert regions for extending the growing season of vegetables. In North-East, greenhouses are constructed essentially as rain shelters to permit off-season production. The coastal and southern regions require naturally ventilated polyhouses for efficient and round the year cultivation of vegetables and flowers. The northern region with hot and cold climate needs cooling (fan and pad) and heating devices. Net houses of at-least 40 mesh were successfully utilized in summer to grow capsicum, tomato, beans *etc.* in the Indian plains. Various cucurbits (cucumber, muskmelon and water melon) can be planted in plains in plastic low tunnels/poly tunnels, and by the time flowering starts, the plastic film is removed mechanically or manually.

Efficient designing and construction of polyhouses (location specific) is the pre requisite for success of greenhouse technology by making use of appropriate and need based cladding material that combines the desirable characteristics of various materials (Castrilla *et al.*, 2004). The general trends of improving greenhouse designs are by increasing roof slope to enhance reduction in transmission and to evacuate condensed water (Baille, 2001), and building higher structure suited to local conditions for increased ventilation (La Malfa and Leonardi, 2001). An ideal polyhouse or greenhouse may be standardized as per agro-ecological situation for harnessing real potential of greenhouse technology. A number of new crops can be introduced, but their possibility to effect significantly the biological structure of protected cultivation appears moderate. The demand for varieties suitable for green house cultivation and having tolerance to various biotic and abiotic stresses is progressively increasing thereby, ensuring higher productivity, better quality, lesser pesticides use as well as environment pollution than open cultivation.

Scope of Protected Cultivation in Kerala-A Typical Tropics

Kerala 'Gods Own Country' is bestowed with rich diversity of flora and fauna and is one among the rich biodiversity hotspots of the world. The State is situated at 8°18' and 12°48' North Latitude and 74°52' and 77°22' East Longitude. Kerala's climate is mainly wet and maritime tropical, heavily influenced by the seasonal heavy rains. Annual rain fall of the state is 3107 mm. The total geographical area of the state is 3.88 M ha and total area under Horticulture crops is 1.66 million ha.

Kerala is always branded as a consumer state though there is potential to be self sufficient. Out of the total horticultural area, only 9 % is under vegetables and the share of vegetable production from this area is about 33 % of demand *i.e.,* 8.25 lakh tones. As per the report of Director of Research, KAU (Peter, 2012), the estimated production of vegetables in Kerala is 8.25 lakh tones, which is grossly inadequate (20-25%) to meet the requirement of the state. Vegetable cultivation is confined to a mere 3-4% of total cropped area (Vijayan, 2012). The requirement of vegetables in Kerala is 36.7 lakh tones *i.e.*, only 22.5 % of vegetables requirement is met by domestic production and remaining imported from neighbouring states like Tamil Nadu, Karnataka and Andhra Pradesh. This gap needs to be bridged. High rainfall, warm humid climate with high relative humidity, acidic soil, increased cost of production resulting from high wage rate, scarcity of suitable land for cultivation etc. make vegetable culture very distinct in Kerala. Due to vagaries of weather, most of the vegetables are infected with various biotic and abiotic stresses which are congenial in this type of climate.

The cultivation of vegetables in tropics is very important as a source of fresh food, vitamins and income for the population. The temperature and the global radiation are very suitable for vegetable production in tropics round the year, but open air cultivation is severely hampered, mainly by high rainfall and high humidity. High rainfall can damage the crop and the infestation by diseases is caused by high humidity and rainfall, and use of pesticides increases. The main question is: how to produce more clean vegetables at better quality with less pesticides, with less water and with less land/space. One answer and one component for integrated production system is: the development and use of efficient but cheap technologies for protected cultivation in adaptive protected shelter/structures.

Greenhouse Construction and Designing

Green house is a frame structure covered with transparent materials like glass, polyethylene and poly carbonates. It consists of two parts frame and covering material. The covering material used for a greenhouse works as a barrier to air flow and traps energy within the greenhouse, which heats both the plants and the ground inside it (Gupta, 2010). This warms the air near the ground, which increases the temperature inside the greenhouse. The greenhouses can be used for growing high value vegetables, flowers, nursery of different crops, ornamental and medicinal plants.

Conventional versus Greenhouse Production

Conventional agricultural practices can only control the nature of root media through tillage, manuring, fertilizer application, irrigation etc (Gupta, 2010). There is no control over light, temperature, air composition, humidity etc. in open field cultivation. Greenhouse cultivation can be used as one of the solutions for these parameters.

Some of the benefits of greenhouse cultivation are

- Efficient use of agro-chemicals by minimizing emission of waste and recycling
- Efficient use of most parts of the phyto- mass produced
- Efficient use of solar, wind and thermal energy
- Off season cultivation
- Better insect and disease control
- Reduced use of pesticides
- Protect plants from environmental hazards like heavy rain, strong wind, excess solar radiation and extremes of temperature and humidity
- Better quality of produces

- Higher productivity
- High-income generation for small land holdings

Design Criteria for Tropical Greenhouses

The climate condition of a region has to be investigated and compared with main requirements for plant growth, to test suitability for crop production under shelters and to develop suitable constructions.

The main influencing characteristics in the tropical humid climate are:

1. High monthly precipitation with heavy rainfall during the wet season
2. High mean air humidity of nearly one hundred per cent on cloudy days, and in the morning and evening
3. Little temperature variation throughout the year and relatively a little temperature difference between day and night
4. Steady mean daily global radiation throughout the year but high irradiation at noon because of the short-day length
5. Temperatures are above the biological minimum of the crop throughout the year
6. Frequent tropical storms and tropical cyclones

Greenhouse shelter/structures for tropical climates should have the following characteristics

1. Crop protection from rain, wind and too high global radiation
2. Very efficient ventilation. Ventilation openings in the side walls and ridge. Ridge ventilation is absolutely necessary
3. The relation of greenhouse volume to the ground-floor area should be as large as possible. The higher the structure with ridge ventilation, higher the ventilation efficiency by the chimney effect
4. Gutters are necessary to drain-off the water and to prevent rain-water from penetrating the greenhouse
5. Resistance of construction for wind and the crop load
6. Foundation construction should guarantee wind resistance and should prevent stanchions from rotting or corrosion
7. If necessary ventilation openings have to be equipped with nets to protect against birds and insects but the ventilation efficiency should not be effected too much

8. Film destruction by fluttering in the wind has to be avoided. The film must not flutter but has to be stretched tightly by simple stretch devices
9. Simple methods to change the film especially in high labour cost areas
10. Water saving irrigation systems should be used to avoid the increase of air-humidity

Crop Production Factors

The crop production inside the greenhouse is affected by micro climate, fertilizers, quality of seeds, seed treatment, pest control, field preparation and irrigation. The plant growth also depends on other factors that are well within one's reach and control. The key environmental factors affecting plant growth are light intensity, temperature, humidity, carbon-dioxide concentration, airflow etc.

Light intensity

Visible light constitutes a source of energy for plants. Light energy, carbon dioxide, water all enter into the process of photosynthesis through which carbohydrates are formed. Considerable energy is required to reduce carbon that is combined with oxygen in carbon dioxide to reach the state in which it exists in carbohydrate. The light energy thus utilized is trapped in the carbohydrate. Later, the carbohydrate can be moved from the green stem and leaf cells where photosynthesis occurs to the other parts of the plant. The carbohydrate can be converted into all other compounds needed in the plant. Energy from the sun is utilized in all of these compounds. These processes result in plant growth. Respiration occurs in all living organisms at all time. In fact, most living organisms are ultimately dependent on light energy. The day length (i.e. photoperiod) also affects the plant growth. The relative day-night lengths may affect the rooting of cuttings, initiation of flowers and fruits, bulb and tubes formation, advent and cessation of dormancy, etc.

The intensity of incoming solar radiation is an important parameter for influencing the photosynthetic activity of plants. The light intensity varies from place to place but it generally varies from zero at the beginning of the day to about 100,000 to 150,000 lux (lumen/m^2) around noontime. Light intensity on cloudy days is quite low which leads to poor photosynthetic process. Light intensity below 3200 lux and above 129,000 lux is not ideal for plant growth since the optimum light intensity for plant is 32,000 lux. In this case, solar radiation transmittance needs utmost attention while considering controlled environment agriculture obtained in greenhouse. It is also influenced by the orientation of structure (greenhouse) and the sun elevation.

Temperature

All crops have certain temperature range in which they grow better. Below this range, the process necessary for growth stops. Under very high or extreme temperatures, enzymes become inactive, causing necessary processes for plant's development to stop. Thus, the temperature may affect the movement of minerals, water and food in roots, stems and leaves. This can affect the photosynthesis process significantly. Prolonged low temperature will result in loss of flowers and fruits; slowing their growth. On the other hand, prolonged high temperature will result in loss of flowers, fruits, burning of leaves and lowering in plant growth. Soil temperature also affects the plant growth because it affects the absorption of water from the soil by the plant.

Humidity

Humidity in the atmosphere has its own importance as it governs most of the metabolic and photosynthesis activities of the plants. It is observed that a relative humidity between 20 and 70 per cent is ideal for plant growth. This is so because, very high relative humidity provides better environment for pathogenic organisms making the plant susceptible to diseases. Low humidity is also harmful for plants since it increases the evaporation rate, and at the same time, enhances the water requirement. In environmentally controlled chambers, relative humidity between 55 to 65 per cent and temperature between 20 to 25°C is maintained.

Carbon dioxide (CO_2)

The amount of CO_2 present in the plant environment affects the plant growth considerably because it is essential for photosynthesis. The amount of CO_2 required for optimum plant growth depends on the plant type, state of development, temperature, light intensity, leaf area, air, velocity, humidity, water stress etc. But the CO_2 requirement differs for various plants.

Research shows that controlled environments, such as closed greenhouse system, offers ample opportunity to improve production through the elevation of CO_2 levels. Carbon dioxide, which comprises about 0.03 percent (300 ppm) of ambient air, is essential for plant growth. This level of carbon dioxide in atmospheric air is sufficient to meet the photosynthetic requirement of open field crops. In closed conditions, the level of carbon dioxide rises to nearly 1000 ppm, because respired carbon dioxide remains trapped overnight. As sunlight becomes available, photosynthesis process begins and carbon dioxide from greenhouse air gets depleted. As a result, the CO_2 level in the greenhouse even goes below 300 ppm before noontime. If the greenhouse air does not receive additional carbon dioxide from any other source, the plant would be deficient of carbon dioxide resulting in poor growth. Reports have shown that crop yield increased by 20-30 per cent when carbon dioxide level was maintained from 1000-1500 ppm inside the greenhouse.

The most common method of CO_2 supplementation is through burning of carbon fuels.

Air Movement

Although air movement affects the plant growth, through its effect on transpiration, evaporation of water from soil, CO_2 availability, tearing of leaves, cooling effect, etc. The air velocity affects the boundary layer thickness of the plant's leaves through which the plant transpires and water vapour diffuses into the outside air while the atmospheric CO_2 required for photosynthesis enters into the plant.

Greenhouse Orientation

The most important and complicated decision that what should be the orientation of greenhouse? The greenhouse orientation should be such that: it receives maximum sunlight in winter and prevailing winds should have the minimum effect on greenhouse structure and other operations.

- The maximum sunlight will be received in the winter if the long side of the greenhouse is oriented from east to west as compared to north south (Figure 1).
- However in case of the multi-span greenhouse the gutter should be oriented east-west to avoid the shadow of structural components.

The long side of the greenhouse should be so oriented that the wind exert minimum pressure on the greenhouse. It should be parallel to wind direction.

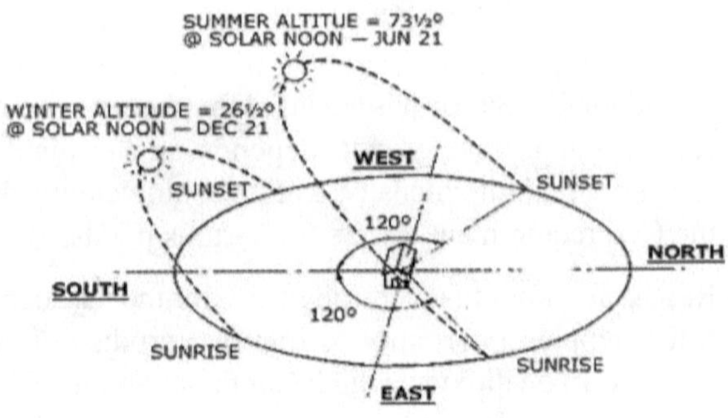

Fig. 1. Summer and winter solar path

Size of Greenhouse

The size of the greenhouse depends upon: production required from greenhouse, land available and the capital investment.

Greenhouse Frameworks

The greenhouse frame can be constructed from different types of material. The selection of right framing material will depend upon capital investment, size of greenhouse, height of greenhouse and availability of material. The most common materials for greenhouse frameworks are bamboo, wood, mild steel (MS) pipes, galvanized pipe and aluminum. The framework should be strong enough to bear wind, dead load including the load required for training of plants as well as for hanging of plant pots. It should allow the maximum amount of light to reach the plants. It should require little maintenance. Avoid the use of MS pipes as it will require frequent painting and its rusting will damage the greenhouse cover.

Greenhouse glazing

The covering of the greenhouse is referred to as the ***glazing***. The considerations in choosing a glazing material include durability, light transmission, cost, and heating and cooling costs.The solar radiation spectrum can be divided in several specific wavebands, which are defined by their range of wavelengths or energy content (e.g., radio and TV radiation, microwave radiation, visible light, etc.). The higher the wavelength, the smaller will be the energy content. Typically, the wavelength of light used by plants is expressed in the units of nanometer (nm). Not all components of sunlight are useful for plant growth and development. In general, ultraviolet (UV; less than 380 nm) and excessive infrared (IR; above 770 nm) or heat radiation can be harmful to plants and should be avoided. Plants use Photosynthetically Active Radiation (PAR; 400--700 nm), as their energy source for the process of photosynthesis. Therefore, greenhouse structures and especially the glazing material should have a high transmittance of PAR radiation.

The common materials for greenhouse glazing (covering) are: glass, plastic films and rigid plastics (poly carbonate and acrylic).

Glass: Glass has the highest light transmission, lasts the longest and is the most expensive. However the single layer glass covering have relatively high heat transfer coefficient.

Plastic films: The use of plastic films as greenhouse glazing is very popular across the globe. The advantages offered by them such as lightweight, translucency, flexibility, toughness, hydrophobic nature and durability, make plastics films the most preferred material for cladding of greenhouses. Different types of plastic materials such as Poly Vinyl Chloride (PVC), Poly Carbonate (PC) and Low Density Polyethylene (LDPE) are commonly used for cladding of greenhouse structures. LDPE remains the most popular of all plastic material because of its features, ease in availability and economics (Gupta, 2010).

Polyethylene films when exposed to outdoor environment for longer duration are prone to degradation as a combined effect of sunlight, heat and oxidation. Interaction with the Sulphur or Halogen based Crop Protection Chemicals (CPC), further accelerates this degradation process. Special chemical additives are incorporated into the polymer matrix to provide a longer lifetime in case of greenhouse films. These additives are known as U.V. Stabilizers, when used along with anti-oxidants, protect the poly film from photo as well as thermal degradation.

The main types of LDPE films used for greenhouse covering are: cross laminated, clear, multi-layered (different layers having different properties). The different variants of multi-layered sheets are available such as: anti-dip, anti-dust, diffused, thermic, anti sulphur, UV blocking, thermic etc.

Different types of chemicals are blended into the material when manufacturing these films. The two main types of U.V. Stabilizer systems being used are: HALS (Hindered Amine Light Stabilizer) + UV absorber Films based on the above package are colorless / translucent in appearance, stabilizes the harmful U.V. radiation form entering in to the Greenhouse and allow maximum transmission of light. For chemical resistance against halogens or sulphur, it is necessary to include a co-stabilizer in this system. Nickel-quencher + UV absorber Films stabilized with above package are greenish / yellowish in appearance. This shade gives a slightly hazy effect thus reducing the intensity of direct light transmission. This stabilization system gives excellent U.V. Stabilization also provides good protection against action of CPC's. This property is also sometimes termed as Anti–Sulphur. Proper light transmission filtering, out the harmful Ultra Violet (U.V.) radiation is very essential for optimized growth of the crops inside the Greenhouse. The photosynthesis activity in plants is directly proportional to the amount of light. Higher light transmission enhances the growth of plants. Therefore it is crucial to select the right combination of polymer and specialty additives, to get maximum light transmission. This is more significant for greenhouses located in colder climates where availability of daylight is limited. The ultra violet radiation is harmful for the plants growth. As well as when the protective polyethylene films are exposed to outdoors environment for longer duration are prone to degradation as a combined effect of sunlight, heat and oxidation. Interaction with the Sulphur or halogen based crop protection chemicals (cpc), further accelerates this degradation process.

Special chemical additives are incorporated into the polymer matrix to provide a longer lifetime in greenhouse films. These additives are known as U.V. stabilizers, when used along with anti-oxidants, protect the film from photo as well as thermal degradation.

The selection of film depends upon the type plants grown inside the greenhouse and properties of glazing cover required. The film thickness is specified in microns or weight density, 200 microns thick or 120-150 GSM (grams per square meter, weight density) sheets are uses for greenhouse covering.

Rigid plastics: The rigid plastics (e.g., polycarbonate and acrylic) are less expensive than glass and last seven to 20 years. They are usually manufactured as twin-walled sheets. The air space between the two walls acts as an insulator. Light transmission through rigid plastics is very good, although it usually decreases over time as the plastics age and turn yellow due to the amount of UV radiation contained in sunlight. The large sheets are much lighter than glass and require fewer support bars to attach them to the greenhouse frame. However, these rigid panels are not so easy to install on curved roofs.

Scope of different polyhouse structures in humid tropics

1. Naturally ventilated saw toothed polyhouse

Naturally ventilated saw toothed polyhouse made of G.I pipe and covered with UV-Stabilized polythene sheet on the roof and insect –proof netting along the sides are suitable for growing vegetables in tropics. Temperature and humidity built up inside the poly house can be controlled by providing sufficient roof and side ventilation and required height to the structure. Optimum height of the poly house depends on the floor area, ambient temperature, relative humidity, solar radiation and wind velocity of the locality Fig. 2.

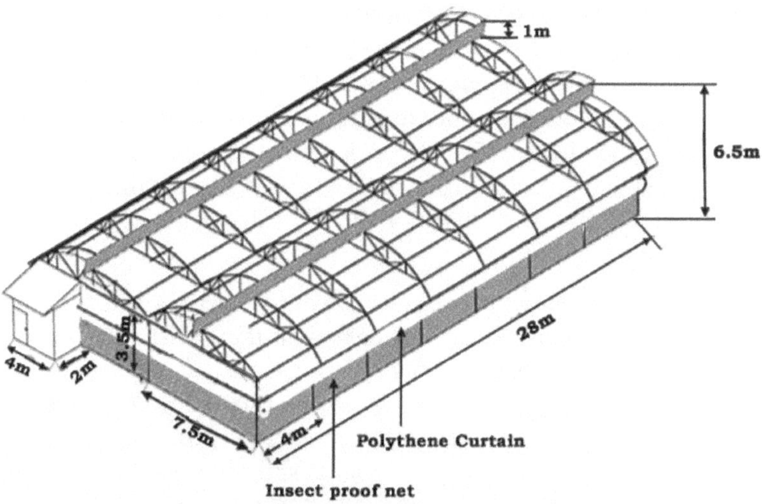

Fig. 2. Natural Ventilated Polyhouse (28 x 15 m = 420 Sq.m (Varghese, 2013)

A saw tooth shaped poly house with sufficient ridge height and gutter height, single-span or multi-span is suitable for growing vegetables in tropics. The structure can be constructed using B-Class GI pipe (2mm thickness). The diameter of the pipes used for construction may vary with the size of the poly house. In any case the structure should be guaranteed to withstand wind velocity up to 140 km/hr. The width of each span should be between 7-9m with roof ventilation not less than 10% of the floor area of the green house. The vertical height of the roof ventilation in each span should not be less than 1m.

Ridge and Gutter height for Green houses in Plains of Tropics

Floor area (m^2)	Gutter height (m)	Ridge height (centre) (m)
40	2.50	4
80	2.85	4.85
120	3-3.25	5
200	3.50	6-6.25
500	4	7
1000 and above	4.50	7.5- 8.0

The roof ventilation and the sides of poly house up to gutter height may be covered with UV stabilized insect proof net (40 mesh). Additional provision may be given to cover the insect proof net provided at the sides with UV stabilized sheet which could be rolled down during night hours, so that the CO_2 given out by the plants during night is trapped inside and used for photo synthesis in the morning hours. The roof of the poly house may be extended up to 1m on all the four sides to prevent entry of rain water into the structure during heavy rains. Provision may be given to collect the runoff from top to a water harvesting tank which could be effectively utilized for irrigating the plants. The green house should be provided with sliding type double door entry to prevent entry of pests inside the structure. The entry point should be sufficiently wide in small poly houses to provide entry of a tiller/ mini tractor inside the poly house for carrying out tilling operations. In large poly houses (500 m^2 & above) a specific entry point may be provided for bringing machinery inside.

The roof of the poly house may be covered with 5 layer UV stabilized polythene sheet (180-200 microns) with at least 85% light transmissibility. The UV sheet and the insect proof net may be fixed by using zigzag spring insert (2-3mm dia with cold galvanization) and aluminum profile (200-220 GSM).Below the UV stabilized polyethylene roof, shade net/ chromatic net 30% may be fitted at gutter height inside the poly house with provision for spreading and folding as and when needed. Provision may be given to control the climate by operating foggers installed at gutter height at a distance of 2.5m c/c. A timer is to be installed to control the operation of foggers. The foggers can bring down the temperature inside the poly house 3-4° C below the ambient temperature it operated during summer months. Inside the poly house, circulating fan may be provided for better ventilation. The foggers should be operated between 11 am to 3.30 pm. It is always better to minimize the duration of operation of foggers at a time and operate the foggers at frequent intervals for effective temperature control. Drip and fertigation system may be operated manually.

Greenhouse Production Systems

Inside the green-house the crop may be raised in the soil or on other media called soilless culture. Growing vegetables directly is the soil easiest way to start green house vegetable production. For this the soil inside the poly house is ploughed well and then brought to a fine tilth. The soil pH is then adjusted to 6-6.5. Raised beds 1 m wide and 8-10 cm height are then taken after mixing the soil well with farmyard manure/ vermicompost, sand/coir-pith and rice hulls (3:2:1). The greenhouse soil is then sterilized either through solarisation or chemical sterilization or a combination of both the two. Before sterilization the prepared beds are moistened well.

Solarisation

Solarisation is effective during summer months. The soil is fully covered with 150-200 gauge transparent polythene sheets. The green house is kept completely closed for 30 days so as to trap maximum heat inside the structure.

Chemical Sterilization

Formaldehyde(35%)solution can be used for chemical sterilization in poly houses. For this dilute the commercial formulation 10-12 times with water and then apply uniformly inside the poly house. Keep it covered for 10 days. Remove the cover and keep the poly house open for 2-3 days. The sterilized poly house is now ready for planting. A combination of solarisation + formaldehyde sterilization is found to be better than either of the two alone. In this case apply formaldehyde commercial formulation at the rate of 2ml/litre of water uniformly inside the poly house and cover it with transparent polythene as in solarisation. Keep the poly house covered for 10 days. Remove the cover and keep the poly house open for 2-3 days for the fumes to escape.

2. Rainshelter-a low cost naturally ventilated greenhouse for Kerala

A greenhouse with sophisticated environmental control is not very essential in Kerala conditions. Such hi-tech structures can not be recomended for the farmers of Kerala whose resource bases are very low. Here comes the application of low cost technology of protected cultivation named Rainshelter.

It is a naturally ventilated low cost greenhouse which facilitates year round production of high value crops like tomato, capsicum, salad cucumber, amaranth and cabbage etc. this technology is acceptable to small scale and marginal farmers of Kerala. The framework can be made of galvanized iron pipes, wooden or bamboo poles. Cladding is provided with UV stabilized low density polyethylene (UVLDPE) film of 200 micron thickness. It is economical, 70-80 per cent transparent with 2-3 years usable life.

The approximate cost of a low cost rainshelter with a floor area of 100 m^2 will be about Rs. 20,000/-. If the framework is made up of G.I. pipes, it will be around Rs. 40,000-60,000/-. The site selected for construction of rainshelter shold be level and properly drained. Sunlight should be available in plenty and natural wind breaks are desirable at a distance of 3-4 metre from the structure. Rainshelter should preferably be oriented in North-South direction. The results of two year experiments revealed that rain shelters without roof ventilation is ideal for rainy season but during summer, structure with roof ventilation is good for tomato, capsicum and amaranth. Capsicum plants could not survive in the open field during rainy season, but at the same time the rain shelter crop recorded a yield of 42 kg/cent. The coefficient of disease infection (CODEX) for leaf spot disease was minimum in the rain shelters without roof ventilation (42.9%) whereas it was 81.5% in the open field for red leaved amaranth (Indira *et al.*, 2004).

Though rain shelter is meant for off-season cultivation, it can be utilized throughout the year. The study revealed that the performance of tomato and capsicum inside the rain shelter is better during october-january. Though this is an on-season for these crops, the rain shelter crops recorded thrice the yield than that of open field crop.

3. Shade house

A shade house is a structure enclosed by agronets or any other woven material to allow required sunlight, moisture and air pass through gaps. It creates an appropriate microclimate conducive to the plant growth. These are used for nursery raising, indoor plants and hardening of tissue culture plantlets.

The shade house structure composed of two basic components *i.e.* frame and cladding material. The shade house frame provides support for cladding material and designed to protect against wind, rain and crop load. The mild steel angle frame shade house lasts upto 20-25 years, if anti-rust treatment is done at regular interval whereas bamboo structure can lasts up to three years. Shade nets are used to provide relief to the plants from the scorching sunlight, high winds, direct rainfall as well as insects and pests. U.V. stabilized shade nets are more preferred. The agro shade nets lasts for 3-5 years depending on the climatic condition. Shade nets are available in different colours with wide range of shade percentages *viz.*, 25%, 30%, 35%, 50%, 60%, 75% and 90%.

In Kerala where tropical conditions prevail, the shade houses are the better choice for commercial production of high value vegetable crops like parthenocarpic cucumber, tomato, capsicum, musk melon, water melon and beans etc. because of the advantage of cooling effect and maintaining the temperature inside the polyhouse which otherwise is not possible in outside/open conditions. It also protects the crops from high precipitation in the rainy season and provides ample scope for off-season vegetable production in Kerala (Mathew *et al.*, 2011).

4. Low tunnels for raising seedlings

Nursery is the place where young seedlings are raised and nurtured before planting them in the main field. Now-a-days the nursery raising of vegetables have become a specialized job with increasing susceptibility of vegetables to biotic and abiotic stresses.

For raising a good crop it is utmost essential that seedling should be healthy, vigorous and disease-free. Seedlings are susceptible to a number of diseases due to its delicate, succulent and highly tender nature. Further high cost of hybrid seeds warrants attention of the grower to produce the seedlings under protected conditions because every hybrid seed has its accountability owing to its high cost which is 20 to 50 times higher than the open pollinated seeds. Raising of nursery under polyhouse/green house and sowing the individual seeds of hybrid in potting plug with artificial culture media is a well established practice in developed countries.

However, in our country most of the nurseries are raised under open conditions and it is one of the main reasons of setback to development of hybrid programme during rainy season in India. Farmers can raise seedlings in low tunnels and low cost polyhouses. Production of early tomato, cauliflower, capsicum, chillies often have problems in high temperature and high rainfall areas.

Low tunnels are miniatures of greenhouse to protect the plants from rain, wind, low temperature, frost and vagaries of weather. The use of transparent plastic sheets as low tunnel provides ideal conditions for successful raising of seedlings than conventional methods. The time taken for seed germination and seedling growth is reduced under low tunnels. These tunnels facilitates entrapment of CO_2, thereby enhancing the photosynthetic activity of the plants and hence the yield. However, under low tunnels, artificial controlling of conditions is not possible. Besides, being inexpensive, they are easy to construct and dismantle. They are usually one meter in width and 0.5 meter in height, though it may also vary depending on the nature of the growth of plants. The seed sown in propagation trays/pro trays or in polybags, show early germination and make faster growth under protected cover.

There are different pro trays available in market with varying number of cavities. Mostly available are 38, 50, 98, 221 and 228 cavities (Mathew *et al.,* 2011). The tray help in proper germination provide independent area of each seed to germinate, eliminate the mortality rate, and maintain uniform and healthy growth of seedlings. These trays are easy in handling and storing, reliable and economical in transportation. The use of shade and agronets has been found very effective in raising of seedlings during high temperature. Shade reduces temperature by 5 ^{0}C to 10 ^{0}C and protects the seedlings from sun scorching and harmful effects of high temperature. Agronet protects the seedlings from insects and reduces the vector-borne viruses' infestation and damage by other insects.

5. Plastic mulching

The process of covering the soil around the plant with dry leaves, straw, hay, stones etc., was practiced for ages. However the introduction of plastic film as mulch increases the efficiency by improved moisture conservation, increased soil temperature and elimination of weed growth and thereby increasing the yield. Covering of soil around the root zone of plant with a plastic film is called plastic mulching. It is an effective practice to restrict weed growth, conserve moisture and reduce the effect of soil borne diseases through soil solarization (Mathew *et al.,* 2011).

Low density polyethylene and linear low density polyethylene plastic films are commonly used for mulching. The plastic mulches mostly available in market are mainly black colour or clear or transparent colour. The double colored sheets like yellow/black, white/black, silver/black, red/brown are also available now. However, LLDPE black colour mulch films are the most popular.

The notable advantage of the use of black plastic as mulch material is its impermeability which prevents direct evaporation of moisture from the soil and thus, reduces the water loss. Moreover, whatever water evaporates from the soil under the plastic film condense on the lower surface of the film and falls back on the soil as droplets, hence conserving about 25 per cent of the available moisture for plant use. Mulching also induces better root growth in addition.

Mulch film thickness for different crops*

Thickness (microns)	Crops recommended
25	Short duration crops, vegetables (3-4 months)
50	Medium duration crops (11-12 months)
100	For long duration crops

*(Mathew *et al.,* 2011)

Under the National Horticulture Mission (NHM), assistance is provided for activities under protected cultivation including greenhouse construction, mulching, shade net and plastic tunnels as indicated below:

Structure	Estimated cost	Pattern of assistance
Greenhouse structure		
Fan and pad system	Rs. 1465/sq. m	50% of the cost limited to 1000 sq.m per beneficiary
Naturally ventilated system		
Tubular structure	Rs. 935/sq.m	50% of the cost limited to 1000 sq.m per beneficiary
Wooden structure	Rs. 515/sq.m	50% of the cost limited to 2 units (each unit not to be exceed 500 sq.m per beneficiary)
Bamboo structure	Rs. 375/sq.m	50% of the cost limited to 5 units (each unit not to exceed 200 sq.m per beneficiary)
Plastic mulching	Rs. 20,000/ha	50% of the total cost subject to a maximum of Rs. 7000/ha limited to 2 ha per beneficiary
Shade net house		
Tubular structure	Rs. 600/sq.m	50% of the cost limited to 1000 sq.m per beneficiary
Wooden structure	Rs. 410/sq.m	50% of the cost limited to 5 units (each unit not to be exceed 200 sq.m per beneficiary)
Bamboo structure	Rs. 300/sq.m	50% of the cost limited to 5 units (each unit not to exceed 200 sq.m per beneficiary)

(Mathew *et al.*, 2011)

Nursery Production Technology of Vegetable Transplants for High Tech Agriculture

Vegetable seedlings are raised in a variety of containers ranging from molded trays to pots. Plastic trays/ Styrofoam trays are the standard containers used in various parts of the world. Plastic pro-trays of uniform size with cells of equal size are most widely used in our country. Plastic trays with cell shapes round, hexagonal and inverted pyramid are used for raising seedlings. The standard cell size is 3cm x 3cm for short cycle vegetable crops and 4.5 com x 4.5cm for longer duration crops. Plastic pro-trays helps in proper germination provide independent area for each seedling to grow, reduce the mortality rate, maintain uniform and healthy growth of seedlings and are easy to handle and store.

Soilless medium for raising seedlings

Three ingredients coco peat, vermiculite and perlite are used as rooting medium in high tech nursery. The above ingredients are mixed in a 3:1:1 ratio to get a homogeneous medium having uniform physical and chemical properties throughout the mix. The medium is light in weight and easy to transport. It is free from insect pests and diseases.

Sowing seeds and raising seedlings

A saw tooth shaped naturally ventilated poly house should be constructed in an open area with a minimum gutter height of 4.0m and 6m at the centre for growing the seedlings. A structure of size 36m x 8m can house 1.5 lakh seedlings at a time. Roof ventilation (1 m wide) should be provided all along the length of structure. The sides and roof ventilation are covered with insect proof 40 mesh UV stabilized net. A 30% movable shade net is also provided inside the poly house at gutter height to regulate excess sunlight and temperature inside poly house. The prepared medium is filled in pro-trays and kept ready for sowing.. Seeds are usually planted at a shallow depth (1 cm) one seed in each plug by gently pressing the media with the finger. After sowing the seeds are covered with a thick layer of vermiculite. In order to obtain healthy seedlings one seed per cell is sufficient. However if more seeds are sown then one seedling may be retained per cell after thinning and the removed seedlings can be used for gap filling in cells where there is no seed germination. This operation can be done at two leaf stage.

Watering and fertigation of transplants

The quantity and frequency of watering depends on cell size, growing media, green house ventilation, crop, age of seedlings and weather conditions. It is important to moisten the entire plug to promote uniform root growth. It is better to irrigate transplants daily in the morning. If the plants are irrigated in the evening the plant remain wet overnight which in turn may invite fungal diseases.

Vegetable seedlings are usually fertigated with water soluble fertilizers which are applied along with irrigation water. The common fertilizers used for vegetable transplant production are 19:19:19, 20:10:20, and 20:20:20. Fertilizer requirement vary with cell size, type of fertilizer and the crop grown in the trays. Tomatoes can be fed once in 5-7 days with 250-350 ppm N. Capsicum seedlings require still higher concentration of fertilizers and hence 350-400ppm N is applied at weekly intervals. Cole crops like cabbage and cauliflower requires fertigation only at the rate of 150-200 ppm N at weekly intervals. Cucurbits need 2-4 applications of fertilizers weekly at 100-150 ppm N concentration.

Pest management

Vegetable transplants are raised in pro-trays and kept in a poly house provided with insect proof netting on all 4 sides. Due to this the seedlings are virtually free of attack by insect pests except for thrips and mites which may often become serious. In such cases the seedlings should be sprayed with jump (Fipronil 80%WG) @ 2g in 5 liters of water. Proper sanitization of green house is highly essential to ward off diseases especially damping off. The used trays should be disinfested with sodium hypochlorite 0.1% before reusing them. All the over grown and unwanted seedlings should be timely removed from the green house. Drenching the rooting media with liquid *Pseudomonas fluorescens* formulation @ 5ml/litre of water is effective in controlling the fungal pathogens and to promote the growth of young seedlings.

Hardening of Seedlings

Hardening of vegetable seedlings before transplanting to the main field is highly essential to withstand the transplanting shock and for better crop stand. Seedling should be gradually hardened by acclimatizing them to the growing conditions at least a week before planting. Withholding water or minimizing irrigation is the best way to harden a seedling. Generally seedlings of all vegetables will be ready for transplanting in 25-30 days. Cucurbit seedlings will be ready for transplanting within 20 days of sowing.

Planting and after care

After green house sterilization, the drip lines are laid over the beds. The beds are then covered with polythene mulch (30 microns). Holes are then made on the mulch according to the drippers on the line and the seedlings are then planted. After planting seedlings are irrigated. Fertigation starts from the 3rd day onwards. The recommended doses of nutrients for various crops are given in the schedule.

Hi Tech Production of Parthenocarpic Cucumber

The parthenocarpic cucumber is a type of cucumber vine that produces only female flowers and produces greenhouse cucumbers without the need for pollinators, such as bees, that provide fertilization by moving pollen from male to female flowers. Because they produce fruit without pollination and thus fertilization, parthenocarpic cucumbers are a popular cucumber for greenhouse production where the growing system is closed for pest and disease control (Sharma *et al.*, 2010). The seed for parthenocarpic cucumbers costs more than ordinary cucumber seed but the germination or sprouting rate is higher and the chance for disease or pest problems is greatly decreased in a closed system. If pollination does occur, the fruit will form seeds, the shape of the fruit will be distorted and a bitter tasting fruit will develop. It is therefore essential to prevent bees and other pollinators from entering the greenhouse and carrying pollen from outdoor gardens.

Cucumber fruit fits into two classes: slicers and picklers. Traditionally, slicers are grown for fresh consumption while picklers are grown for processed consumption. Fresh market greenhouse-grown cucumbers are harvested year-round, while pickling cucumbers are harvested mainly in the spring. About 60 per cent of cucumber consumption is fresh with the remainder in pickled products. Rising fresh-market consumption likely reflects the popularity of salads, the general trend toward more healthful lifestyles, and an increasing interest in greenhouse cucumbers.

Parthenocarpic cucumber is gaining popularity for polyhouse cultivation in India. The productivity and quality of the cucumber grown under protected conditions is better than open condition. Moreover, fruit fly incidence is negligible under protected conditions. In Kerala, cucumber can be grown commercially in modified naturally ventilated greenhouse or rain shelters for almost all the year round supply of high quality fruits, which otherwise, is not feasible under the open conditions. Depending upon location, various structures like plastic low tunnels walk in tunnels and naturally ventilated polyhouses can be used for successful production of both off-season and main season crop of cucumber.

It is a sub-tropical vegetable and grows best in naturally ventilated polyhouses, which provide favorable and stable environmental conditions. When pests are under control, the plants grow rapidly and produce heavily. The crop is, however, more sensitive to temperature, which may cause reduction in both growth and yield. It is relatively a short season crop and cannot tolerate frost. The optimum temperature for speedy development of the fruits and higher fruit yield is 19-24°C. Cucumber requires mild climate and as such does well under greenhouse conditions. Well drained sandy loam soils are preferred. Heavier, wet and poorly drained soils should be avoided. Cucumber can tolerate acidic soil conditions; however, best results are obtained with soil pH of 5.5 to 6.8.

Greenhouse temperature

Air temperature of the polyhouse plays an important role for successful production of parthenocarpic cucumbers. The temperature affects germination, vegetative growth, flower initiation, fruit growth and fruit quality. The greenhouse temperature should not be less than 18 °C and above 35 °C for the sustained production of the cucumbers. The growth rate of the crops depends on the average 24 hour temperature. The wide variation in day-night air temperature results in taller plants having smaller leaf size. Maximum fruit production is achieved with a night temperature of 19-20 °C and a day temperature of 22-24 °C.

Day Length

Parthenocarpic cucumbers are not sensitive to day length.

Light

Light plays an important role in the greenhouse production of cucumber as it regulates the production of the optimum amount of photosynthates and thus helps in obtaining higher yield. Cucumbers develop well in high light conditions (250 to 1000 watts/sq meter). In higher light intensity, there is a need for shading. Lack of light (below 100) severely decreases plant development. Cucumber cannot tolerate shade for longer duration. In case of parthenocarpic cucumbers, shading is required for the development of fruits inside the greenhouse in summer months because fruit formation and development seizes in high light intensity conditions.

Relative humidity

Cucumber plants develop well in relatively high humidity (65-85%). They suffer in extremely dry weather. It controls the growth of the plants, affects evaporation, transpiration etc. and incidence of insect-pests and diseases. Generally the high relative humidity favors the growth but leads to condensation of vapors over the leaves favoring high disease incidence. High humidity also reduces transpiration resulting in inadequate absorption and translocation of certain nutrients especially calcium to the leaf margin and fruit. On the other hand, low relative humidity favors certain insect-pests and diseases like powdery mildew and mites. Irrigation requirements also get enhanced with the decrease in the relative humidity. Thus optimum humidity should be maintained in the polyhouse by adequate aeration and by installing the various misting devices like misters and foggers etc. so as to ensure near optimum growing conditions for cucumber in the polyhouse.

CO_2 Enrichment

Addition of carbon dioxide to greenhouse has been demonstrated to improve vegetable yields. Carbon dioxide is normally present in the atmosphere at a concentration of 330 parts per million. Concentrations of 1,000 to 1,500 ppm in greenhouse atmospheres have given the best results. Increases of 20 to 40 percent in yield have been reported for various vegetables. Generating and monitoring equipment are readily available.

Parthenocarpic varieties

The varieties/hybrids grown in the polyhouse are usually gynoecious i.e. these varieties bear the female flowers only. However, such varieties can become bitter and club shaped in case pollination occurs with the wild cucumber by the bees. Therefore, immense care is to be taken to avoid entry of pollinizers in the polyhouse. Parthenocarpic fruit setting results in an earlier and more regular production rhythm since it circumvents the inhibitory effect of seed formation on subsequent fruit development.

A few important parthenocarpic varieties of cucumber from Europe and Israel are Hasan and Sarig (for summer crop), Muhasan and Dinar (for winter season), Mustang and Bronco (for spring crop), Kian, Alamir, Isatis, Multilina, Mansaur and Nun-9729 etc. Downy mildew, powdery mildew, mite and nematode tolerant/ resistant should be preferred for cultivation under protected conditions. The growers are advised to evaluate the performance of Parthenocarpic hybrids/ varieties under prevailing agro-ecological situations before going for its commercial cultivation in polyhouses. The Parthenocarpic varieties should be grown in protected structures only and not in the open environment.

For Kerala , the perfect parthenocarpic cucumber varieties are Kian, Hilton, WS-556, Pant Parthenocarpic cucumber-2 and Pant Parthenocarpic cucumber-3.

Moreover, the PhD research work is under progress for the development of parthenocarpic hybrids of cucumber in Department of Olericulture, Kerala Agricultural University, Vellanikkara, Thrissur. Parthenocarpic fruit development is achieved under high temperature conditions in Kerala as well.

Nursery raising and planting

For greenhouse cultivation, the nursery is usually raised in soil-less media in plastic pro-trays. Cocopeat, vermiculite and perlite is used in 3:1:1 ratio (volume basis). About 600 g seed is sufficient to raise the nursery for one hectare area. Alternatively, the cucumber seedlings can also be grown in the poly-tubes by using soil mixture with one part each of sand, soil and farmyard manure or vermi-compost. The seedlings are usually ready for transplanting in 28-30 days during winter and 15-18 days during the summer months. The root system remains intact in case of soil-less media as all the roots come as a plug, while in case of the seedling raised in poly-tubes, the plant is replanted with whole of the earth ball intact. The cucumber nursery should be transplanted at two true leaf stage.

Plant geometry

Proper spacing, training and pruning is essential for growing of the cucumbers in the polyhouse, which ensures sufficient sunlight, proper nutrition and ease of various intercultural operations. The cucumber seedlings are planted on the raised beds with two rows in a bed with row to row spacing of 70 cm and plant to plant spacing of 30 cm. Sometimes, one row is planted in one bed by keeping 140-150 cm spacing between rows and 30 cm within row. Factors such as cultivar, season and growing media influence the plant spacing. Recommended and proper spacing, as per cultivar characteristics, should be followed so that there is efficient use of polyhouse vertical space by minimizing competition for light and ensuring sufficient ventilation. For Kerala conditions, the optimum space requirement for a cucumber plant is 70-80 cm^2 and spacing between rows is kept at 1.5m and 50 cm between plants in a row.

Depending upon size and design of polyhouse, there are different training systems for production of greenhouse cucumber such as upright system and V-cordon system etc. The basic purpose of the training and pruning cucumbers is to uniformly maximize the leaf interception of sunlight in the polyhouse and to avoid competition for nutrients. The selection of a particular training system will depend upon the height of the greenhouse, training facilities and choice of the grower. In greenhouse, plants are trained upward and the main stem is allowed to climb to the over head wire along a polythene twine. The wires are fixed 8-9 feet above the ground. The twine of each plant is alternatively tied to the horizontal overhead wires or steel cables running along the length of the rows. The base of the twine is tied/anchored to the base of the stem with a non-slip nose. After training the wine on the horizontal wire upto 8-9 feet, it is again trained to downward direction.Trellising of the plants should be done carefully to avoid damage to the female flower buds on the main stem. In this system of training, avoid fruits upto 1.5 to 2.0 feet above the ground.

Pruning of the plant is based on the plant vigor and fruit load. Extensive vegetative growth should be discouraged to allow sufficient light and nutrition. In single stem training, female flowers should be removed or pinched upto 5th node to ensure sufficient vegetative growth of the plants to sustain the fruit load. The multiple fruits in the axils of the leaves should be avoided and only one fruit should be retained in the leaf axil at every node. The lateral shoots should be pruned and only one or two main stems should be allowed to grow depending upon training system and height of the polyhouse. Maintaining the proper balance of foliage and fruits in each in each plant is essential for maximizing productivity and improving fruit quality through-out the crop period.

Fertigation

Fertigation refers to applying fertilizers along with the irrigation water. Cucumber crop requires constant water supply for higher yield and improved quality. The crop is very sensitive to water stress during the period of growth, flowering and fruit enlargement. The greenhouse parthenocarpic cucumbers grow very fast and are very demanding in terms of the nutritive requirements and hence constant fertigation is done usually through drip irrigation system by mixing various micronutrients and water soluble fertilizers like Samadhan or polyfeed (19:19:19) with the irrigation water. The highest demand for nutrients is during the peak fruit production period. Generally, nitrogen and potassium requirement is higher for greenhouse cucumbers in addition to other nutrients.

Cucumber is quite sensitive to N deficiency and grows slowly and produces less flowers and smaller fruits in low nitrogen regimes. In addition to NPK, calcium is also required for enhancing fruit quality and the yield of cucumber grown under protected conditions. The fertigation and irrigation will depend upon the stage of the crop, weather conditions and the fertility status of the soil. It responds to heavy doses of fertilizers and nutrients. Therefore, a basal application of 50-100 kg NPK/ha is recommended. The polyhouse soils should be analyzed for the nutritional status for ensuring judicious and economic use of liquid fertilizers.

The optimum irrigation regime of 20-25 kpa is ideal. However, over watering and water stress adversely affect the crop growth and fruit quality, and increase the incidence of insect pests and diseases. Drip irrigation system with fogger/mister facilities ensure near optimum growing conditions for healthy crop.

In general green house cucumbers are irrigated using in line drippers with discharge rate of 2 litres / hour. Under Kerala conditions the crop can be irrigated daily with one litre of water. Fertigation is carried out once in 3 days.

Ad-hoc fertigation schedule for cucumber*

Water/fertilizer	Early stage	Fruit bearing stage
Nitrogen (ppm/m^3 of water)	80-100	100-150
Phosphorous (ppm/m^3 of water)	60-70	80-100
Potash (ppm/m^3 of water)	100-120	120-150
Calcium (ppm/m^3 of water), if required	30-40	40-50

*(Singh, 2005)

There are reports to fertilize the crop even once, two times and three times daily, and two times every other day to increase the marketable yield .

Fruit Development

A fruit may be developed at each node and more than one fruit may begin to develop at some nodes. It is usually best to thin these multiple-fruit clusters to a single fruit, however, vigorous plants can sometimes mature more than one fruit at a node. Miniature and beit-alpha types will support several fruit per node. Any distorted fruit should be removed immediately. The greatest growth of the fruit occurs between day 6 and 14 after the bloom opens (anthesis). Maximum fruit length occurs at day 14 followed by diameter increase. The shape of the fruit is somewhat tapered being largest at the stem end prior to day 10 after bloom, however, the fruit becomes uniformly cylindrical by day 14. During the spring season, commercially acceptable fruit size is usually reached by the 11th day after the bloom opens.

Harvesting and storage

Gynoecious cucumbers usually come in bearing within 30-40 days after transplanting. The fruits are harvested when they are more or less cylindrical and attractive. Fruits keep on developing as the plants grow and reach harvestable stage in 10-14 days after flower opening depending upon the variety. Fruit picking is usually done two to three times within a week. The harvested fruits must be protected from desiccation and bruising as the gynoecious cucumbers have very tender skin. Under good management conditions, parthenocarpic hybrids can yield one tonne of fruit from 200m^2 area in 3-3.5 months.

The cucumber fruits should be carefully harvested and then stored in the plastic or padded containers to avoid bruising and damage. The cucumber fruits should be harvested at the optimum stage with knife or scissor in the morning or evening hours, when the fruits attain the optimum size. After harvesting, the fruits should be prevented from direct sunlight exposure and each fruit should be individually wrapped in shrink wrap film before packaging. Grading of the produce should be practiced for higher returns. During transportation, minimize heat and store produce in cold storage (12° C) as soon as it arrives at destination. The cucumber fruits should not be stored with the active ethylene producer like tomato as the quality and appearance of the cucumber will deteriorate after exposure to ethylene, Fruits stored under ideal conditions i.e. 10-13°C temperature and 90-95% relative humidity has a shelf life of 10-14 days.

Grading of fruit

Defective fruit, insects, etc should be removed before packing. Size grading can be done by machine in the packing house, and washed by water at 55 to 58 centigrade for 20 seconds, and then dried by cool air to improve appearance and shelf life.

Packing

Packing should be done according to needs of wholesalers. It is common to place a foam pad on the bottom of each box to maintain freshness and reduce shrinkage. Cucumbers can be held in coolers at 10-12 centigrade and relative humidity of 90-95% for up to 10 days.

Plant protection

The incidence of various insect-pests are less in modified naturally ventilated polyhouses having annex/double door and insect proof nylon mesh (40-50 mesh size) on the side and top vents, which helps to discourage the entry of the various insects inside the polyhouse. However, occasionally, in-spite of all the precautions, a few insects like mite, aphids and white fly enter the polyhouse and cause severe damage to the crop. Besides, diseases like powdery mildew and downy mildew also infect the crop. List of a few important diseases and insect pests of cucumber along with their control measures are as under:

Important diseases of cucumber under protected cultivation

Downy mildew: This disease is caused by a fungus-*Pseudoperonospora cubensis* and an optimum temperature range of 15-23 °C is congenial for spread of the disease. The initial symptoms appear as pale green areas separated by dark green patches. The spots are angular, yellow colored and often restricted by the veins on the upper surface. Purplish downy growth appears on the lower side during moist weather. Infected plants show wilting symptoms. The disease cannot be controlled easily by spraying Mencozeb only @ 3 g in one liter of water. However, alternate sprays of Ridomil MZ (0.2 %) and Mencozeb(0.3 %) as soon as the symptoms appear on the vines. Repeat the spray after 10-12 days interval. In Kerala, spray of Equatium Pro @1ml/litre of water or Sectin @ 2g/litre are also very effective for controlling downy mildew, another chemical for spraying is Melody Duo @ 2g/litre can also control the spread and control of this serious disease. Irrigation should be done with drip irrigation system especially during the monsoon season. Maintain sanitation inside the polyhouse by removing the infected leaves, plant debris etc. Avoid high humidity inside the polyhouses during rainy and autumn seasons.

Powdery mildew: This disease is caused by *Erysiphe cichoracearum* and *Sphaerotheca fuliginia*. The fungus forms white floury patches on leaves, stems and other succulent parts of the plants. Fruits remain poor in quality and flavor. The optimum temperature for disease development is 20-25°C. The crop should be sprayed with triadimefon or bitertanol or dinocap or triadimenol (0.5 %). *Ampelomyces quisqualis* and *Verticillium lecanii* are effective bio-control agents. Disease resistant varieties like Condesa, Alamir, Tenor, Ilas, Parad, Jinyou No 12 etc. should be grown.

Anthracnose: The causal organism of this disease is *Colletotrichum orbiculare*. It causes lesions on leaves and fruits in cucumber. On leaves, small yellow water soaked areas are formed that enlarge rapidly. The spots commonly start on the vein and extend into brown spots, which are angular or roughly circular. The spots on fruits are roughly circular, sunken, water soaked with dark borders. In the centre of these spots, pin head sized black fruiting bodies are seen. The fungus is soil and seed borne. The germination and growth of the fungus occurs best at 22-27° C. The disease development occurs at 20-30° C. In order to control the disease use healthy seed and treat the seed with Thiram or Captan @ 2.5g/kg seed. Spray the crop with hexacap (0.25%) at an interval of 15 days. Destroy all the wild hosts nearby.

Fusarium: The causal organism for this disease is *Fusarium* species. Mainly monoculture is responsible for higher incidence of this disease under polyhouse conditions. If possible, crop rotation should be followed in polyhouse to control the disease. Besides this, grafting of the cucumber seedlings over the resistant root stocks such as pumpkin can be effective.

Important insect pests of cucumber under protected cultivation

White fly: These are small fly like in their adult stage. The wings lie fairly flat over its abdomen, almost parallel with the leaf surface. Damage by the pest is caused by inserting the stylet like mouth parts through plant tissue into the phloem thus remove the plant sap resulting in yellowing of leaves which fade and dry away. The nymphs also secrete honey dew which makes the leaf sticky and shiny thus lead to the growth of sooty mould which interferes with the photosynthetic activity of the plant. White fly is one of the serious insect pests in modified naturally ventilated polyhouses. It can be controlled by using yellow sticky traps at a distance of 4 m^2 and by placing yellow painted water filled trays inside the polyhouse. Spray of *Boveria* is also affective in controlling the white fly population. All the unwanted plants in the polyhouse should be removed because these harbour the immature and the adult stages of the pest. Prevent white fly entry in the polyhouse by using the double door/annex door in the polyhouse. It can be controlled by spray of imidaclorpid (0.007%) followed by another spray of Nuvon (Dichlorvos) (0.008%) at an interval of 10 days. Avoid spray of imidachlorpid during fruit harvesting stage.

Mites: Mites cause lot of damage in cucumber particularly during the warm and dry conditions. Mites feed on the plant sap and prefer to live under the underside of the leaves. The adult mite is red in color measuring about 0.52 mm in length and 0.30 mm in width. All the stages have eight legs except for the larval stage which has six legs. Damage is caused both by nymphs and adults who rupture the plant cells by piercing, stylet like mouth parts and suck the cell material. Due to extraction of chlorophyll, the leaves show characteristic blotching and bronzing. Each feeding site is very small but soon the feeding marks become so numerous that they group together causing the leaf to turn yellow and to appear dusty. Severe damage to foliage reduces photosynthesis and the plant dies regardless of the presence of adequate water and nutrition. As the number of spider mites increase, they move to the upper surface of the leaves. The mites continue upward to the apical or the terminal leaves where they produce the silken webs and the use these as leaf bridge to cross to other plants. Entry of the pathogen should be prevented inside the polyhouse by following adequate sanitation etc. Mite problems are encouraged by high temperature and dry conditions, so sprinkler irrigation aid in control and in delaying build up of mite population. Dicofol (0.05%) or Avamectin (0.05%) or Magister @ 2ml/litre of water can be sprayed to control the mites, but it should be ensured that the spray must be directed to the lower side of the leaves. Application of neem based formulations at 12-14 days interval also help in keeping the crop free from mites.

Root knot nematodes: Nematodes heavily infest the cucumber inside the polyhouse when it is grown intensively. These are microscopic organisms which feed on or in plant roots causing gall formation in the roots. The root knot nematode (*Meloidogyne* spp.) infected plants show day time wilting and the plant growth is stunted. On uprooting the plants the roots with gall formation and egg massses are found. Various soil amendments like oil seed cakes and *Trichoderma* spp., *Paecilomyces lilacinus, Pseudomonas fluorescens* and *Arthrobotrys oligospora* help in reducing the nematode population. Steam sterilization of the soil of polyhouse and the growing media is another alternative. Carbofuran (Furadan 3G) granules @ 3 Kg a.i./ha are effective in controlling the nematodes however, the waiting period of 45 days is required.

Aphids: Aphids are generally small (1-3 mm) and soft bodied insects which vary in color from light green to rose. Some species of aphids are light yellow to dark green or almost black. Aphids cause damage in cucumber by feeding on phloem through their needle like stylet thus removing plant sap thereby reducing the plant vigour. They produce honey dew which encourages the growth of the sooty mould which in turn hampers photosynthetic activity of the plant. Aphids also act as vectors for the transmission of the various viruses. These can be controlled by spraying confidor @ 0.5ml/litre of water or Acetamiprid @ 0.5g/litre of water. Sanitation inside the polyhouse should be maintained and all the weeds which serve as alternate hosts should be removed near the polyhouse. Door should be properly closed and any unwanted crevices in polyhouse should be plugged especially during fall and spring to prevent the migration of the pest inside the polyhouse.

Leaf miner: The adults are minute (2 mm) grayish black fly with yellow spot on the thorax. The larvae are orange yellow without any leg while, pupae are oval shaped with yellow color. Females make small punctures on upper leaf surface with their ovipositor and lay eggs singly beneath the epidermis. The larvae cause the maximum damage by mining of the leaves, which result in the destruction of the leaf mesophyll. The mine becomes noticeable about 3-4 days after oviposition and becomes larger in size as the larva matures. The pattern of mining is irregular. Both leaf mining and stippling can greatly depress the level of photosynthesis in the plant. Extensive mining also causes premature leaf drop. Wounding of foliage also allows entry of bacterial and fungal growth. In order to control the pest, ensure that healthy seedlings free from infestation are planted in the polyhouse. Yellow sticky traps can be used to lower the adult population. Spray of contact insecticides at weekly intervals can help in checking the adults that will continue to emerge from puparia. Insecticides such as acephate (0.15%) or deltamethrin (0.0056%) may be used to control the pest. Spray of Fipronil @ 2ml/litre of water or neem based formulations can also help in controlling the pest inside the polyhouse.

Tobacco caterpillar: The fore wings of adult moths have beautiful grayish brown pattern with grey wavy lines. A single female lays about 300 eggs in clusters covered by brown hair. The full fed larva, upto 5 cm in length, is velvety black with yellowish green dorsal stripes and lateral white bands. They pupate in soil.

They are mainly foliage feeders, but occasionally damage fruits. Eggs are deposited by female moths that have made their way into the greenhouse and lay eggs in large masses. Heavy infestations may occur in isolated areas within the greenhouse, and timely spot treatments can often provide effective control. It is a very destructive pest if not controlled at early stages. Its control can be achieved by spraying specific NPV (Sl NPV) @ 2ml/litre or Ecalux @ 1.5 ml/2litre of water.

Screen mesh sizes needed to exclude major greenhouse pest species*:

Insect-Pest	Hole size (micron)	Mesh (Number of threads per linear inch)
Leafminer (*L. trifoli)*	610	34
Cotton whitefly (*B. tabaci*)	462	42
Aphid (*M. persicae)*	340	52
Greenhouse whitefly (*T. vaporariorum*)	290	58
Thrips (*Thrips* spp.)	192	76

*(Sood, 2010)

References

Baile, A. 2001. Trends in green house technology for improved climate control in mild winter climates. *Acta Hort.* **559:**161-67.

Castilla, N. J., Hernandez, A. F. and Abou-Hadid. 2004. Strategic crop and greenhouse management in mild winter climate areas. *Acta Hort*. **486:**301-305.

Goncharova N A, Van der Vlist J A A M and Verstegen. 2004. Changes in horticulture sector in the Nethetlands. *Acta Hort* **655**: 319–31.

Gupta, R. K. 2010. *Green house construction and designing.* Compandium of Summer school on Protected cultivation, Dep. of Vegetable Science and Floriculture, CSKHPKV, Palampur, India. p 13-19.

Hickman G W. 2011. A review of current data on international production of vegetables in greenhouses. wwwcuestaroblecom, 73 p.

Indira, P., Gopalakrishnan, T. R. and Sreelatha, U. 2004. *Rain Shelter for off season vegetable production.* National Agricultural Technology Project. Dep. of Olericulture, Kerala Agricultural University, Vellanikkara, Thrissur, India. p-8.

Jensen M H. 2002. Controlled environment agriculture in deserts tropics and temperate regions – A world review. *Acta Hort* **578**: 19–25.

La Malfa G. and Leonardi, C. 2001. Crop practices and techniques: Trends and needs. *Acta Hort.* **559:**31-42.

Mathew, E. K., Varghese, A., Nair, P. G. and Menon, J. S. 2011. *Protected cultivation.* Precision farming development centre, Kelappaji College of Agricultural Engineering and Technology, Kerala Agricultural University, Tavanur, India. p-18.

Pathania, N. K. 2010. *Protected cultivation - an overview.* Compandium of Summer school on Protected cultivation, Dep. of Vegetable Science and Floriculture, CSKHPKV, Palampur, India. p 1-5.

Peter, K.V. 2012. Final report of task force on vegetables. In: *Proceedings of the member secretary State planning board;* 11 september, 2012. Goverment of Kerala, India, pp 4-6.

Sabir, N. and Singh, B. 2013. Protected cultivation of vegetables in global arena: A review. *Indian J Agric Sci.* **83**(2):123-35.

Sharma, P., Pathania, N. K. and Pardeep-Kumar. 2010. *Hi tech production of parthenocarpic cucumber.* Compandium of Summer school on Protected cultivation, Dep. Of vegetable science and floriculture, CSKHPKV, Palampur, India. p 85-92.

Singh, B. 2005. *Protected cultivation of vegetable crops.* Kalyani Publishers. Ludhiana 1st ed. p 168.

Sood, A.K. 2010. *Integrated Pest Management Under Protected Environment: Principles and Practices.* Compandium of Summer school on Protected cultivation, Deptt. of Vegetable Science and Floriculture, CSKHPKV, Palampur, India. p 216-226.

Varghese, A. 2013. Personal communication

Vijayan, N. 2012. Vegetable scenario in Kerala (oral presentation). In: *Proceedings of the Member Secretary State Planning Board;* 26 September, 2012. Goverment of Kerala, India.

Wittwer, S. H. 1993.World-wise use of plastics in Horticulture. *Hort. Tech.* **3**:6-27.

Wittwer, S.and Castilla, N. 1995. Protected cultivation of horticultural crops worldwide. *Hort Technology,* **5**:6-23.

14

Protected Cultivation of Spices: Realities and Challenges

A.K. Sadanandan and K.V. Peter

Abstract

Protected cultivation technology in black pepper and cardamom offers distinct advantages over conventional farming. Production of quality planting materials irrespective of the season was made possible. Such propagating materials can be well maintained under polyhouse, irrespective of the seasons and can be used judiciously for replanting/ gap filling. In this era of climate change, protected cultivation assumes significance. In black pepper and cardamom, rapid multiplication techniques of planting materials were made possible. A novel idea of growing pepper, different from conventional system as "bush" that can be grown under potted and field conditions were also developed. Nutrition and water management studies could be undertaken with precision. In cardamom, quality planting material production and high quality capsule production are made possible all throughout the year by the adoption of protected cultivation. For optimizing production, prescriptions of fertilizers, using DRIS norms were developed for black pepper and cardamom based on critical soil and foliar nutrient concentrations. Typical deficiency symptoms of major and micro nutrients on black pepper were characterized to rectify the deficiency. Technology for propagation and cultivation of bush pepper in pots, under protected cultivation was also developed and popularized. Technology for using organic manures, non edible oil cakes, biofertilizers towards production of high quality zero defect black pepper and cardamom was also developed and popularized.

Keywords: Protected cultivation, Pepper and Cardamom, Organic Spices

Introduction

World over there is growing interest in controlled environment agriculture and horticulture. In view of the consistent population growth and current poor economic situation; better systems of intensive farming assume significance to meet the growing demand of food, fruits, vegetables and spices. In this context protected cultivation offers tremendous scope of augmenting production, productivity and quality of spices. The aim and objectives of protected cultivation technology are to provide a good growing environment akin to the natural condition or far better for successful growing with minimal use of harmful chemicals for production of high quality plants and produces round the year. In this era of climate change, protected cultivation ensures continuous supply of fruits, vegetables flowers and spices all throughout the year. Controlled environment/techniques would therefore be essential to promote spices as backyard/rooftop cultivation to get zero-defect fruits, vegetables and spices particularly in the context of prevailing varied range of agro-climatic conditions in India (Sadanandan, 2012).

The greenhouses are structures covered with transparent to translucent materials such as LDPE, FRP, Polycarbonate sheets, agro-net etc. which allow the solar radiation to pass through, but trap the thermal radiation emitted from the objects within. This creates a congenial environment for the plants to growth. Energy received from the sun is converted to heat within the greenhouse during the day as well as used to evaporate water during the normal transpiration of plants. The various parameters influencing plant growth like light, heat, humidity, CO_2 and moisture could be controlled in these structures. Studies carried out so far showed ample scope for adoption of protected cultivation technology for production of zero-defect spices and maintaining its intrinsic quality.

Principles of Protected Cultivation

Protected cultivation is a system of farming in which crop plants are grown under controlled environmental conditions. It also gives socio-economic gains. Greenhouse cultivation and other modes of controlled environment cultivation were evolved to create and/or stimulate favorable micro-climates, which favour crop growth, development and production which could be made possible all through the year or part of the year. Greenhouses and other associated technologies for controlled environment and plant production system are associated with the off-season production of ornamentals and/or foods of high value, particularly in cold climate areas where outdoor production is not possible owing to adverse condition.

Environmental parameters controlled are temperature and humidity by providing heat to overcome extreme conditions. Environment control includes: cooling to mitigate excessive temperatures, light control either shading or adding supplemental light, carbon dioxide levels, relative humidity, water, plant nutrients and pest and disease control. The scientific principles of design of greenhouse should be based upon, controlled environment for plant growth. This requires far more capital investment per unit area than field agriculture and thus must essentially be correspondingly more intensive to justify investment costs.

Greenhouses are structures covered with a transparent material for admitting natural light for plant growth. The main components of greenhouse like structure, covering/glazing and temperature control systems need proper design for healthy growth of plants.

Ultra violet resistant low density polyethylene (UVLDPE) single film cladding of 200 micron thickness is sufficient for Naturally Ventilated (NV) greenhouse and fan and pad (FP) greenhouses.

The location and orientation of the greenhouse should avoid falling of shadow on the adjacent greenhouses. To avoid shading effect from one green house to another greenhouse, that should be oriented east to west. Wind direction and latitude are also to be considered while locating green house. If the greenhouse is naturally ventilated, the advantage of natural wind direction has to be taken to the maximum extend possible. The maximum dimension (length) of greenhouse should be perpendicular to the wind direction especially in summer. For fan and pad greenhouse, the natural wind direction should be same as the air blown by fan.

The plant is required to take up large amounts of water and nutrients, with a relatively small root system, and manufacture photosynthesis for a large amount of produce per unit area with a foliar system relatively small in relation to required production. **Water needs shall be met by drip and nutrient need by fertigation.**

As regards socio-economic aspect, agriculture as a profession is not attractive for the educated youth, which is attributed to the drudgeries associated with field work. To motivate the educated youth, agriculture has to be developed to be a remunerative and drudgery-less industry as competitive as any other industry using improved agro-technologies. Then only a sense of pride will be associated with agriculture. In this context protected agriculture assumes significance.

Status and Strategies of Indian Spices

Indian spices are a multibillion dollar industry. India can boast as the monopoly supplier of spices, spice oils and oleoresins the world over. In respect of spice oils and oleoresins, curry powders, spice powders, spice mixtures and packed spices too, the country is in a formidable position. India exports more than 0.40 million tons of spices annually and accounts for around 48% of the global export volume and 44% of the export value. This includes Chilli, Black Pepper, Cardamom, Ginger, Turmeric, Tree spices like Clove, Nutmeg & Mace, Cinnamon, Cassia, Asafoetida, and Tamarind. Seed spices like Coriander, Cumin, Fennel, Fenugreek, Celery, Garlic, Vanilla, Saffron, Oils & Oleoresins, Mint products and Curry Powder are also exported from the country.During the year 2007-08, India's export earnings from spices surpassed the $ one billion mark for the first time in the history and registered an all time high both in terms of quantity and value in spice exports. The export of spices from India during the year has been 4, 44250 ton valued at $1101.80 million registering an increase of 39% in value over 2006-07.Kerala, which is known as the Spices Garden of India, is the home to all major items of export like Pepper, Cardamom, Ginger, Turmeric, Curry powder, Spice oils and Oleoresins, Vanilla, Nutmeg and mace.During 2006-07, Kerala's share to total Pepper export from the country was above 88% both in quantity and value. The export of Vanilla from Kerala increased from 22.18 tones in 2005-06 to 56.75 tones in 2007-06 and earned Rs.635.39 lakhs against Rs.405.35 lakhs in the previous year, achieving 56.75% increase in value and 155.9% increase in quantity.The major destination of spice exports is USA followed by the European Union, Malaysia, China, Singapore, Sri Lanka, Japan and the Middle East.

Realities and Challenges

i. **Global warming:** Global warming fed by greenhouse gases from industry, transport and shifting land-use poses a long-term threat to India's prosperity, health and food output. Country faces extremely grim ecological and environmental conditions under the impact of continued global warming that changes to India's regional environment. But that possible fall can be held in check by improved crop choice and farming practices, protected farming, as well as increased irrigation and fertiliser use. Climate change will lead to severe imbalances in India's water resources within each year and across the years. In most areas, precipitation will be increasingly concentrated in the summer and autumn and floods and droughts will become increasingly frequent. India's efforts to protect vulnerable coastal areas with embankments are inadequate.

   There are sure to be a shift in India's cropping patterns as well. Shrinking water availability could lead to a marked decline in agricultural production, productivity and quality.

ii. **Food availability:** Amid the current low level of per capita food availability, India faces the daunting challenge of feeding its large and growing population with rising purchasing power. With a population of over 1200 million expanding at about 1.5 per cent a year and, thereby, adding roughly 18 million mouths to be fed every year, the demand for food can only keep rising. The demand for Spices, oleoresins and curry power for food preparation will atomically be increased. The vantage position of agriculture to deliver 'growth with equity' is by now well recognized.

iii. **Food prices rise:** With population growth and rising incomes, demand for food growing higher, rapidly changing food habits especially among the middle class are set to transform the composition of the traditional food basket. While supply growth trailing, demand growth and production costs escalating, food prices will continue to rise significantly. This definitely has adverse consequences for the poor, whose food and nutrition intake already stand diluted in the stake of nutritional security. With higher incomes, the emerging middleclass can afford to consume more food, fruits, vegetables, meat and spices which require much more land and water to produce.

iv. **Food safety risks:** As people's demand for more processed foods increases, food safety risks along the supply chain also increases. These risks may also increase with more intensive crop and livestock farming through contamination with chemicals and/or pathogens. Interestingly, while intensifying food production can boost the country's food security and serve the poor, it can also cause land degradation, depletion of water resources, water pollution, and new pests and disease problems. We have to work towards economic development, environmental protection and greater social welfare. India is already facing the serious challenge of provision of adequate and nutritious food and water at affordable prices. In these contexts, protected farming assumes significance to produce zero-defect food.

v. **Experimental methods:** The research work pertaining to Spices in the capacity as Project Leader/ Associates carried out at IISR / KAU/ NGOs by the authors in relation to Protected Cultivation are reported and discussed briefly.

A. Nursery techniques of raising plants for field planting

Black pepper (*Piper nigrum)*

In Spices, raising rooted cuttings in commercial scale is possible only under protected/ greenhouse condition, having structures covered with transparent to translucent materials such as LDPE, FRP and Polycarbonate sheets etc. which allow the solar radiation to pass through, but traps the thermal radiation emitted from the objects within. To the net house, fitting Foggers provide a superior and consistent controlled temperature and humidity for accelerated growth of saplings. Under this system, four types of nurseries have been standardised

i. **Three noded rooted pepper cutting:** Pepper is propagated vegetative from cuttings made from vines. Runner shoots emerged at the base of mother vine is selected and coiled to prevent them from striking roots in the soil. Such runner shoots are separated from the vines in Feb-Mar. The middle one third portions of runner shoots are preferred. Very tender and too hard portions of vine are avoided. Leaves are clipped off leaving a small portion of the petiole on the stem. The shoots are cut into pieces of 2-3 nodes in each. Dipping the Lower cut end (up to 2 cm) of the cuttings in 1000 ppm solution of 3 Indol Butyric Acid (IBA) or IAA for 45 seconds enhanced root formation and development.

Polybags 20x10cm are arranged in lots of 200-300 in raised beds and protected airtight with transparent poly sheet 200gauge for 30-40days to retain the humidity inside (Fig 1). In large nurseries, bio-*pandals* are made using bamboo, coconut plated leaf for housing poly bags nursery. The cuttings are well protected from direct sun. Filtered light and frequent watering are needed in the nursery to maintain a humid and cool atmosphere around the cuttings. Watering is done two or three times a day. Saplings will be ready for field planting in 3-4 months.

ii. **Poly-bag nursery technique:** This is a system of raising rooted black pepper plants continuously under protected green house condition. In this system, 2-3 months old poly bag pepper vines having 3-5 nodes are allowed to strike roots at every node in potting mixture, taken in poly bags, arranged underneath, at every node of the parent vine (Fig 2). The vines are allowed to grow for about 3months by which time 12-15 single rooted single noded saplings can be raised in arrow. This process is allowed to continue till about 20 nodes get rooted. The first 10 poly bags with rooted nodes are separated by cutting the internodes and the rooted nodes are pressed into the bag and allowed to root profusely. The auxiliary bud from each node in the bag sprouts and develops into 4-5 noded cutting in 2-3 months. After a gap of another 20-30 days, the rest of the bags are also separated. A protected greenhouse of about 140 sq metre dimensions can generate 20-25 thousand well rooted saplings annually.

iii. **Rapid multiplication technique.** This system of raising nursery also requires overhead protection against adverse weather conditions prevalent during growing season. Split halves of bamboos with septa having 8-10cm diameter and 1.25m length are fixed at 45degree angle on a strong support (Fig. 3). In areas where bamboo is not freely available, split halves of PVC fitted with septa at 30cm intervals can be used.

Rooted pepper vines are planted in the trench at the rate of one cutting each for one bamboo/PVC splits. The lower portions of the bamboo/PVC splits are filled with a rooting medium (coir dust-FYM mixture 1:1) and tied the vines with the bamboo. As the vines grow this process is continued. But each node is pressed to the rooting media and tied with the bamboo. Once the vine reaches the top, the terminal bud is nipped off. The vines at third node above the base are crushed. This is done to activate the auxiliary bud.

Fig. 1.

Fig. 2.

Fig. 3.

Fig. 1, 2 & 3: Bush black pepper (cv. Panniyur-1 and Karimunda) cultivation in green house with mist irrigation

After about 10 days the vines are cut at the crushed point and removed from the rooting media and each node is cut and separated. Each node cutting with the bunch of roots intact is planted in individual poly bag filled with potting mixture (red soil: sand: FYM). A protected greenhouse of about 140 sq metre dimensions can generate 25-30 thousand well rooted saplings annually for field planting.

iv. **Bush pepper sapling production technique:** This system involves planting lateral branches (Plagiotropes) of pepper; the vine can be grown as a bush. One year old lateral branches with 3-5 nodes are planted in poly bag 20x10cm in potting mixture in the nursery during March-April for pre rooting after treating the cut ends with 1000 ppm solution of IBA/IAA for 45 seconds. Rooting percentage of laterals is less than 20%.

Rooting in coir pith taken in 40x30cm poly bag will give 40-50% germination in 45 days (Fig 4). Only well rooted and established plants are to be used for transplanting (Sadanandan 2005). Rooted cuttings are planted at 1-2 per pot/pit. The bush nature of the plant will have to be ensured by pruning off the runner/ hanging shoots if any. The potted plants are to be kept preferably under partial shade. Under good management, bush pepper can be retained in 30cm pots for 10-15 years.

B. Foliar diagnosis and nutrient deficiencies

A study to develop nutrient deficiency symptoms for black pepper was undertaken in green house condition. Typical characteristic symptoms were drawn out. This eases the optimum use of nutrients to correct the deficiencies, to increase production and quality up gradation. (Sadanandan *et al.*, 1996).

C. Nutrition of bush pepper

Technologies for growing black pepper as a bush in pots under protected cultivation, and growing in fields under agro-net and their nutritional requirements were standardized. (Sadanandan et *al.*, 1997)

D. Bush pepper in pots

Bimonthly application of NPK fertilizers @ 1.0: 0.5: 2.0 g (10 kg of soil) increased the spiking intensity, yield and berry volume. The yield level was around 150 - 200 g pot^{-1} $year^{-1}$ (Sadanandan and Hamza, 1996). Studies on the foliar nutrition of Zn showed that spraying ZnSO4 @ 0.5% on the 60th day after flowering increased the yield by 19%. The oleoresin production was also enhanced due to $ZnSO_4$ spray (Sadanandan and Hamza, 1996). Studies on the use of biofertilizers for black pepper revealed that a combination of Azosprillum with Phosphobacteria and VAM increased the biomass, nutrient uptake and dry matter production. (IISR Annual Report 1996-97).

E. Protected Cultivation of bush pepper

Studies on the protected cultivation of bush pepper revealed that in 140 sq.m. 350 numbers of bush pepper can be fited. The light intensity under the system was around 50-60%. The investment can be realized during the fifth year of planting with a net profit of Rs. 5,000 per year. There was no *Phytophthora* disease incidence. (IISR Research Highlight 1997-98).

F. Organic farming

Technology for organic farming in black pepper was developed. Studies both under green house and field conditions revealed that bi monthly application of chicken/goat manure was superior to FYM regarding the yield (57%) and quality of black pepper. Applications of 30g Neem cake or 15g Groundnut cake at bimonthly intervals was comparable to NPK fertilizers at the recommended rate, regarding nutrient uptake and yield. There was no *Phytophthora* disease incidence. (IISR Annual Report 96-97). Introduction of earth worms (*Phretima heterochaeta* and *Eudrillus eugenia*) *in situ* in bush black pepper pots supplemented with FYM and *Glyricidia* leaves increased the fertility, and yield response (IISR Annual Report 1996-97). Coir compost (*Terra care*) in combination with FYM, vermicompost and biofertilizers retained the soil moisture and increased the yield of black pepper by 176%.

i) Cardamom (*Elettaria cardamomum*)

A. Methods of propagation

Cardamom is propagated both by seedlings and vegetative methods under protected conditions like pandal or green house.

a. **Vegetative propagation techniques of Cardamom:** For this method, rhizomes with not less than 3 shoots are used. Plants propagated vegetatively come to bearing one year earlier than the seedling propagated plants. But this method has the disadvantage of spreading the '*katt*' disease to widespread diseased areas. It would be safer to use seedlings for propagation.

   Studies carried out at Kalpatta in Wyanad distric of Kerala showed that for rapid vegetative multiplication of Cardamom clones for field planting is the most economic (fig. In this technique elite clones of cardamom with 2-3 shoots are planted at one metre spacing following square system of planting. Under good management, a green house of 140 Sq metre can produce enough quality planting material for planting one hectare (Sadanandan 2005).

b. **Nursery:** Ripe capsules are collected from high yielding plants during Sept-Oct. The seeds are extracted by gently pressing the capsules. The extracted seeds are washed in cold water four times to remove the mucilaginous coating on them. The washed seeds are drained and mixed with ash and allowed to dry in shade for 2 or 3 days. The seeds should be sown in nursery having overhead protection with shade net.

c. **Primary nursery:** The nursery site is selected in open, near a source of water. Beds of size 6x1x0.3m are then prepared. Seeds are sown on the bed in lines. Ten grams of seeds are required for one sqm area. Sixty grams of seeds will be required for a nursery bed of 6sqm. The nursery bed is mulched with dry grass to a thickness of about 2 cm. These have to be watered daily in the morning and evening. The mulch should be removed on commencement of germination. The seedlings have to be protected by providing shade pandals. During June-July, the primary seedlings are transplanted to the secondary nursery.

d. **Secondary nursery:** During June-July, the seedlings from the primary nursery are transplanted at a spacing of 25-30cm in the secondary nursery provided with overhead Shade pandal. Bio-mulching the bed with dry leaves is a must to conserve soil moisture. Regular watering during dry months, weeding, application of fertilizer, control of pests and diseases and mulching are the essential operations for the maintenance of the secondary nursery. One month before uprooting seedlings, the overhead pandal is removed in 2-3 stages wherein encourage better tillering of saplings.

e. **Polybags nursery:** Polybags 20x10cm can be used for raising secondary seedlings. For such nurseries, seeds are to be sown in beds in primary nurseries in Sept and transplanted into Polybags in Dec-Jan. These seedlings are moved to green house and maintained till it is ready for field planting in June-July. In this case, the nursery period could be reduced to 10 months.

B. Rapid multiplication

For rapid multiplication of cardamom clones for field planting, Mudigere (Karnataka) centre has developed a technique. This gives a multiplication ratio of 1:15 in 10-12 months of planting. The mothers' clones are planted in trenches of 45cm^3 at a row spacing of 0.9m and fertilized with organic (15 t ha^{-1}) and inorganic fertilizers @ 75:75:150 kg ha^{-1} at bimonthly intervals (AICRPS Research Highlights -1997-98).

C. Protected cultivation

Studies on the agro-net protected cultivation of cardamom revealed that in 140 sq.m, 140 numbers of Cardamom clumps can be planted. The light intensity under the system was restricted to around 50-60% with agro-net. The investment can be realized during the fifth year of planting with an annual net profit of Rs. 5,000/-. Interestingly disease and pest incidence were negligible

Organic Spices Production

Organic Farming is receiving attention world over in the recent past. Globally 26.5 million ha in 5.6 lakh farms are currently managed organically. Area under organic farming in India is likely to grow over ten-fold to one crore ha in the next five years on buoyant domestic market and increased farmers' interest to ensure zero-defect food, at sustained yield level at moderate costs. The Indian organic product market is currently pegged at Rs 1,500/- crore and this would likely to go up to Rs 10,000/- crore in the next five years. Organic Agriculture (OA) is expanding globally due to increased consumer interest in organic food, environmental protection and EU polices. Global trade in organic products is projected to increase to $102 billion by 2020, due to increasing demand in Australia/Oceania, Europe, Latin America, USA, Asia and Japan.

In India, Projects on Organic Farming in Spices were conceptualized and implemented at ICAR Institutes, SAUs and Research Centres of Commodity Boards and NGOs intensively since 1990s. Guideline in organic production was followed wherever possible.

The issues related to OA with reference to inputs such as organic fertilizers (plant residues-compost/ vermin compost, animal manures/ de-oiled cakes which contains nutrients in complex organic forms, bio-fertilizers etc), their impact on soil quality indicators, quality of major spices and its economics in important spice crops under humid tropical conditions of South India were studied:

In black pepper the effect of commercial organic manures was studied (based on nutrient equivalents) and compared with conventional NPK chemical fertilizers and farmers practices. Investigations revealed that irrespective of the sources, application of organics altered soil pH, nutrient availability in the soil and crop uptake. Among the organic manures poultry manure followed by goat manure was significantly superior with regard to nutrient uptake yield and enhanced piperine and oleoresin content of black pepper (Sadanandan 1999) The OA technology was further extended to farmer's fields in three important black pepper growing States of South India viz., Kerala, Karnataka and Tamil Nadu during 2000- '03 under the Emeritus Scientist Project (Sadanandan2003). Application of the recommended dose of N as organic (FYM + de-oiled cake of neem and pea-nut, P as phosphate rock + bone meal and K as wood ash, in conjunction with bio-fertilizers Azospirillum and Phosphobacteria each applied @ 40kg/ ha per year enhanced the spiking intensity of the pepper vine, yield and nutrient uptake. The FYM application significantly improved physico-chemical properties of soil (Sadanandan, 2003). In Organic Agriculture, organic manures play a significant role in improving soil and crop quality. There is, however 15-25 per cent reduction in crop yield, while adopting organic agriculture that should be compensated by premium pricing of organic produces. Further research is called for in understanding soil health quality- plant relationship under protected cultivation.

Experimental evidences showed that there is ample scope, under protected cultivation system, for fine tuning production technology for organic farming including planting materials production.

Summary, Conclusion and Outlook

India's Spices is a multibillion industry. The country has an exciting future in production and marketing of spices to meet the emerging global needs and domestic consumption. The export market is becoming ever more competitive and stringent and favors 'zero' defect Spices for marketing. To meet the emerging scenario, development of appropriate technology is imperative.

Protected cultivation technology is an area that has immense potential to be exploited by spice industry. The results obtained from the studies suggest that protected cultivation is a feasible technology and is a handy tool particularly for small and marginal farmers for production of quality planting materials, growing off season crops viz. Pepper, cardamom, coriander etc.

In major spices growing tracts, there is a sharp decline in soil quality. Recurrence of drought is yet another natural phenomenon in the major producing centers. Rejuvenation of the existing plantation, replanting senile gardens for which sound nursery techniques assume importance. Development of appropriate technology for sustainable spices production and integrated nutrient and water management techniques are warranted so that the delicate ecosystem is not disturbed anymore (Peter and Sadanandan, 1998). The authors bagged the prestigious Rafi Ahamed Kiwai award in 1998 for the innovation.

In the above background the significant technologies were developed *viz.* sound nursery practices for the development of quality pest and disease free planting materials of major spices likes pepper, cardamom, tree spices like clove, nutmeg etc. under protected cultivation system. Prescription of optimum nutrient level for sustainable production of spices is yet another area of significance. Green house studies have shown that, coir pith as compost- a byproduct of coir industry, when used as a source of organic input enhanced the water holding capacity of the soil, reduced the bulk density and enhanced root growth and crop yield. The technology for production of zero- defect Spices through IPNM/ Organic farming was developed. The use of bioagents viz., *Phosphobacterium, Azosprillum* and VAM as nutrient supplementary in the system, helps in scaling down the use of chemical fertilizers which improved the soil quality indicators that optimise crop yield. To meet the requirement of urban households and for homestead cultivation, techniques of growing black pepper as a bush using plagiotrophic shoots were developed. The nutrition and water requirement for such a system were standardized under protected cultivation system.

Protected cultivation of bush pepper for adoption in commercial scale is yet another technological attainment. The Science and Technology Entrepreneurs Development (STED) Govt of Kerala has come forward through a project "Bush pepper for urban women" costing Rs. 7.0 lakhs in all the 14 districts of Kerala. Bush pepper programmes are also propagated through the Dept. of Agriculture, Govt. of Kerala/ Pondicherry (Mahe). The techniques of raising bush pepper are being adopted by a number of nursery men.

Bioprotection

To attain sustained production, productivity and quality up-gradation in Pepper and cardamom, and to make farming a viable vocation and livelihood support to the farmers, biotic and abiotic protection play key roles.

While abiotic protection is the negative impact of non-living factors on the living organisms in a specific environment, biotic protection includes such living organisms which cause disturbances such as virus/ fungi or harmful insect pests.

Both pepper and cardamom are very sensitive to a number of physical environmental factors which include altitude, latitude, radiation, temperature, rainfall, humidity, wind, seasons etc. Recent studies have shown that these factors have their unique influence not only on the absorption of nutrients and water but also affect production, productivity and quality of produce (Sadanandan, 1998).

Amongst the factors, sunlight is an important factor that influences growth of pepper and cardamom. The radiant energy is converted into chemical energy by the plant during photosynthesis. This is the reason why shade loping is scrupulously followed both in pepper and cardamom just before the onset of monsoon, and the crop is exposed to the sky during the monsoon (Sadanandan 1998). Photoperiod also influences pigmentation. Sunshine is also essential for pollination, development and green colour to cardamom capsules. The trade variety Alleppy Green Extra Bold (AGEB) cardamom capsule fetches premium price in the market. Rainfall is an important factor that indirectly affects flowering and fruit setting by triggering the process of germination of pollen grains and even fertilization of black pepper.

Biotic protection includes protection against living disturbances caused by virus, fungi, harmful insect pests etc. Biotic protection, both for crop nursery and plantations towards sustainable crop production is inevitable. Recent studies have shown judicious adoption of bio- and/or chemical protection has distinct advantage over conventional protection.

In the nursery, spraying/drenching the plants at fortnightly with 2% *Pseudomonas fluoresceneas* suspension, phyto-sanitation, changing the nursery site once in two/ three years to avoid population built up are measures which give ample protection.

Spraying and drenching with potassium phosphonate 0.3% or metalaxyl-macozeb (0.01%) at monthly interval is unavoidable. Alternatively spraying with (1%) Bordeaux mixture and drenching the soil with copper oxychloride (0.2%) at monthly interval will control Anthracnose, leaf rot and blight. Viral infection, characterised by mosaic/ yellow mottling of leaves are checked by periodic removal and destruction of affected plants. Plants are protected against insect vectors such as aphids/ mealy bugs by spraying with dimethoate (0.05%) recent studies have shown that biotic and a biotic factors are dreadful if they occur together, and lead to total crop loss.

References

Anomyanone Growing soils of South India. Paper accepted for presentation at PLACROSYM - XIII Coimbatore, 16-18 Dec. 1998.

Peter, K.V. and Sadanandan, A.K. 1998. Report for the award of Rafi Ahamed Kiwai Award 120pp.

Peter, K.V., Sadanandan, A.K. and Prabhakaran Nair, K.P. 1997. Thrust areas of Spices Research & Development in Keralam during IX th Five Year Plan. Report Presented at the Task Force on Research, Education and Training, State Planning Board, Kerala. Jan. 1, 1997.

Prabhakaran Nair, K.P. Sadanandan, A.K., Hamas. and Jose Abraham. 1997. The importance of Potassium Buffer Power in the growth and yield of cardamom. *J. Plant Nutrition*. **20**(7 & 8):987-997.

Sadanandan A. K. 1998. Scientific cultivation practice in pepper towards improving productivity. Invited paper presented at 105th UPASI Annual Conference, Coonoor, and 7th to 8th Sep. 1998. 6p.

Sadanandan A.K. 1998. Project Document/EFC Memo of AICRP on Spices. 35 p.

Sadanandan A.K., 1998. QRT Report of IISR, Calicut 1992-1996. 108 p.

Sadanandan AK 2013. Roof-top Farming of Spices Home stead farming of Spice. Paper presented at State conference on Jan 12, 2013.

Sadanandan AK 2002 Project Report of Emeritus Scientist, IISR Calicut 198p.

Sadanandan AK 2006. Organic Farming in spices DASD, Calicut pp 32.

Sadanandan, A. K. 1998. Research Accomplishments in Spices. Paper presented at the Rural convntsion of KarshakaSree 1998. Agriculture Seminar organized at Kannur by Malayala Manorama during 20th Nov. 1998.5p

Sadanandan, A.K. 1998. Black Pepper Production Technology. Paper presented at UPASI congress at Bangalore on 3rd Nov. 1998. 8p.

Sadanandan, A.K. 1998. (In) Descriptors for Black pepper *(Piper nirgrum .L) IPGRI*, Rome, Italy, 39 p.

Sadanandan, A.K. 1998. Production and Processing Technology of Vanilla (*Vanilla planifolia*). In: *Natural Resources of Kerala.* (Eds.) K. Balachandran Thampi, N.M. Nair and C.S. Nair. PP: 309-311.

Sadanandan, A.K. and Hamza, S. 1996. Organic and inorganic fertilizer management for sustainable spices production in India. Paper presented at the Summer Institute on "Nutrient and Pesticide residue management in horticulture crops for yield and quality for domestic and export market." IIHR, Bangalore, 30 Aug., 1996.

Sadanandan, A.K. and Hamza, S. 1996. Studies on nutritional requirement of Bush pepper (*Piper nigrum L.*) for yield and quality. (Abs.) Paper presented at PLACROSYM-XII, RRII, Kottayam. Nov. 27-29, 1996. p. 48.

Sadanandan, A.K. and Peter, K.V. 1997. Technology for Spices Production. Paper presented at the Working group on Horticulture development for the formulation of IXth Five Year Plan. Pondicherry. 19 April 1997.

Sadanandan, A.K. Hamza, S. and Johny, A.K. 1998. A decade of potassium nutrition research of spice crops in India. *J. Potassium Research* (In press).

Sadanandan, A.K. Krishnamoorthy, K.S., Kandiannan, K and Korikanthimath V.S 1998. *Water and Nutrient management for sustainable production and quality of Spices*. (Proceedings of National Seminar, 5-6 Oct. 1998 Madikeri) Indian Society for Spices Calicut, Kerala

Sadanandan, A.K., Bhargava, B.S. and Hamza, S. 1996. Leaf nutrient norms for black pepper (*Piper nigrum L.*) using DRIS. *J. Plant. Crops.* **24** (Suppl.): 53-59.

Sadanandan, A.K. and Hamza, S. 1998. Organic farming on yield and quality of Spices in India. (Abs) 16th World Congress on Soil Science, Montpellier, France during 20-26 August, 1998 pp-738.

Sadanandan, A.K. and Sarma, Y.R. 1996. Report on the Off - Campus Training Programme on Spices held at Gangtok from 23-29 May, 1996. 6 p.

Sadanandan. A.K. 1998. AICRP on Spices. In 50 years of Horticulture Research (Eds.) Ghosh S.P., Bhatnagar, P.S. and Sukumaran. N.P. Division of Horticulture. ICAR KAB-II New Delhi, pp 207-214.

Sadanandan. A.K., Vasugi, C. and Johny A.K. 1998. AICRP in Spices. Annual Report 1997-98. 98 p.

15

Protected Cultivation of Floricultural Crops

S.S. Sindhu

Abstract

Floriculture is the fast emerging sector in horticulture sector and people are changing very fast from the traditional field crops to flowers because of high returns per unit of area and better price. The flowers are being used during all the occasions. Though the art of growing flowers is not new to India, large scale commercial cultivation and protected cultivation in polyhouses is relatively new in India. With the opening up of world market in the WTO regime, there is a free movement of floriculture products worldwide. India is having a better scope in future as there is a shift in trend towards tropical flowers and this can be gainfully exploited by countries like India with high amount of diversity in indigenous flora. The major flowers grown under protected cultivation are rose, chrysanthemum, gerbera, carnation and lilium. Even some of the flowers are location specific and made a dent on the state economy, the best examples are carnation in Himachal Pradesh and gerbera in Maharashtra. Now with market trend is changing and biotechnological tools must be exploited for the crop improvement to evolve novel varieties with desirable colour traits. The other major area of work is on protected cultivation of cut flowers which is coming up on a large scale and value addition in floriculture i.e. extraction of pigments, essential oils, pharmaceutical and nutraceutical compounds have emerged as frontline avenues in floriculture. We have many challenges ahead like trained HRD, quality planting material, technical knowhow, limited choice of the varieties and products in the market to compete in global market.

Keywords: Protected cultivation, Rose, Chrysanthemum, Gerbera, Carnation, Lilium

Introduction

The floriculture industry, at present, comprises of cultivation and trade of cut and loose flowers, potted plants, foliage & bedding plants and dried flowers under open field as well as under protected environment conditions. Globally, more than 145 countries are involved in cultivation of Floriculture crops. Though the art of growing flowers is not new to India, large scale commercial cultivation and protected cultivation in polyhouses are relatively new in India. Enormous genetic diversity, varied agroclimatic conditions and versatile human resources offer India a unique scope for diversification into new avenues which were not explored to a greater extent. India also has a strong dry flower industry, which contributes a major share to the overall trade. Other segments like fillers, potted plants, seeds and planting materials, turf grass industry and value added products also contribute a share in the overall growth of floriculture sector.

Cut flower production is a growing business in most Asian countries as the result of higher per capita income of people. Technology for production is well known and has been utilized to the maximum in many countries. As a consequence, cut flowers are commercially produced in larger quantities in these countries resulting in surpluses during the normal periods of blooming, the end result of which is a lower price for cut flowers. On the other hand, such prices are much higher during the periods of needs such as certain special occasions or off-season.

Specific-Date Production

The demand for cut flowers is quite high during certain occasions, such as Christmas Celebration, New Year's Day, Mother's Day, Memorial Day, Valentine's Day, Graduation Exercise Day, Buddhist Lent's Day, etc. Thus, it will be profitable to the farmers if they could produce cut flowers on these occasions.

What is Greenhouse Technology?

Today about 92% of plants, raised by man, are grown in the open field. Since the beginning of agriculture, farmers have had to cope with the growing conditions given to them by Mother Nature. In some of the temperate regions where the climatic conditions are extremely adverse and no crops can be grown, man has developed technological methods of growing some high value crops by providing protection from the excessive cold and excessive heat. This is called Greenhouse Technology. Greenhouse Technology is the science of providing favourable environment conditions to the plants. It also protects the plants from the adverse climatic conditions such as wind, cold, precipitation, excessive radiation, extreme temperature, insects and diseases. An ideal micro climate can be created around the plants. Now the area under protected cultivation has reached from 4000 ha to 20,000 ha which include the production of high value flowers. Greenhouses are framed or inflated structures covered with transparent or translucent material large enough to grow crops under partial or fully controlled environmental conditions to get optimum growth and productivity.

Advantage

- Crops can be grown throughout the year
- Quality of the produce is superior
- Higher input use efficiency
- Higher income per unit
- Easy to manage insect-pests and diseases
- Water requirement is minimum
- Less labour requirement

Limitations

- High cost of initial installation
- Non-availability of various components
- Needs a special postharvest facility
- All the operations are very intensive

Present Status

The domestic industry is growing at annual rate of 7-10% per annum. The area under production of flowers is around 1.91 lakh hectares concentrated mostly in West Bengal, Karnataka, Tamilnadu, Andhra Pradesh, and Maharashtra with a production of 10,31,000 tones of loose and 69.03 million of cut flowers.

Table 1. Area and Production of Flower

Year	Area (in, 000 ha)	Production	
		Loose flower (in 000 tones)	Cut flower (In million numbers)
1993-94	53.3	232.5	555.6
2000-01	98.4	556.4	803.5
2001-02	106	535	2565
2002-03	70	735	2060
2003-04	101	580	1793
2004-05	116	655	1952
2005-06	129	654	2150
2006-07	143.9	880.2	37175
2007-08	166.2	868.4	43654
2008-09	166.5	987.4	47942
2009-10	183	1021	66671
2010-11	191	1031	69027

Source: Indian Horticulture Database, NHB 2010-11

Table 2. Area and production status of major flower growing states (as on 2010-11)

States	Area (in 000 ha)	Loose flower (in 000 tones)	Cut flower (in 000 numbered)
Tamilnadu	32.0	247.3	-
Karnataka	27.0	203.9	5860.0
Andhra Pradesh	21.8	133.7	6202.0
Maharashtra	17.5	91.1	7914.0
Punjab	1.7	82.0	-
Haryana	6.2	60.3	1084.0
West Bengal	23.1	59.2	23919.0
Uttar Pradesh	10.4	17.6	2958.0
Delhi	5.5	5.7	1038.0
Rajasthan	5.4	9.6	-
Himachal Pradesh	0.7	0.6	605.0
Uttarakhand	1.3	59.2	3416.0

Source: Indian Horticulture Database, NHB (2010-11)

Indian exports mostly target the major floricultural important events like Christmas (December) New Year eve (January) Valentine day (February) and Mother's day (May). The major factor being the unfavorable weather conditions during winter in the major production centers in the Northern hemisphere that limits the production. Therefore, the markets are open to produce that comes from more favourable climates from the Southern hemisphere. India therefore finds itself competing with other equally favourable countries like Kenya, Ecuador, and Morocco etc. during such events.

Table 3. Cut flower export from India

Sl No.	Country	Share (%)
1.	EU	36
2.	Japan	40
3.	Middle-east	10
4.	Australia	6
5.	S.E. Asia	5
6.	Others	3

The market price if Floricultural products are highly volatile. The production strategy should be such that growers are not at loss. The price determination in floricultural products is based on following parameters:

- Season and demand
- Variety and quality
- Free from pest, disease and mechanical damage
- Uniformity in opening
- Bright colour
- Packaging
- Regularity in export

Institutional Inventions

During the last two decade the ICAR and SAU's have made a tremendous significant growth in Research and development. The comprehensive research programmes on crop improvement using conventional and biotech tools, production in open and protected environments, technology dissemination for the benefit of farming fraternity. The IARI has made a distinction of evolving the largest number of rose varieties in the country possessing desirable traits. 75 varieties comprising of Hybrid teas, Floribundas, Grandifloras, Miniatures and Climbers were released. Induced mutants viz., Abhisarika, Striped Christian Dior and Madhosh: natural mutants Chitra, Pusa Abhishake, Climbing Sadabahar, Climbing Dr. Homi Bhabha, Delhi Pink Pearl, Pusa Gaurav and Nav Sadabahar exhibit a unique spectrum of patterns and are well received by the garden enthusiasts.

Many new cultivars have been developed in chrysanthemum viz., Baggi , Birbal Sahni (at NBRI) 'Arka Ravi' (at Hessaraghatta, Solan, Udaipur and UAS Bangalore), 'Punjab Gold' (at Hessaraghatta and Pune), Mother Teresa' (at Hessaraghatta), CO_2 and Indira (at Coimbatore), 'Yellow Gold' (at Solan), 'Ratlam Selection' and 'Basanti Local' (at Hyderabad), 'Sonali Tara', 'Sunil' 'Baggi' and 'IIHR Hybrid-11' (at Pune).

Crop Specific Production Technology of Different Crops

A. Crop Improvement

B. Production Technology

1. Rose

Rosa indica var. *odorata* rootstock was found the best for Coimbatore, Delhi, Lucknow, Ludhiana and Pune conditions; *R. multiflora* and IIHR - Thornless for Bangalore, Hyderabad and Kalyani conditions; and *R. chinensis* (Titri) for Udaipur conditions. Depending upon the varieties used, 30 x 20 cm spacing was found the best for Super Star and Gladiator under Pune and Udaipur conditions; 30 x 30 cm spacing for Montezuma at Kalyani conditions; and 30 x 40 cm for Super Star under Delhi, and for Sonia Meilland under Udaipur conditions, as quality flower production was maximum there.

For commercial cultivation of rose, NPK 400:200:200 ppm/plant/week at 'run off' stage is recommended. Coimbatore centre recommended NPK [$(NH_4)_2$ NO_3, H_3PO_4 and KNO_3] 150:40:150 for cv. First Red under protected conditions. At Ludhiana, 60:40:40 g/m^2 NPK with cv. Super Star is recommended. Delhi centre found 10 cm pruning from Dept. 23 to Oct. 1, in rose cvs. Queen Elizabeth and Happiness, and 45 cm leaving 4 shoots, from Oct. 13 to Nov. 17 in cv. Super Star as the optimum. Against weeds, Oxyflurofen at 0.5 kg a.i./ha drenching at Kalyani, and 1 kg a.i./ha at Ludhiana; Oxyflurofen (1 kg a.i./ha) or Diuron (2 kg a.i./ha) at Delhi in cvs. Super Star and Raktagandha, and Diuron (2.5 kg a.i/ha) or Simazine (3 kg a.i./ha) or Round Up' (1 kg a.i./ha) at Yercaud and Bangalore conditions were quite effective.

Rose is one of the natures beautiful creations and is universally acclaimed as the "Queen of Flowers". It is a top ranking cut flower in the flower trade on the basis of average production and consumption. Great diversity in plant growth, flower colour, flower shape, fragrance, slow opening of buds and good keeping quality made rose so popular that it is grown commercially to meet the demand of cut blooms. In India, roses are grown for cut flowers, making essential oil, rose water and gulkand. Whereas it's dried petals are used for making incense sticks. People are not aware of different propagation techniques of rose, therefore they have to purchase rose planting materials from nurseries at very high cost. So the main propagation methods, through which we can produce plants at high level, are as below:

Propagation

Roses can be propagated both by seeds and by various vegetative methods. But now-a-days tissue culture techniques are being used for the production of disease free plants in large number.

1. Seed propagation

This method of propagation is generally adopted by breeders for developing new cultivars with desirable characters. Seedlings of rose species are also used as stock for grafting or budding. The rose fruits (hips) are harvested when fully ripe and thoroughly dried before extraction of seeds. In rose, seed germination is very poor because most of the seeds when mature are in resting conditions requiring an after ripening period before germination. So stratification of seeds at 1.6-44.4°C for 6 weeks improves germination. Germination can also be improved with scarification with different chemicals like sulphuric acid for 1-2 hours.

Care of seedlings

For raising seedlings, clean seeds are sown about 5 cm apart in pairs containing pot compost with adequate organic matter or in small pots. The flower buds appeared on new seedlings should be pinched off until plants become reasonably large and sturdy. A continuous moisture and nutrient supply and control of weeds, diseases and pests contribute to successful seedling growth.

2. Vegetative Propagation

Roses are propagated by various vegetative methods like cutting, layering, budding and grafting.

a. Cutting

Propagation by cutting is normally done to raise stocks for grafting or budding and also adopted for multiplying vigorous types of cultivar, climbers, rambler, polyantha and miniatures. Cuttings are always prepared from shoot. Root cuttings have also been found to form shoots and ultimately a new plant.

i) Stem cutting

It is one of the least expensive and the easiest method of rose multiplication. Cuttings may be single, double or triple eyed, the later being preferred as there is greater chance for more shoot and better root formation. Cutting should normally be taken during the monsoon or spring. The cuttings are made from pencil size thickness (8-9 inch) shoots just below the node and each cutting should have 2-3 buds. Sand is the most ideal medium for rooting of cuttings.

ii). Root cutting

Some rose species like *Rosa blanda, R. nitida, R. virginiana* can be propagated by root cuttings. The cuttings should be taken from young and healthy plants and it is important to maintain polarity when planting. The cuttings should be inserted vertically with the upper most ends at level of the rooting media.

b) Layering

This type of vegetative propagation is usually limited to climbing and rambling roses and usually practiced during the monsoon or early spring and leads to own rooted plants. Mainly 2 types, ground layering and air layering, are used.

i) Ground Layering

It is performed by bending the shoot to the ground and covers it with soil leaving the terminal end exposed. Root promoting substance can also be used for improving root formation in layers. Rooting usually takes place in a month or so and the layered shoot is detached 15-20 days after root formation.

ii) Air Layering

It consists of removing a ring of bark (about 2.5 cm long) from stem, around the shoot and wrap sphagnum moss around the ringed portion and covering with polythene film. For the improvement of root formation in air-layer, some root promoting substances can be used.

c) Budding

Budding is the most popular and successful method for multiplying roses. It provides larger number of plants than cuttings, layering or grafting as a single shoot of the desired scion furnishing a number of buds for budding. Shield or T-bud is the method ordinarily used. On the selected rootstock, the buds are inserted into a T-shaped incision and then tied with suitable wrapping material. Several other budding techniques like forkert, patch, chip, skin etc. have also been tried in rose with varying degree of success.

The time of budding varies from place to place and the right stage of budding is when plants have adequate sap flow and the cambium tissue is highly active. The best time to bud roses in eastern India is from January to March, while in northern India it is from November to February. In places with mild climate, like Bangalore and Pune, budding can be done throughout the year.

3. Some Common Rootstocks in India are as follows

Common Rootstocks

a) Rosa multiflora

This rootstock is tolerant to nematode and frost. It performs well in Bihar, Bengal and in the hills.

b) Rosa bourboniana (Edward Rose)

Susceptible to powdery mildew and die back diseases. Plants provide straight stems and mostly used in northern plains of India for budding standard roses.

c) Rosa indica (var odorata)

Well adapted to both excessively dry and wet soil. Plant raised on this rootstock is vigorous and produces flower with longer stems. It is resistant to powdery mildew and insect pests and recommended for northern plains of India.

d) Rosa canina* or *Rosa manetti: *R.canina* or *R.manetti* is the most ideal for greenhouse cultivation.

Production

Planting time

Planting time depends mainly on the climatic conditions of the region. In general, roses can be planted any time except hot sunny days and during heavy rains when the soil is wet. But winter months are the best and safest. The suitable time for planting roses in the plains of India is during September-October and in the hills during October- November or February-March under green house.

Planting Depth and Distance

The position of the bud union should be looked into. In temperate season, the bud union should be slightly below the soil. In India, better results are obtained if the Planting is done with bud union 2-3 cm above the soil level.

Under green house, roses are planted at distance of 30 × 25 cm (Row × Plant) as a common distance and total 7 to 10 plants per sq. m. are accommodated in the beds. Two rows are preferred on one meter wide bed and a passage is kept of 85 cm between two beds.

Pruning/Bending

It is an important operation for maintenance of floriferousness and flower quality along with vigour of rose plants. The main objectives of pruning are to remove weak and unproductive shoots which allow air and light to reach plant centre. It keeps the plant in proper shape and size and also forces the plant buds to break and grow to produce quality flowers. It also helps in maintenance of proper balance between vegetative growth and flowering.

Pruning should be done when plants are less active or dormant. In northern plains of India, it is mainly done during October-November. The basic rule of pruning is to cut half centimeter above the vigorous, outward directed bud by giving a slanting cut because it does not retain moisture and prevents fungal growth.

In H.T. roses, all dead, weak, damaged, diseased and crossing shoots are removed from the base retaining only 4-5 healthy basal shoots. Moderately severe pruning appears to be ideal. Under the green house conditions, a bending practice is common to regulate the production as well to have better quality flowers.

Manuring and Fertilization

Rose is a nutrient loving plant and about 16 elements are known to play important role for its growth and development. Rose should be fed with both organic and inorganic sources. One hundred gram of mixture containing Groundnut cake, bonemeal, amorphous, Ammonium sulphate are used after pruning.

Single super phosphate (2kg) and potassium sulphate (1kg) should be applied per bush for better result. It is advisable to apply the fertilizers in three split doses and application should be made when there is sufficient moisture in the soil. The first application of the fertilizers should be given 15 days after pruning when the new growth starts, i.e. in October then in December and third dose in February. Under green house 300: 250 and 200 ppm of NPK is recommended through fertigation system for quality production. Normally, water soluble fertilizers are used to feed the crop.

Disbudding

It is also an important operation by removing buds and retaining only central resulting in large flowers with better quality.

De-shooting

Removal of lateral shoots and to allow terminal shoots only are advised. It is necessary, otherwise plants get heavily branched with large number of small flowers.

Cultivars

Golden Gate, Grand Gala, First Red, Kiss, Konfetti, Mercedez, Ravel, Noblesse, Vivaldi, Golden Gate, Hollywood and Starlite.

Planting

Optimal planting time is October-December. The planting density should be between 7-13 plants/m^2 (50-60 thousand bushes/ha area). Planting should be done preferably in 2-row system.

Temperature requirement

The greenhouse temperature is generally maintained at 20^{0}C or 21^{0}C on cloudy days and 24–28°C on sunny days. However, the temperature range of 15-27^{0}C is optimum.

Diseases

Die-back

It is one of the major diseases of rose, particularly in India. The older plants are more prone to the attack of this disease as compared to younger ones. The stem and roots of the dead plants show browning of internal tissues. *Diplodia rosarum* is the causal organism. It may occur through improper use of fertilizers, improper watering, borer damage, poor drainage and lack of light. Proper cultural practice is the prevention for die back in roses. Some systemic fungicides like carbendazim (1g/l) may also be used as soil drench.

Powdery-mildew

This major disease of rose causes much damage to roses all over the world. It is caused by the fungus *Sphaerotheca pannosa*. It can be controlled by applying sulphur but most of the roses are susceptible to sulphur injury.

Black-Spot

This is a fungal disease caused by *Diplocarpon rosae.* Dark brown, circular spots with fringe borders are present on both sides of leaflets. Benlate can be used as control measure.

Harvesting and Yield

Red and pink cultivars are harvested when first 2 petals are beginning to unfold and calyx is reflexed below the horizontal lane. Yellow cultivars are harvested slightly earlier and white cultivars slightly later than red and pink types.

After cutting the stem ends are dipped in 200-500 ppm citric acid or aluminium sulphate and placed in cold storage at 1°C till grading. Stem ends are re-cut, bunched in 20's and placed in preservative solution with sucrose up to 2%.

Flower yield

Flower yield of 250-350 stems/m^2 is considered ideal. Flower yield can be increased by spraying BAP 50-100 ppm before flowering flush. Burning of saw-dust during winter months in the morning hours (7-10 AM) increases flower yield and quality significantly.

Carnation

Carnation (*Dianthus caryophyllus* L) is an herbaceous perennial flowering plant, belonging to the family Caryophyllaceae. It has been under cultivation by man for over 2000 years. The common name Carnation (Latin, *Carnation*, from caro; *carnis*, flesh) is believed from a Greek word 'Coronation' because these flowers were used in decorating the crowns of Greek athletes.

Carnations are indigenous to the Mediterranean areas. They are grown on a large scale in Italy, Spain, Columbia, Kenya, Sri Lanka, Canary Islands, France, Holland Germany, Israel and USA. Due to the high cost of production in greenhouses in Europe and USA, the cultivation of Carnations is shifting to more naturally growing areas where they are being produced at lesser cost. Owing to its ever increasing demand in the world market, high quality Carnations are also being cultivated in India. The major production centers are located around Pune and Bangalore. Carnations are now being commercially grown in Solan, Shimla, Mandi, Kullu, Chamba and Bilaspur districts in Himachal Pradesh. It has ample scope of its production in Uttarakhand as well.

Under polyhouse conditions in Solan, cvs Impala, Super Star, Veleta and Fantasia were supplied with 4 hour extended lighting which induced early flowering with maximum floral stalk. April planting resulted in earlier flowering (160 days) while October planting delayed it (213 days) but gives better stalk length and floral diameter. 'Pinch and a half method' resulted in earlier and regulated flowering. Planting at 15 x 15 cm spacing (4,44,444 plants/ha), double pinching (40 and 60 days after transplanting) and application of NPK @ 30-40:20:10 g/m^2 gave the highest yield with nitrogen in 3 splits (basal, one month after and two months after transplanting).

There are about 250-300 species of *Dianthus* comprising annuals and evergreen perennials but only a few are cultivated. The most common cultivated species are *D. caryophyllus, D. barbatus* and *D. chinensis.*

The present day perpetual Carnation was developed from cross between *D. caryophyllus* (female parent) and *D. chinensis* (male parent). They are grouped into standard and spray types on the basis of flower size. The standard type produces large single flowers on a longer sturdy stem. They suffer from diseases when grown under warm humid climates. The spray or miniature types produce many flowers of a smaller size on a short stem and are suitable to warm climates. The most common varieties of carnation are: Arthur Sim, Sunrise, Lemon drop, Pink Dona, Master, Liberty, Tempo, Lavender etc.

2. Climate

Carnations perform the best in relatively cool climate. It is a long day plant and requires approximately 21.5 K lux sunshine for at least 8 hours a day. Temperature is a major factor that influences the growth and flowering, colour, grade and the quality of cut flowers. The location having a day temperature of 10° C is good for high quality Carnations. The optimum temperature for production of standard Carnation is 18°-23°C. High temperature and high humidity produce inferior quality flowers. The optimum humidity for growing Carnation is between 50-65%.When humidity is higher; plants become susceptible to fungal diseases like *Alternaria* and *Botrytis*. High light intensity and very low humidity are detrimental to a young crop and may cause leaf burning and weak foliage. It leads to infection of parasites such as red spider mites.

2.1 Soil

Selection of site is very important for successful Carnation cultivation. Since roots of Carnation are highly susceptible to poor drainage conditions, it needs well drained and aerated soil. The growth of roots in heavy soil is very much restricted and adversely affects growth of plants and quality of flowers.

A rich sandy loam or loam soil is ideal for its cultivation. Soils having higher silt or clay content should be amended by incorporating organic matter or compost. The pH 6-7 and EC 0.8-1.6 ms/cm at 25°C are ideal. Both higher and lower pH are found to show adverse effects on flower production. The site selected for plantation should be free from weeds, nematodes or soil borne pathogens. Pre-plant sterilization with steam, sun or 5% formalin is beneficial. Reused soils may contain many pathogens especially *Fusarium*, which makes some, beds unsuitable for Carnation cultivation even after disinfection.

Perpetual flowering Carnations is multiplied by vegetative means, soft terminal cuttings while seed propagation is normally practiced for the purpose of hybridization. For successful cut flower production, choose Carnations with well developed roots having minimum four leaves and not developed any flower bud. Cuttings should be of about 10-15 cm long for standard cultivars and 8-10 cm long for spray types, disease free and of best commercial variety. Leaves should be free from any fungal spot.

2.2 Planting

Soft terminal cuttings from healthy mother plant having vigorous root should be planted at shallow depth. Deep planting of cuttings may cause root and stem rot. During planting, keep the base of the cutting above the soil level as they were in the rooting medium to ensure rapid development. To obtain the maximum no. of flowers, planting time and plant density are very important. When planted for commercial scale, raised beds of 30 cm height, 1.05 m width and convenient length are preferred. A planting density of around 32 plants /m^2 (15x20 cm spacing) is generally recommended for 2 years production. When it is planted in pots, a single plant can be planted in 7.5 cm pots containing a potting mixture of equal parts of soil & well rotten F.Y.M. and half part of sand. Normally planting distance is 15 × 15 cm

2.3 Season and Cycle

Planting schedule is important to produce the maximum number of excellent quality flowers. Carnation produces heavy flush of flowers during a short period of time, which is not desirable for commercial production. Therefore, it is regulated through planting time, planting density and method of pinching of plants. In Northern plains of India, planting is done during mid September- November, which produces flower during February to April. In hilly areas where there is no snowfall, October-February planting is suitable to obtain flowers during mid-April to mid July and in hilly areas having seasonal snowfall, planting is done during February- April to get flowers from August- October. In protected conditions, Carnations can be planted round the year.

2.4 Carnation Growing Period

Carnation flowers take approx.150-180 days from planting to flower under open conditions and 120-150 days under protected conditions depending upon the season. The main propagating season of Carnation extends from November to March. Perpetual Carnations are mostly planted on a 2 year cycle after which their yield decreases.

2.5 Irrigation and Fertilization

Rooted cuttings should be watered immediately after planting. Spraying of water through misting is more effective than the surface watering for early establishment of cuttings. After three weeks of planting, drip irrigation has to be adopted. Frequency of watering depends on the weather conditions and the soil type. Stagnation of water should be avoided to minimize the incidence of diseases.

Carnation plants need regular supply of nutrients for continuous production of flowers. Therefore, the application of nutrients in smaller doses but at regular interval favours better growth and yield. Low level of nitrogen and boron may cause calyx splitting. In the first three weeks after planting, no chemical fertilizers are applied. A total nutrition of 5-10 kg of FYM/m^2, 30 gm N/m^2, 20 gm P_2O_5 and 10 gm K/m^2 is applied for standard Carnation while same dose is applied for spray types except the Nitrogen @ 40 gm/ m^2. Half dose of nitrogen is applied before planting and remaining half N one month after planting. Spray of 0.1% urea and potassium nitrate (KNO3) and 1 ppm boron at 10 days interval till flowering produces good quality flowers.

Some micronutrients like Fe, Zn, Mn, Mo and B may also be required depending upon the soil type, mineral content of irrigation water and leaf analysis of plant sample.

2.6 Support Nets

In order to obtain straight stems and to prevent lodging of plants, Carnation needs supporting nets of 4-5 layers. These plants are planted within the net of mesh size 7.5x7.5 cm. Nets are gradually raised with plant growth. For every 2.5 to 3.0 m, the metal wires or nylon rope should be supported with poles. The first net should be fixed at 12.0 cm above the bed. The remaining nets, whose squares should be 12.5x12.5 cm or 15x15 cm, are placed over the first net. These nets are placed at 15 cm apart.

i) Pinching

For successful production of top quality Carnations, pinching is an important operation. During pinching, the tip of the stem is removed to encourage the growth of the standard or sim Carnations. Only 4-6 well grown lateral shoots will be allowed to grow. Pinching is done after 30-35 days of planting leaving only 6 pairs of basal leaves. If too many leaf pairs remain, stem and flower quality may be reduced considerably. Similarly, production will decrease if a few leaf pairs will be left. There are three types of pinching methods generally followed:

Single pinching: In single pinch, apical portion measuring 5-7 cm is pinched off to give about 4-5 lateral shoots, which produces flower. This is done for early crop.

Pinch and-a- half: This method involves single pinching of the main stem to give 4-5 side shoots. Only half of total numbers of lateral shoots are pinched when they are about 5 cm long. This method reduces the load of first crop and provides a steady supply of flowers throughout the year.

Double pinch: In this method, all the lateral shoots arising from the first pinch are again pinched when they are 8-10 cm in length at 2-3 nodes. It is done approximately

5-6 weeks after first pinch. This is generally done for late harvesting or delaying the flowering period. Approximately 8-10 shoots are retained. This method produces large number of flower bearing shoots and poor quality flowers hence not preferred.

ii) Disbudding

Disbudding is the removal of unwanted buds. It is practiced in Carnation to obtain good quality flowers. Buds are taken away before they get too big, as they will considerably reduce the flower size. In case of standard Carnation, the lateral flower buds are removed leaving only the terminal or main flower bud. In spray or miniature Carnation, the terminal or main flower buds are removed, to encourage lateral flower buds. Disbudding helps to produce quality bloom in standard carnation.

2.7 Post harvest handling and Plant protection

Harvesting of Carnation flowers depends on size of bud and growth of petals. Standard Carnations are harvested when they are half open and the first petal lays horizontally or at paint brush stage. Spray types are harvested when at least two flowers have opened and remaining buds are showing colour.

Flowers are the best harvested during the morning hours when they are turgid. Keep the flowers in fresh, clean water or preservative solution for about 4-6 hours and then place in cold store at 4-6°C. Do not cut the flowers too deeply as this may adversely affect production during the next season.

Grading of flowers is based on stem length; flower diameter and physical conditions like stem sturdiness, freedom from pest, diseases and physiological abnormality. They are grouped in bundles of 20 and kept at 0-2°C for 12-14 hours and then packed in corrugated cardboard boxes.

Vase life and storability of flowers can be increased by placing flowers in 10 % sucrose + 1 mm STS for 8-10 hours. Carnation flowers are very sensitive to ethylene gas. Now day's anti-ethylene compound 1-MCP (1- methyl cyclopropene) is being used as floral preservative.

2.8 Insect, Pest, Disease and Physiological disorders:

1. **Insect Control**: The most common insects attacking the Carnation are aphids, thrips and red spider mite.

   i. **Aphids** *(Myzus persicae)* suck the sap from new shoots and leaves. They leave sticky substances on leaves and flower buds. Foliar spray of Rogor (2ml/ litre) or 0.25% demeton-methyl is effective.

   ii. **Thrips** *(Thrips tabaci)* also suck the sap of leaves and turn them yellow and patchy. Under severe attack, they also cause streaks and spots on flowers and stems. Pink and red cultivars are more susceptible. Malathion or Rogor (dimethoate) or dichlorovos, all at 0.1-0.2 % gave the best control of Carnation thrips.

iii. **Red spider mite** *(Tetranychus urticae)* is the most serious pest of Carnation. They are very small red coloured insects mostly feeding on lower sides of leaf and suck the sap. Leaves become yellow with dusty coating and fine webs on their surface. The most effective chemical against mite is vertimac, which is effective against adults, eggs and larvae.

2. **Disease**: Carnations are attacked by a number of diseases like *Fusarium, Alternaria*, Bacterial wilt, Rust, *Rhizoctonia*, and stem rot.

   i. **Fusarium wilt** caused by *Fusarium oxysporum* f. *Sp.dianthi*, also known as vascular wilt, is one of the most serious diseases of Carnation. It is soil born, invades the plant through roots and damages stem tissues resulting in wilting and death of plant. Soil fumigation and treating the beds with Bavistin (0.1%) and Dithan M-45 (0.1%) at regular interval is effective.

   ii. **Leaf spot** caused by *Alternaria dianthi* is another serious air-born disease. It is characterized by grayish- brown spots on leaf and stem. High temperature and humidity favour the disease spread. Removal of infected leaves and foliar spray of Mancozeb (Dithano M-45) @ 0.2% or Bavistin (0.1%) controls this disease.

   iii. **Bacterial wilt** *(Pseudomonas caryophylli)* is prevalent at high temperatures. It is a soil born disease attacking older plants. Production of grayish green foliage, rotting of roots and deep longitudinal cracks on internodes are the major symptoms. Soil sterilization, treating the plant material with 0.01 % streptocycline and using disease free plants help in reducing the incidence.

3. **Physiological disorders**: Splitting of calyx is a serious disorder affecting the quality of Carnations. In this disorder, sepals beneath the flower are unable to form a cylindrical calyx tube which supports the base of petals. As the bud opens and size of petals increases, calyx may split down either half or completely because it is unable to contain the load of extra petals. The primary cause of calyx splitting is unknown but may be due to genetical, environmental like low temperature and fluctuation in day-night temperature or nutritional like low level of nitrogen and boron. The problem of calyx splitting can be reduced by placing of rubber band around the calyx of the flower bud, when they have just started opening. Increased level of potassium and boron, choosing the cultivars which are less prone to splitting and uniformity in day night temperature can reduce the problem.

3. Chrysanthemum

Chrysanthemums, often called 'mums', are a genus Chrysanthemum of about 30 species of perennial flowering plants in the family Asteraceae, native to Asia and North-eastern Europe. Now, botanically it is *Dendranthema grandiflora*.

3.1 Uses

Modern chrysanthemums are much more showy than their wild relatives. The flowers occur in various forms, and can be daisy-like, decorative, pompons or buttons. This genus contains many hybrids and thousands of cultivars developed for horticultural purposes. In addition to the traditional yellow, other colors are available, such as white, purple and red. The most important hybrid is *Chrysanthemum × morifolium* (syn. C. × grandiflorum), derived primarily from *Chrysanthemum indicum* but also involving other species.

Solan centre has developed the technology for year round flower production through staggered planting, pinching, growth retardants, photoperiod regulation and cultural practices, as following:

PRC	Pinching	Retardant	Lighting	Blackout (days)	Blooming period
June 2	June 24	July 28	NLD upto Aug. 4	48-70	Oct.2-Nov.4
Aug. 5	Aug. 27	Sept. 30	NLD upto Aug. 26 + ALD from Aug. 27 - Oct.6	48-73	Dec. 11-Jan. 1
Oct. 11	Nov. 2	Dec. 5	ALD upto Dec. 12	50-74	Feb.11-Mar.4
Dec. 1	Dec. 20	Jan. 23	ALD upto Jan. 30	94-115	May 24-June 6
Feb.2	Feb.24	April 1	ALD upto April 8	48-73	June 8-29
April 3	April 22	May 28	ALD upto April 22 + NLD from April 23-June 4	52-72	Aug.7-29

- Blackout means natural/artificial short days after termination of light to the stage till 60-70% flower buds showed colour
- NLD = natural long days and ALD = artificial long days
- The range under blackout column denotes variation with the treatment
- PRC = planting of rooted cuttings

ALD treatments at later stages of growth (1 and 2 months after planting) extend vegetative growth than LD treatments immediately after planting. GA (50 ppm) advances the flowering while NAA at (50 ppm) delays it. At Ludhiana, by providing natural short day conditions from Dec. to Feb. and artificial short days from March, flowering was induced in May in the cv. Punjab Anuradha. Black polythene covering of the plants of cvs Birbal Sahni, Jubilee, Flirt, Sunil, Jayanti and Vijay for 14 h daily at night till 60-70% buds show their colour, induced earlier flowering.

Rooting was more in suckers than the terminal cuttings though reverse was true with respect to flowering and yield. Spacing of 30x30 cm with single pinching at 4th week or 2 pinchings at 4th and 7th week of planting provide higher yield. *Azospirillum*, VAM and phosphobacterium along with 200 ppm N, 40 ppm P and 200 ppm K provide better growth, yield and post harvest life in cvs. Bronze Spray and Red Spray.

Chrysanthemums are divided into two basic groups, Garden Hardy and Exhibition. Garden hardy mums are perennials capable of being wintered over in the ground in most northern latitudes. Exhibition varieties are not usually as sturdy. Garden hardies are defined by their ability to produce an abundance of small blooms with little if any mechanical assistance (*i.e.* staking) and withstanding wind and rain. Exhibition varieties on the other hand require staking, over-wintering in a relatively dry cool environment, sometimes with the addition of night lights.

The Exhibition varieties can be used to create many amazing plant forms; large disbudded blooms, spray forms, as well as many artistically trained forms, such as: Thousand Bloom, Standard (trees), Fans, Hanging Baskets, Topiary, Bonsai, and Cascades. Chrysanthemum blooms are divided into 13 different bloom forms by the US National Chrysanthemum Society, Inc., which is in keeping with the international classification system. The bloom forms are defined by the way in which the ray and disk florets are arranged.

Chrysanthemum blooms are composed of many individual flowers (florets), each one capable of producing a seed. The disk florets are in the center of the bloom head, and the ray florets are on the perimeter. The ray florets are considered imperfect flowers, as they only possess the female productive organs, while the disk florets are considered perfect flowers as they possess both male and female reproductive organs.

3.2 Cultivars and Pot Culture

i. **Large Flower:** Beauty, Snow Ball, William Turner, Innocence, White, Green Godess, Chandrama, Super Giant, Sonar Bangla, Pink Cloud, Shirley, Perfection, Classic Beauty, Pink Turner, Brave, Distinction, Alfred, Autumn Blaze etc.

ii. **Small Flower:** Mercury, Jyotsna, Birbal Sahni, Himani, Topaz, Archana, Basanti, Nanako, Sujata, Kundan, Nilima, Alison, Jaya, Gem, Jean etc.

Chrysanthemum culture in pot requires very strict time bound monthly intercultural practices with precise operation. Selection of disease free and high quality planting material is ideal for commercial growers. Pot culture includes soil preparation and selection of proper size pot for it.

i) Soil Preparation for Pots

Ideal soil composition for growing healthy chrysanthemum plants in pot includes one part of clay soil, one part of leaf mold and 2 parts of rotten cow dung manure. Fertilizer is also mixed at this stage which includes Bone meal one part, Super phosphate one part, Di -ammonium phosphate one part and Neem Cake one part i.e. in the ratio of (1:1:1:1). The mixture should be kept as a heap and spread for solar treatment and for proper sun drying prior to filling of pots. For filling 100 pots 5kg bone meal, neem cake and fertilizer mixture is required. Soil is prepared well in advance i.e. a month prior to planting. A small amount of Malathion powder is added to the soil to get rid from insects.

ii) Preparation of Cuttings

Cuttings are prepared from mother plants during June to August. In case of early rains, cuttings can be prepared in the end of June otherwise July to mid August is the ideal time for making cuttings. Stem cuttings from terminal portion are selected and its length should be restricted to 10-12cm. Bavistine or Indofil M-45 (2% solution) is prepared for the treatment of cuttings. After quick dipping for 2-4 seconds in fungicide solution, hormones like Rootex No.1 or alcoholic solution of IBA (quick dip method) are applied at the base of the cuttings. Cuttings are planted in 25- 30 cm pot filled by sterilized coarse sand. For large flowered chrysanthemum especially for spider type cultivar, sand filled pots are ideal as the roots developed from stem cuttings are fine and its development are best in sand. The rooting takes place within 15-20 days of planting. Fifty cuttings of small flowered chrysanthemum can be accommodated in a 30cm pots while forty cuttings of large flowered chrysanthemum can be accommodated in 25cm pots.

Planting and Month Wise Operation

Planting and month wise operation include watering, fertigation, pinching, dis-budding, removal of old and diseased leaf, protection from hot winds and excessive rains and staking. Intercultural practices from January to December for small and large flowered chrysanthemum are briefly explained.

Suckers are planted in earthen big size pots or beds in the months of January and February and large flowered chrysanthemum is transplanted in 10 cm pots. Small flowered chrysanthemum can be transplanted in well manured beds. Similarly, a few plants of large flowered chrysanthemum should be planted in beds to have cuttings from them in the month of July. In large flowered chrysanthemum beds, regular watering is required in the month of April and in potted plants, watering is restricted to check the growth. Small flowered chrysanthemum in beds requires more branching by pinching. This is the ideal time to pinch the plants back to 10cm from ground level. In the month of May, large flowered chrysanthemum should be protected from hot wind and repotting in 15cm pot gives better results.

Small flowered chrysanthemum requires second pinching in the month of May. In June, large flowered chrysanthemum requires protection from hot wind in the month of June. Small flowered chrysanthemum requires protection from hot winds during the month of June and this is the proper time for third pinching. The pinching should be done in first week of June. In the month of July Small flowered chrysanthemum should be lifted from beds and it should be planted in 25cm pots. Large flowered chrysanthemum requires repotting in 20cm pots in the month of July. This is the high time for taking cuttings for raising dwarf plant with superior blooms. In August, chrysanthemum plants need protection from excessive rains. Disease foliage should be removed and fourth pinching is required in small flowered chrysanthemum.

Nutrition and Propagation

i. **Dry Fertigation:** For dry fertigation Di- ammonium phosphate, super phosphate, bone meal and neem cake at ratio of (1:1:1:1) are mixed thoroughly. One tea spoon (10g) of this mixture is applied at fifteen days interval in chrysanthemum pots for better growth and flowering.

ii. **Liquid fertigation:** For better growth and quality flowering in chrysanthemum, liquid fertigation is very essential. For making 100lt. of liquid fertigation following components are mixed in a big tank.

- Unripe cow dung..................................5kg
- DAP.. 5kg
- Super phosphate...............................5kg
- Neem Cake....................................5kg
- Calatropis leaf (Madar)........................10kg

The solution should be stirred well prior to use and kept for a 15-20 days before use. Liquid fertigation should start from October and continued up to November at fortnight interval. This liquid should be diluted ten times before use.

3.3 Propagation

Chrysanthemum is propagated both by vegetative and sexual methods. The latter method is adopted by the breeder for development of new varieties. The vegetative methods include multiplication through suckers, cuttings and micro-propagation.

i. **Sucker Method:** Profuse suckering occurs about one month after blooms fade. After separating the suckers from mother plants, they are planted directly in the field or for root formation in pots.

ii. **Cutting Method:** Propagation through stem cuttings is the common method followed by the commercial growers. An ideal cutting is supposed to be

4-6 cm long with a diameter of 3.2 to 4.8 mm. The terminal cuttings are made late in June and July. Seradix B-1 or 2000 ppm IBA to 0.5 cm basal portion. The cuttings root between 3 and 4 weeks.

iii. **Micro-Propagation Method:** Chrysanthemum can be multiplied faster by culturing shoots apex, somatic callus, from peduncle and other floral parts and from leaves *in vitro*. The terminal cuttings 2-3 cm long with at least one node is taken.

3.4 Growth and flowering

Environment has great influence on growth and flowering of chrysanthemum. The major factors are light, temperature, relative humidity and air composition.

1. **Light:** It is well known that chrysanthemum plants continue to grow vegetative till the nights are shorter than critical level of $6^{1/2}$ hrs in most of the traditional varieties. Flourescent light of very high intensity given only for one minute every night completely inhibits the flowering. It has been reported that a minimum of 3-5 days are necessary for bud initiation. It is been observed that day-length of $14\text{-}14^{1/2}$ hrs could initiate flower buds, but their development occurred only when day-length was of $13\text{-}13^{1/2}$ hrs.
2. **Temperature:** The growth rate in chrysanthemum remains at low level at or below 10°C. When the temperature is raised to 15°C, the growth rate increase rapidly. A constant temperature of 16°C is found the best for rooting of cuttings. On the basis of their response to temperature regarding flowering, the varieties are grouped into 3 classes by Cathey.
    - **Thermozoro:** Flowering at any temperature between 10°C and 27°C temperature.
    - **Thermopositive:** At higher temperature at 27°C bud initiation is rapid but flowering is delayed.
    - **Thermonegative:** Bud initiation occurs at low to high temperature between 10°C and 27°C, but continuous high temperature delays bud development
3. **Relative Humidity:** A relative humidity between 70 and 90 is optimum for growth of chrysanthemum.
4. **Air Composition:** In temperate regions, chrysanthemum is grown in glass-houses. During the day time, the required amount of O_2 is available as a result of photosynthesis. During night, proper ventilation is necessary for ensuring sufficient oxygen supply. Production increases significantly by artificially enriching the greenhouse atmosphere with CO_2. Supplementary CO_2 (700-900 ppm) produces longer stem length and early flowering.

5. **Growth Substances:** Growth substances and various other chemicals regulate growth and flowering in chrysanthemum.

   5.1 *Growth promotion:* GA_3 shows marked effect on growth and development depending mainly on the time of application and stage of growth. Treatment with GA_3 during 3rd week of short days caused elongation of stem and of peduncle when applied in 4th week of SD. Spraying chrysanthemums at pinching time with BA (Benzyeadenine) (40 ppm) increases branching.

   5.2 *Growth retardation:* For developing well-shaped pot plants, growth retardants have proved very effective. SADH and phospon are being used commercially. The application of SADH (2000-4000 ppm) after disbudding reduces stem length below the flower head. The liquid form is used as soil drench. CCC (Chlormequat) at 1.25 per cent concentration suppresses the plant height.

4. Gerbera

Gerbera originated in South Africa and belongs to family Asteraceae. It is an important commercial cut flower crop grown throughout the world under a wide range of climatic conditions. During 2001, at Aalsmeer Flower Auction Centre (Holland) it remained at fourth place among the top ten cut flowers of the world floriculture trade in sales volume. In India, its commercial cultivation under low cost polyhouse environment is on rise and has made its cut flowers availability round the year.

Gerbera (*Gerbera jamesonii*) is commonly known as Transvaal Daisy, Berbrton Daisy or African daisy and produces very attractive long stalk cut flowers. Its flowers are of various colours and forms, suitable for different flower arrangements. Its cut blooms have 7-12 days vase life. Apart from upcoming huge domestic consumption, it has got great potential in export also.

4.1 Growing Environment

Gerbera is normally grown under open field condition, but with the introduction of new exotic varieties, which require more care in their management, it has become difficult to get production of superior quality blooms under open field conditions or even agro shade nets. Therefore, gerbera has to be grown under protected environment. Different growing environments, viz., open field, 50 % shade net, insect proof net, naturally ventilated polyhouse (low cost) and fan and pad evaporative cooled polyhouse (medium cost) were compared for growing exotic varieties of the gerbera under Bangalore condition. The yield of superior quality flowers was reported higher under naturally ventilated polyhouse. The ideal temperature for gerbera flower initiation is 23°C and for unfolding are 25-27°C. Flowering is harmed considerably at temperature below 12°Ce and above 35°C. The optimum relative humidity inside the greenhouse should be 80-85%.

4.2 *Commercial varieties:* Several exotic varieties of gerbera in Single, Semi-double and Double types are available with nurserymen. Gerbera varieties are also classified into two groups i.e. , large flowered and mini gerbera with desirable characters like – large flower diameter, long and stout flower stalk, double type of flower arrangement, black center , increased vase life and ability to withstand in long distance transportation, which have demand in the market are significant. Some of the suitable varieties under Indian conditions are: Cabana, Daikiri, Ellymay, Essandre, Evening Bells, Feugo, Gold Spot, Golden Gate, Ibiza, Jankfrau, Lindessa, Lyonella, Optima, Ornella, Paganini Piton, Polar, Red Monarch, Rosabella, Ruby Red, Sangria, Sunway, Tiramisu, Twiggy, etc.

4.3 *Growing medium:* Well drained loamy soil, rich in organic matter and has adequate moisture holding capacity, is ideal for successful gerbera cultivation. A soil pH of 5.5 –6.5 and E.C. 0.5-2.0 ds/cm^2 are ideal. A lower pH may result in the lack of manganese or iron and a high pH affects the soil structure. The land should be ploughed/dug deep enough 3-4 times and brought to a fine tilth. Gerbera is grown on 1 meter wide and 30 cm high raised beds of convenient length, leaving 30-40 cm space between two beds, should be prepared by mixing farm yard manure (8-10 kg/m^2) and sand (3-4 kg/m^2) into the prepared beds or beds may be prepared by mixing farm yard manure, sand and coir pith in 2:1:1 ratio.

1. *Disinfection of Soil*

Before taking up transplanting, soil sterilization or disinfection of the prepared beds is absolutely necessary. It would minimize the infestation of the soil borne pathogens, which would otherwise destroy the crop partially or completely. The raised beds should be drenched by 4 % solution of formaldehyde (1-1.2 litre/m^2) and immediately covered with air tight polyethylene film for 4-5 days. Soon after, the treated beds should be watered thoroughly to drain off the chemical after disinfestations and subsequent washing out of the soil; it is advised to wait for two weeks before taking up transplanting. Soil disinfection may also be done with methyl bromide (@ 30g/m^2 area). Growing medium may also be sterilized by application of Basamid @40g/m^2 area. The treated beds have to be covered gas proof by polyethylene film for 6-7 days. A few days before transplanting neem cake @ 1kg/m^2 can be incorporated into the prepared beds. Along with this, Furadan @ 10g/m^2 can be applied as a preventive measure against the nematodes.

2. *Transplanting*

Gerbera is commercially propagated either through suckers or tissue culture plantlets. Its propagation through tissue culture plantlets has become the predominant method to provide uniform flowering, free branching and pathogen free plants. Faster the delivery of the young plants, it is necessary to transplant them as soon as possible. It is good thing to keep them in a room without direct sun, arrange the boxes one by one without placing above each other.

In this way, it is possible to guarantee a good aeration. Proper planting is important. If the plants are planted too deep, the crown rot is encouraged and too shallow, the exposed roots are easily disturbed and plants may wilt and die. The young plants should be planted so that the crown is slightly 2-3 cm above the surrounding soil level. The crown must be dried out between two waterings; otherwise condition will be ideal for infection of *Phytophthora cryptogea*. Planting can be adopted at a spacing of 30 X 30 cm, in three rows per bed of one-meter width. Plantlets are transplanted without disturbing the root ball. After planting, the relative humidity inside the greenhouse should be kept 70-80 % or shade nets should remain closed for 4-6 weeks to avoid the desiccation of the plants.

3. *Watering*

Gerbera requires a constant supply of water but in a smaller quantity. The soil surface should dry out between watering. The deep roots draw water from the base of the beds. Excessive dampness around the leaves and crowns encourage the incidence of Botrytis, powdery mildew and crown rot diseases. Immediately after transplanting, watering should be done using overhead sprinklers, watering cane and watering hose fitted with a 'rain' nozzle up to one month. Thereafter, gradually change to drip irrigation. Generally, one dripper per plant is required. The average water requirement may be 700 ml / plant /day (4.5 to 6.0 litre/day / m^2), depending on the season, soil texture, light intensity and crop stage. It is necessary to avoid over watering, even though plants tend to dry out rapidly due to their large leaf area. Watering can be done in 2-3 shifts a day. A watering shift should not be too short to ensure vertical flushing of excessive salts. On soils with insufficient drainage, irrigation should be limited to only one large watering before, 10 a.m., in order to let the soil dry out during the day.

4. *Nutritional requirement*

Gerbera requires a plenty of macro and micronutrients for proper growth and development. The first dose of fertilizers should be applied 10-15 days prior to the transplanting. Application of 10:15: 20 g NPK/m^2, respectively during first three months of transplanting will be helpful in better establishments and growth of the plants. From fourth month onwards, application of 15: 10: 30 g NPK/m^2 at two months intervals is desirable. Fertilization with 80 % of the recommended dose (1: 10: 30 g / m2) two months through straight fertilizers and remaining 20 % through the water soluble fertilizers reported cost effective to obtain the higher returns and good quality of cut flowers. Improved flower yield, stalk length and flower diameter can be obtained by application of 15 tone / ha vermicompost plus 75% recommended dose of NPK (15:10: 30 g/m^2/two months). Spraying of micronutrients like boron, calcium, iron, copper and magnesium @ 0.15 % each, at monthly intervals is also desirable. Concentrated complex of micro-nutrients mixtures may also be sprayed (0.2 %) at monthly intervals for obtaining increased flower production.

Application of fertilizers through drip irrigation (fertigation) preferably in the morning around 8.00 a.m., is advisable. Fertilizers should be given 10 days after the transplanting. Initially 19:19:19 NPK should be given at 0.5 to 1.0 g /l of water on daily bases upto flowering stage. After appearing of the flowes, the fertigation should be given as per the following ratio: viz., 1:1:3:1: 0.2 NPK, Ca and Mg. Application of micro-nutrients through drip irrigation is also beneficial. However, it is suggested to analyze the soil and water samples once in two months regularly to fine tune the fertigation programme according to the soil and water analysis report.

5. *Harvesting of Flowers*

The first flowering occurs in 7-12 weeks after transplanting. Harvesting of flowers should be done when outer 3-4 rows of the disc florets have fully developed. However, some growers remove the first flower as soon as feasible. They believe this results in more rapid and uniform elongation of the remaining flowers escapes. In a young plantation, the plants may be pulled loose while picking. Therefore, the first flowers have to be picked very carefully. The cut flowers should be pulled rather than cut, because cutting will have a stem stub on the growing plant that encourages development of diseases. The base of the flower stem should be pushed slightly down and to one side and then separated from the crown with gentle upward tug. There is absolutely no need to apply knife, scissors, etc. while harvesting of flowers. Water movement through the basal portion of the stem is slower than through the segment 5-6 cm higher. Therefore, the basal 5-6 cm portion should be cut using very sharp knife removed before placing in the harvesting bucket. Gerbera has a fairly soft stem and walls of the vascular bundles are also soft so that they are prone to damage if cut with pruning shears.

Yield and Plant Protection

a) Yield: The flower yield of gerbera crop depends upon various factors like growing environment, varieties planted, spacing adopted, etc. The average yield under low cost naturally ventilated polyhouse is 30-55 flowers / plant / year (200-400 flowers/m^2/year). If properly maintained, healthy plants give economical yield for a period of $2^{1/2}$-3 years, after that plantation will have to be replaced.

b) Insect-Pests

i. **White fly** (*Bemisia tabaci*) lays eggs on the lower surface of the leaf. Nymphs and adults suck cell sap from the leaves resulting in discolouration, reduced plant vigour and development of the shooty mould on leaves. Removal and burring of heavily infested leaves check white flies build up. Chemical control includes sprays of Confidor/Admior or Dicofol @ 1.5 -2 ml / 1 of water at regular intervals.

ii. **Leaf miner** (Limomyza *trifolii*) causes considerable damage to the foliages. Adults lay eggs in the leaves by punching. Maggots feed inside the leaves by leaving characteristic mines which turn brown , brittle and dry in case of heavy infestation. Removal and destruction of the heavily infested leaves reduce further damage. Spray during morning hours of pyrazophos, metasystox, or dimethoate @ 1.5- 2 ml/l of water are quite effective measures.

iii. **Aphids** (*Myzus persicae*) attack growing flowers. Both nymphs and adults colonize on lower sides of petals and spread to entire flower in heavy infestation. Aphids suck the sap from the petals leading to disclouration, wilting and distortion of the flowers. Spray of monocrotophos or dimethoate @ 1.5 – 2 ml /l of water at regular intervals is quite effective to control aphids.

c) Diseases

Crown rot or root rot is a major disease and severe infection can destroy the entire gerbera crop. The crown of the plant turns dark and decays. The leaves of the infected plants wilt gradually, become dry and turn to radish brown colour. The infected roots develop a water brown colour. The disease can be effectively controlled by drenching of the root zone of infected plants with 0.2 % solution of mataxyl, captan or carbendazim or copper oxychloride at fortnightly intervals.

i. **Powdery mildew** (*Erysiphae cichoracearum*) infection causes white powdery coating on the leaves. Spraying with Thiophanale methyl @ 1.5 %, Carbendazim @ 0.2 %, Hexaconazole @ 1ml/l, Triadimefon @ 1g/l or Dipotassium Hydrogen Orthophosphate (K_2HPO_4) @ 4.3 g/l are effective against this disease.

ii. **Wilt** (*Fusarium oxysporum*) infected plants show yellowing of lower leaves. The collar portion shows vascular discoloration. Rouging and soil application of Benzamidazoles (0.2%) help in reducing the disease spread.

iii. **Root Rot** is caused by *Pythium.* Root skin is easily removed and finally wilting of the plants. Avoid excess watering. Drenching with Benlate @ 2 gm/l or with Bavistin @ 2 gm/l is helpful to control the root rot.

Prospects of Floriculture Industry Under Protected Environment

Worldwide consumption of floriculture products is estimated to be 60 billion US dollars. Cut flowers contribute about 60 per cent of the world trade and the remaining are live plants, cut foliage, dry flowers, dried plants etc. Since India is situated comparatively closer to major flower consuming countries than its Asian counterparts it has good scope and potential in flower trade. The severe winter in major flower producing European Countries is also an advantageous factor to

India, especially areas like Bangalore, Pune, Hyderabad, Nasik, North East (for Orchids and Anthurium) which enjoy moderate climate all throughout the year. The prospect of an auction market for floricultural products coming up shortly at Singapore give an added advantage to India. (Sindhu, 2010).

In India, nearly 98.5% of flowers are grown under open cultivation and hardly 1.5% flowers are grown under greenhouse. Besides the export growth and the domestic demand for cut flowers are expanding at an impressive rate of 7-8%, which open up new avenues for greater investments in this sector. Until this time, the floriculture industry in India is overdependent on cultivation of roses and now its time for us to diversify into other crops like orchids, anthurium, lilium, alstroemeria, carnation, etc. (Sindhu and Misra 1997).

The global floriculture market is around 70 billion US dollars of which 3940 million dollars value is traded in cut flowers only. The Aalsmeer in Holland is the largest market for cut flowers. To ensure quality flower production of cut flowers, greenhouse technology is required for uniform production. Before taking cut flowers under greenhouse certain consideration should be kept in mind for flourishing business.

- Greenhouse is to be constructed according to the location, climate and crop to be grown
- Availability of labour, quality water and planting material
- Development of Post harvest handling techniques and efficient marketing channel

Table 4. Existing production centres in India

Flowers	Existing Location	Future Location
Rose	Bangalore, Pune	Kodaikanal, North -East
Carnation	Sangali, Satara, Bangalore, Ooty, Kullu, Shimla	Uttarakhand, North -East, H.P
Gerbera	Sangali,Satara, Pune, Bangalore	New Delhi, Kolkata, North -East
Orchids	Tamil Nadu. Kerala, Maharashtra, Mangalore, Guwhati	Goa, North-East
Anthurium	Coorg, Kerala, Bangalore, Alibagh	North East, Bangalore
Lilium	Ooty, Bhimtal, Kullu	North-East, Uttarakhand
Bird of Paradise	Bangalore	North-East, Pune, Uttarakhand

Table 5. Crops Suitable for Cut Flower Production Under Greenhouse

Crops	pH	EC (mMhos/cm)	Temperature (D/N)	Photo-period(Hr)	CO_2 (ppm)	Yield/m^2/year	Varieties
Rose	6.2-6.8	0.7	28/18	14	1000-1500	150-300	First Red, Grand Gala, Konfetti, RavalVivaldi, Golden Gate
Carnation (Standard)	6.0-7.0	1.2	18/13	16	500-1500	200-300	Scania, Garbo, Tanga,Raggio-de-Sole,Calypso
Chrysanthemum	6.2-6.7	1.0	16/10	10	700-900	250-450	Snowball,Snowdon, Thiachen Queen
Lilium (Asiatic & Oriental)	65-7.0	1.0	18/12	50% shade	1000-2000	100-200	Connecticut, King, London, Pollayana, Stargazer, Nepal
Gerbera	6-7	1.4	16/12	14	800-1200	200-300	Diana, Nettle, Sangria, Rosula, Gloria, Sun Way
Alstroemeria	6-7	1.0	18/15	16	1000-1500	-	Pluto, No.-14, Alladdin, Tiara, Cardinal

Allied Activities other then Cut Flowers

Cut Green – An Evergreen Tool for Diversification

Flowers and foliage fit perfectly in the changing social sector of the world over. The billion-dollar floriculture industry worldwide has projected itself as worldwide phenomena in the recent decade. In the sphere of agro-export, cut foliage trade is now identified as an extreme focus segment for rejuvenating the stagnant economy. Cut foliage is used as a filler, lining and background material in various flower arrangements. They also used for bringing life to the bouquets, wreaths and garlands. Cut foliage has the following advantages: -

- Lesser cost of production because most of them can be grown in open conditions
- Year round production of a particular crop
- No strict time limit for storage
- Lesser risk of damage to quality while transport
- Longer shelf life

1. Flower seed production

Seed production of seasonal flower crops is a lucrative business and is practiced in considerable areas in Punjab and Haryana. This offers higher returns from unit area. Of late, demand is increasing in the domestic market also. Research work is required on the development of high yielding varieties including F_1 hybrids, agro-techniques for producing uniform seed with higher certification standards.

2. Pot Plant Production

The demands of potted plants for use of various purposes in India and for the export to other countries are increasing. At present, India exports potted plants nearly 50% and it is 10% higher than the cut flowers. Many nurserymen in our country are exporting foliage plants to West Asian countries. Flower/foliage leaves have to take advantages of the changed socio-economical situation in various parts of the country. The per capita consumption is on rise worldwide including India, but proportion of spreading is not so high in India. The trends indicate that production is steadily increasing which is essential for the growth of pot plant industry in India. The pot plants are grouped into four broad categories

- Flowering pot plant
- Green or foliage pot plant
- Hanging basket plants
- Bulbous pot plants

3. Nursery as Industry

A nursery can be a part of a diversification strategy to make a farm more profitable, or it may only be an enterprise. Nursery production usually involves reduced levels of synthetic fertilizers and pesticides, use of IPM systems and focus on building the soil to promote plant health. The plant material available in most of our nurseries today unfortunately is outdated or substandard. Lack of quality planting material is the major hindrance for not realizing the full potential of floriculture in India. Plant material of various kinds (seedlings, budded plants, rooted cuttings, bulbs, tubers, corms, annual seed, etc.) are required for commercial flower production, pot plant production (and their rentals), for adding to home garden and for landscaping (corporate landscaping, bio-aesthetic planting, etc.).

Most of the nurseries still follow the same old practices of production, management and marketing of nursery materials. An attempt was made to focus some of the strategies for Hi-tech nurseries for future growth of nursery industry in our country which includes –

- Production of quality planting materials i.e. through ***satellite nurseries***
- Development of infrastructure
- Container plant production and rentals
- Plug plant production
- Corporate landscaping
- Plant tissue culture
- Seed production

This segment may not require much research but requires adequate extension for transferring the already developed technologies. On the research side, the use of PGR's can further be exploited in propagation and seed production of ornamentals.

4. Value Addition

Value addition in floriculture trade takes place at every step to create a novel product to cater the demands of the heterogeneous consumers. The trends in floriculture keep pace with the trends in the fashion industry and hence the evolution of the floriculture is much more rapid than other industries. The flower trade across the continents is valued at US$ 6 billion but in reality the trade is much more in magnitude as the value added products and their sale start after the sale of the fresh flowers at the auction houses. Value addition in floriculture includes extraction of essential oils. The major flower crops important for essential oil extraction are rose, jasmine, tuberose, vanilla etc. The major producers of essential oils are Brazil, China, USA, Egypt, India, Mexico, Guatemala and Indonesia. The demands for the essential oils progress at a rate of 7-9% per annum and offer unique scope for large scale cultivation of essential oil bearing crops and their processing using both conventional and ultra modern vacuum distillation/ solvent extraction/ liquid CO_2 methods. Essential oils and perfumery from natural sources are in great demand. In India, flower crops grown for essential oil production are limited and include mainly rose, jasmine, tuberose, etc. *Rosa damascene* is exclusively cultivated for extraction of essential oils, rose water, attar, gulkand, etc. in certain pockets of Rajasthan and Uttara Pradesh.

Research should be focused on development of varieties with higher oil content and standardizing distillation methods for higher oil recovery. Further, identification of more crops and standardization of production technology need to be included in the research agenda. Promotion of this sector encourages ancillary industries like steam distillation and use of indigenous technical knowledge (ITK) for making value added products.

References

Anonymous 2006. Indian Horticulture Database 2009. NHB, Ministry of Agriculture: 18-20.

Barreto, M.S. and Jagtap, K.B. 2007. Assessment of substrates for economical production of gerbera flowers under protected cultivation. *J. Orn. Hort.*, **9**(2):136-138.

Bhattacharjee, S.K. and De, L.C. 2003. Advanced Commercial Floriculture. Aaviskar Publishers, Jaipur, Vol -1, pp 4-5.

Bose, T. K. and Yadav, L. P. (Ed.) 1989. Commercial Floriculture. Naya Prokash, Calcutta.

Chadha, K. L. and Choudhury, B. 1986. Ornamental Horticulture in India, ICAR, New Delhi.

Chaudhary M.L. 2000. A Flower City Banglore special issue on high-tech horticulture Fhal fool –Hindi (17 -21).

Das, P., Samanta, P.K.S. and Parthasarathy, V.A. 2002. Gerbera. In. Commercial Floriculture Vol. I. Eds. T.K. Bose, and others. Naya Prokash. pp 163.

Dutt, M., Patil, M.T. and Sonawane, P.C. 2002. Effect of various substrates on growth and flowering of chrysanthemum. *Ind. J. Hort.* **59**(2): 191-195.

Kumar S. 2000. Floriculture industry. Ready for takeoff. Agriculture today, pp 15-17

Mishra R.L. and P. Naveen Kumar 2007. Biennial progress report of AICRP, ICAR New Delhi

Ramesh Kumar 2010. The Genesis of Directorate of Floriculture Research, A Bulletin, ICAR, New Delhi

Sindhu, S. S. 2009. National Conference on Recent Trends and Future Prospect in Floriculture 5-8 March 2011 held at Meerut. A Souvenir. 173.

Sindhu, S.S. and R.L. Misra 1997. Problems and Prospects of flowers trade in India. Indian Horticulture Vol. (2) : Golden Jubilee issue (50 Years of India's Independence)

Sindhu, S.S. and M. L. Choudhary 2001. Commercial Flower Production, Division of Floriculture & Landscaping, IARI, New Delhi-12.

Singh H.P. 2000. Current status of Floriculture in India, Floriculture Today, Sept., 6 -14.

16

Scenario of Flower Breeding for Protected Cultivation

Ramesh Kumar, Ganesh B. Kadam and Prabhat Kumar

Abstract

Domestic market both for cut flowers as well as loose flowers have increased significantly owing to rapid urbanization, changes in social attitude, increase in income level and the increasing habit of 'saying it with flowers'. The per capita consumption of flowers is increasingly growing up due to which the production has been increased and non-traditional areas have emerged as important players in floricultural production. In this scenario of intensive production system, protected cultivation can be a viable option. Protected cultivation offers unique opportunities to produce high quality flowers round the year and insulate from weather vagaries thus, reduceing risk from the climatic variations. Protected structure comprises of high cost, medium cost and low cost polyhouses and each structure needs specific varieties and technologies to achieve maximum yield. Hence, breeding program should be based on structure and climate as it has been observed that varieties bred in Northern hemisphere under high cost greenhouses are not performing alike in medium and low cost polyhouses of Southern hemisphere. Breeding work in India has been carried out in wide range of floriculture crops. Breeding of many floricultural crops are in fact not always simple and straightforward due to polyploidy, complex ancestry involving one or more species, aneuploidy, and in some families self and cross sterility barriers, etc. It appears that breeders have vast opportunities to incorporate disease and insect screening and development of varieties for protected cultivation in routine breeding programs during the evaluation of seedlings. To develop varieties for such adversities, our research institutes should have enough infrastructures to carry out required research. In a few areas, protected cultivation has succeeded in India due to efforts of private companies and enterprising farmers. Recently, with greater efforts of ICAR a few public institutions have started breeding of flower varieties suited for protected cultivation.

Keywors: Breeding, Protected cultivation, Rose, Gerbera, Carnation and Lilium.

Introduction

Floriculture in India has flourished in last couple of decades. The area under flower crops has increased several fold and at present it is estimated around 1, 91,000 ha with a production of 10, 31,000 tones of loose flowers and 67,027 lakh cut-flowers annually (NHB, 2010-11). There is, however, great scope to increase productivity of flower crops. In view of the growing international demand for floriculture products, especially cut flowers which have been registering an annual growth of over 11 percent (Anonymous, 1997), floriculture was given a priority status for export market development in India. Cut roses accounted for 19 percent of the international cut flower trade; therefore, special attention was paid to develop cut rose production as the major crop for export in India. However, in order to meet international standards for cut flowers, special protected structures are required for development of the floriculture industry in India. The cut flowers, which are presently exported from India, are from the hi-tech floricultural units. Protected cultivation although covers only 5% of total area under cultivation of flower crops which contribute significantly to total floricultural export. The broad term "protected cultivation" refers to any activity to protect crop plants against biotic and/or abiotic stress using physical, chemical or biological means. This opinion has its focus on the physical means of climatic control aimed at: i) eliminating the effects of meteorological events such as rain, hail and snow; ii) alleviating the severity of environmental stress, such as low temperature (e.g. freezing, chilling or just sub-optimal temperature), heat, excessive or poor irradiance, in particular regions and/or seasons and iii) modifying one or more environmental parameters (e.g. temperature, carbon dioxide concentration, light intensity etc.). Hereafter, protected crops are defined as cultivation carried out under any kind of permanent or temporary shelter covering the entire crop with the aim of enhancing its productivity.

In India, more than 90 % of farmers have land holdings with an average farm size of about 1.57 hectare comprising nearly 55 % of the arable land. These farmers are resource-poor and often do not have access to new technologies. But this huge human resource can sustain an intense and efficient agro-production system if provided with critical inputs like technology, marketing and credit. The most important aspect is to ensure greater participation of small and marginal farmers and convert their inherent weaknesses into opportunities. Growing flower crops under protected structures has huge potential to increase the acreage under intensive flower production thus, increasing the floricultural export and thereby income generation.

Since, protected cultivation is not a common practice in India, the greenhouse designs and structures have been imported from countries like Netherlands, Israel, France, and the United States. Hi-tech greenhouse rose production in India, like other countries, is highly capital intensive, with wide variations in both capital cost and return. Despite of all the efforts, Indian greenhouse industry contributes less than 1% to the total global cut flower trade. Being a nascent industry, Indian growers have yet to achieve the necessary volume and quality demanded by the international trade. Therefore, they are at a competitive disadvantage in the global market and often receive low prices due to weak marketing power and low product quality. Reasons lack of suitable varieties, poor infrastructure facility, lack of trained man power to handle hi-tech greenhouses, poor marketing facilities and other limiting aspects like cost, infrastructures, etc. However, with new upcoming areas and intensified research programs at ICAR level, the floriculture industry is expected to do better in development of greenhouse technology in India.

World Scenario

The greenhouse technology is not new and Europeans were considered the pioneers in this field. Later, with the advent of plastics during the World War II, a new phase in the greenhouse technology emerged. At present nearly 90 per cent of the new greenhouse are being constructed by using ultra violet (UV) stabilized polythene sheets as the glazing material. Total area under greenhouse in the world was 275,000 hectare as reported for the year 1999-2000.

In international terms, the consumption of cut flowers is concentrated in three regions: Western Europe, North America and Japan. The highest growth is expected in Japan and the USA. The West European market is becoming saturated. As a result, its cut flower consumption share is declining. New markets are emerging in the Eastern European countries. In one segment of the market, cut flowers are becoming part of the consumption patterns of people with high incomes. In Japan, domestic consumption of cut flowers will become more regular than in the past. In the past, consumption was mainly based on special occasions and institutional consumption (hotels, parties, etc.). This pattern is changing to a more Western style. Consumption is rising not only in Japan but also in other Asian countries with rising expendable incomes and a flower-minded culture. If economic development keeps accelerating in Latin America, a strong increase demand is also to be expected here.

There are more than 50 countries now in the world where cultivation of crops is undertaken on a commercial scale under cover. United States of America has a total area of about 5000 ha under greenhouses, mostly used for floriculture with a turnover of more than 2.8 billion US $ per annum. According to EUROSTAT (http://epp.eurostat.ec.europa.eu), about 136,000 hectares were grown under cover in 2007 in Europe (excluding France). Other sources indicate that the area cultivated in France was about 11,400 ha in 2005 (Sas-Paszt 2010; Sigrimis 2010). Thus, the area devoted to protected crop cultivation in European Community is roughly 150,000 ha. The countries with the higher areas under protected cultivation are: Spain (66,000 ha); Italy (34,600 ha); France (11,400 ha); the Netherlands (10,200 ha); Poland (6,300 ha) and Greece (4,900 ha). Protected crop cultivation is mostly based on plastic tunnels and greenhouses except in the Netherlands where glasshouse production is the most common.

The Netherlands is the traditional exporter of greenhouse grown flowers all over the world. With about 10,800 ha under cover, the Dutch greenhouse industry is probably the most advanced in the world. Dutch greenhouse industry however, relies heavily on glass framed greenhouses, to cope up with very cloudy conditions prevalent all the year round. A very strong research and development component has kept the Dutch industry in the forefront. The development of greenhouses in Gulf countries is primarily due to the extremity in the prevailing climatic conditions. Israel is the largest exporter of cut flowers and has wide range of crops under greenhouses (15,000 ha) and Turkey has an area of 10,000 ha under cover for cultivation of cut flowers and vegetables.

Egypt has about 1000 ha greenhouses consisting mainly of plastic covered tunnel type structures. In Asia, China and Japan are the largest users of greenhouses. Along with the Netherlands, the US and Italy, Japan is a major producer of both flowers and potted plants, but its production is mainly for the domestic market. In 1994, the value of production was about the same as the Netherlands and the USA. The production of both potted plants and cut flowers is growing rapidly. Both have doubled in space in ten years to about 18,000 ha of flowers and about 2,000 hectares of potted plants. The limited but rapidly growing international trade provides a strong stimulus to domestic cut flower consumption. Domestic consumption is also benefiting from this.

Only one third of the cut flowers are grown under glass or plastic. The importance of protected cultivation is increasing. A shift has taken place from seasonal to year-round production. The main cut flower is chrysanthemum, followed by carnation and rose. In Israel, total area increased in the early nineties. During the past four years, however, the total area has remained stable at about 1,900 ha. About 70% of the cultivation is done under glass, plastic or other protective cover. Israel is well known for its product innovation and knowledge transfer. It is the fourth largest exporter, mainly to the EU. Reduction of import duties by the

EU gives Israel extra export possibilities. Due to economic situation and high exchange rate of the dollar, its export position is under pressure.

Throughout Asia, production of ornamental products is rising. In traditional production areas such as Thailand and Malaysia, production is more and more orientated for the Japanese and (for special products like orchids) European markets. In Thailand, more than 7,000 ha is under cultivation for cut flower production. Production is increasing in the newly developing countries like Vietnam and China. In China, almost 60,000 ha area is under flower and foliage crops. In a period of three years, production increased from about 100 million stems to about 400 million stems. Development of the sector is just in an early stage. For example, standard quality, post-harvest technology and management skills are not at the required level at the moment.

Table 1. Area under protected cultivation of flower crops in world

Country	Area (ha)	Source
China (2010)	27,60,000	Yang, 2011
Korea (2009)	57,444	Lee, 2011
Spain	52,170	EuroStat, 2005
Japan	49,049	MAFF, 2011
Turkey	33,515	TurkStat, 2007
Italy	26,500	EuroStat, 2007
Mexico	11,759	SAGARPA, 2010
Netherlands	10,370	EuroStat, 2007
France	9,620	EuroStat, 2005
United States	8,425	US Census Hort Spec 2010

Indian Scenario

Application of greenhouse technology for commercial purpose is however, in its infancy in India. Greenhouses which existed in 1960s were used for commercial purposes but the impetus and the much required thrust came only in the mid 1980 s with the emergence of the industries manufacturing U.V. stabilized Low Density Poly Ethylene (L.D.P.E.) and development of indigenous technology for low cost greenhouses. In 1985, Indian Agricultural Research Institute (I.A.R.I) designed and set up first green house in Leh (J&K). Today there are more than 1840 small and medium sized greenhouses at Leh.

Farmers are also utilizing low and medium cost greenhouses for raising potted plants and seedlings in the nursery. In the Northern Gangetic plains, especially in Punjab, Haryana and Uttar Pradesh, farmers are using this technology to raise healthy seedlings of high yielding crop varieties so that they can be transplanted early in the fields during the onset of spring season to capture the early markets and thus, reap higher returns. South Indian states like Karnataka and Andra Pradesh have made considerable progress in development of nursery industries under protected conditions. In North Eastern states, especially in Assam, efforts are made to raise vegetable crops in the greenhouse-cum-rain shelter structures during the long South West monsoon periods. In these regions, stress is being given for the development of low cost greenhouses using bamboo frame structures, as these construction materials are readily available in these regions.

The State Seed Corporation of India is utilizing at present, the medium and the high cost greenhouse technology in their seed production units to maintain higher degree of purity of parental lineage and genotype. A large number of private companies are utilizing the high cost multispan greenhouse technology which has provisions for microprocessor controlled environment control system, like operation of motorized shading system, on line drip irrigation system, supplementary heating and lighting system and the likes, for vegatative propagation of chrysanthemums, roses, carnations, etc.

Since implementation of the new policies in 1991, 103 projects with foreign investment of more than Rs.80 crores have been approved to be set up in the country at an estimated cost of more than Rs.1000 crores around Pune, Bengaluru, Hyderabad and Delhi. At present, approximately 1850 ha area is under protected cultivation and many new areas like North East and Himachal Pradesh are coming up, with area likely to increase in the future.

Table 2. Area under protected cultivation of flower crops in India

States	Area (ha)	Crops
Maharashtra	450	Rose, Gerbera, Carnation
Tamil Nadu	380	Rose, Carnation, Gerbera, Lilium
Karnataka	370	Rose, Gerbera, Carnation, Orchid, Anthurium, Calla Lily
Andhra Pradesh	160	Rose, Gerbera
Haryana	150	Rose, Gerbera
West Bengal	150	Gerbera, Carnation
Uttrakhand	100	Gerbera, Carnation, Lilium, Rose
Himachal Pradesh	70	Carnation, Rose, Gerbera, Lilium
Uttar Pradesh	20	Gerbera, Rose

Performance of Flower Crops Under Different Protected Structures

Polyhouse

Polyhouse is a framed structure cladded with polyethylene film which can provide the favourable conditions for growth of the plants in several ways, viz., favourable environmental condition, protection from heavy winds, pests, diseases and other climatic conditions (Khan, 1995). Nagaraj *et al.* (1999) found better growth, quality and yield in exotic rose cultivars under polyhouse when compared to open condition. In an evaluation of chrysanthemum varieties under open and polyhouse conditions, an increase in plant height and branching (50% and 28%, respectively over open condition) was noticed and consequently yielded 35 to 40 per cent more flowers under polyhouse condition (Gaikwad and Dumbrepatil, 2001). Singh and Ramachandran (2002) reported that two types of greenhouses viz., naturally ventilated (NV) and fan and pad cooled system (FP) were compared for growing nine exotic cultivars (Diablo, Lyonella, Ornella, Sunset, Tara, Thalassa, Tiramisu, Twiggy and Whitsun) of gerbera (*Gerbera jamesonii* Bolus ex Hook f.) at IIHR, Bengaluru during 1989-99. The results revealed that NV greenhouse gave better vegetative growth and higher yield of flowers without affecting their quality. The cultivars Diablo, Lyonella and Tiramisu gave higher yield than the others. An experiment conducted to evaluate 20 different spray chrysanthemum cultivars to assess their growth, flowering and yield parameters under open and polyhouse conditions revealed that open planting had significantly better growth, early flowering (74.39 days), flower number (71.25), flower size (7.59 cm2) and yield (2.24 kg/m2) compared to polyhouse planting. However, the blooming period (41.68 days) and vase life (15.08 days) of the flowers were found significantly better under polyhouse condition (Talukdar *et al.*, 2003).

Mahanta and Paswan (2003) studied the comparative performance of eleven gerbera cultivars under open conditions and plastic rain shelter. The cultivars exhibited significantly better performance under plastic rain shelter as compared to the open condition with respect to growth, physiological and flower characters. Patil et al. (2003) while comparing the performance of rose cv. Gladiator under fan – pad cooled greenhouse with that of open field cultivation, found that the greenhouse crop out performed the open field crop in terms of total number of flower production (3897), weight per flower (16.71 g), number of flowers per plant (25.47), number of flowers per unit area (152.82/m^2) and average stalk length (53.00 cm). Gerbera grown under naturally ventilated polyhouse produced maximum plant height, plant spread and number of leaves (Gupta et al., 2004). Singh *et al.,* (2006) reported that six gerbera cultivars gave better yield in naturally ventilated polyhouse and three cultivars gave better result in fan and pad greenhouse. More yield was observed in cv. Diablo (39.20%) followed by Lyonella (37.00%), Sunset (15.00%), Thalassa (8.80%), Twiggy (1.60%) and Whitsun (6.20%) in NVGH in comparison to FPGH. In FPGH more yield was observed in cv. Tara (36.50%) followed by cvs. Ornella (2.00%) and Tiramisu (6.00%) in comparison to NVGH. The quality of flowers was better in NVGH as compared to FPGH. Guttal and Takte (1993) studied the effect of low cost polyhouse on production of chrysanthemum and obtained significant results with respect to yield (20.19t/ha), average weight of flowers per plant (389g), diameter of the flower (6.5 cm), vase life (12 days) and attractive colour as compared to cultivation under open conditions. The cultivar Kiss produced the highest flower yield (80.96/m2) followed by Yunone (64.16/

m^2) and Vivaldi (57.79/m^2), as compared to Gladiator, which produced the least number of flowers (44.13/m^2). Quality parameters viz., bud length, bud diameter and total stem length, superior in Gladiator followed by Yunone and Vivaldi. The Cv. Kiss had better vase life (9.37 days). Gladiator recorded 9.0 days and Yunone 8.25 days' vase life whereas, Eterna recorded the least vase life (7.25 days) (Nagaraja,1996). Sindhu and Kumar (2004) laid out an experiment to study the performance of four commercial cut flower rose varieties viz., Konfetti, Novajo, Grandprin and Rakathagandha under unheated polyhouse, Maximum number of flowers/m^2 (10.42) was produced by Novajo followed by Konfetti (8.60) and Rokatgandha (8.22). Length of the cut stem was minimum (47.64cm) in Novajo. Konfetti, Gradprin and Rakthagandha were at par with each other. Size of the flower and number of petals/flower were maximum in Grandprin, resulting in the longest vase life.

Chandragiri *et al.* (2004) reported that 20 exotic spray varieties of Chrysanthemum were evaluated for their performance under naturally ventilated greenhouse. Among them, Solomon Impala recorded the highest plant height (132.16 cm) at harvesting stage. Rafel recorded the least number of flowers per plant (6.93), while varieties Shuttle and Require recorded maximum flower diameter (7.39 cm). Caymen recorded the highest number of ray florets per flower head (314.27). Maximum vase life was observed in varieties Linekar and Linekar Amber. The study found that all exotic varieties produced good quality cut flowers and can be grown successfully around Bangaluru under protected structure. Malhotra and Kumar (2000) conducted experiment in Department of Floriculture and Landscaping, PAU, Ludhiana and reported the effect of different treatments overhead shading and polythene covering on rose Cv. Raktagandha. The maximum plant height and plant spread were observed in plants, which were given 25 per cent summer shading fellowed by winter covering. However, maximum number of flowers per plant was observed in plants maintained under 50 per cent shade and polythene covering. The plants produced blooms about 10 days earlier under 25 per cent shading than 50 per cent shade plants.

Other Requirements for Protected Cultivation

Physical properties of soil, moisture and groundwater

The moisture content in greenhouses is essential if one has to obtain the highest possible yield from the production. In addition, permeability of the soil and groundwater level also play a major role. Humus, in particular and clay retain water and are thus important constituents of the soil. Soil without any humus dries out rapidly and needs additional and regular watering. The greenhouse grower must drain excess water to avoid water logging. Good permeable soil is therefore, the best. Furthermore, the groundwater level should be low (60-80 cm minimum below surface level) so that excess water can easily subside. If groundwater level is high then water will need to be able to flow to lower areas via drains. The depth of main rooting zone (topsoil) depends on plants need for optimum root development. Generally, a topsoil of 40 to 50 cm in depth is adequate. It has to aerate and drain well. There are dozens of crops which can root deeper. Fertility can be increased by adding nutrients to the irrigation water for quicker uptake by the plants. If the soil has been under greenhouse cover for a long time the content of unabsorbable salts will increase. Furthermore, salination of the topsoil will also occur if frequently stagnant surface water and ground water disappear from the ground through evaporation. The degree of salination can be measured using an EC-meter. If leaching does not occur naturally by rainfall or is insufficient, then the soil must be leached with neutral water. Leaching is usually done with large quantities of water (200 to 300 mm), preferably portion by portion, to allow the salts to dissolve gradually. Next, it is important that water rapidly drains off deep via the ground water, taking with it the excess salts in solution. In shallow soil (with a high ground water level) an intensive drainage system is indispensable for effective leaching and discharging of drain water. If leaching of the soil present insurmountable difficulties, use of glass wool or rock wool as growing medium could be the answer. However, this would mean a more complex form of greenhouse cultivation that would necessitate the use of drip irrigation and added artificial fertilizer. Laying beds on the ground, consisting of coarse sand wrapped in plastic film, could be a compromise. Then any excess water containing salts could be drained off via the soil.

Current Status of Cut Flower Breeding in India

Plant breeding is both an art and science and the most important tool to improve ornamental crops. Research is conducted by both public and private organizations and in public sector various central and state agricultural universities are carrying out most of the fundamental research for improvement. Private sector like seed companies is also doing the varietal development. In addition to this, amateur horticulturists are also engaged in development of cultivars in many flowers.

Cut flowers grown under protected conditions in India are rose, carnation, gerbera, lilum, chrysanthemum, orchids, bird of paradise, anthurium, etc. In India, breeding of flowers for garden and home decoration was started years ago. Scientific rose breeding in India, through various breeding techniques, started in early thirties when P. Bhattacharji of Bawan Bigha, Deoghar, Bihar made first cross. This has got impetus with the work of Dr B.P. Pal, former Director, I.A.R.I., New Delhi, when he released first hybrid cultivar 'Rose Sherbat' in 1959. Later on, work got prominence in the Division of Horticulture. There are about 20 other amateur Rose breeders who too have contributed in evolving new rose varieties suited to the varied agroclimatic conditions through conventional breeding methods. Today more than 500 Indian bred roses are available. In I.A.R.I. New Delhi, during the past three and half decades, around 200 cultivars were developed. These include those bred by Dr B.P. Pal in his experimental plot at I.A.R.I., and are well described in his book 'Rose in India'. We are well known in the development of rose's varieties for garden purpose which are floriferous and highly fragrant but characteristics required in varieties for cut flower production under greenhouse conditions were lacking.

IARI has the distinction of evolving the largest number of rose varieties in the country possessing desirable traits. 75 varieties comprising of Hybrid teas, Floribundas, Grandifloras, Miniatures and Climbers were released. Varieties with unique colours and hues top the long list of hybrids / varieties. A chocolate brown coloured rose hybrid, Mohini attracted the attention of rose breeders across the globe. Induced mutants *viz.*, Abhisarika, Striped Christian Dior, Pusa Christina and Madhosh: natural mutants Chitra, Pusa Abhishek, Climbing Sadabahar, Climbing Dr. Homi Bhabha, Delhi Pink Pearl and Nav Sadabahar exhibit unique spectrum of pattern and are well received by the garden enthusiasts.

Hybrids like Raktagandha, Arjun, Raktima, Pusa Gaurav, Pusa Pitambar, Mother Teresa, Priya and Dr. B. P. Pal evolved through vigorous breeding programmes possess desirable traits comparable to exotic cut flower varieties suitable for greenhouse cultivation. Rose varieties namely Bhim, Chambe Di Kali, Ganga, Jawahar, Mridula, Mrinalini, Nehru Centenary, Nurjehan, Priyadarshini, Raja Surender Singh of Nalagarh, Banjaran, Jantar Mantar etc., evolved at IARI, have become very popular among commercial growers and garden hobbyists for their eternal beauty and higher yield.

During the early stages of development in protected cultivation of flower crops, cultivars developed for open field conditions were grown. Soon, it was found that many of them did not grow well under protected conditions. The protected crop industry required cultivars which are early flowering, resistance to diseases, high tolerance to low temperature and low light conditions, low CO_2 compensative point, high quality and yield per unit area. It is very difficult for field cultivars to meet all these conditions. Therefore, in 1990s some work has been initiated by research institutions to improve the varietal wealth for protected cultivation. After some initial work, very a few cultivars like, Raktgandha, Arjun, Bahadur, Pusa Gaurav, were developed by the Indian Agricultural Research Institute for cut flower production but quality flowers were not up to the mark. The tropical background of local cultivars leads to poor flower quality and hence, import of the temperate cultivars which are highly suitable for the high-tech greenhouses were started. Indian farmers started importing planting material from countries like The Netherlands, USA, Israel and UK for which initially they charged very less but after the protection of plant varieties and farmers right act (PPV and FR), breeders have given right to royalty for their variety and hence, prices of importing material started increasing.

Research work on protected cultivation technology was carried out in some parts of the country. Flower crops like gerbera, carnation, Bird of Paradise, etc. are being sensitive to environmental conditions. It is difficult to get superior quality blooms under open field conditions. Therefore, it is recommended to grow them under protected environment to utilize the genetic potential to maximum. Recently, Gerbera has emerged as an important commercial flower crop under protected cultivation. The performance of cultivars varies with season, indicating clearly the interaction between genotype and environment. Photoperiod and temperature influence growth and flower production. Recently IIHR, Bangaluru developed Arka Krishika and YCD-1 cultivars of gerbera and Arka Fleme of carnation for protected condition and are performing well near Bangaluru conditions.

Breeding of Flower Crops under Greenhouse

Breeding Objectives

Breeding is the purposeful manipulation of plant species to create desired genotypes and phenotypes for specific purposes. Breeding of new varieties of ornamental crops is one of the most fascinating subjects to all floriculturists. Floriculturist, however, faces challenges which require both new and innovative approaches, as well as utilization of trusted classical breeding methods. Although, new breeding methods, such as marker assisted selection and genetic modification, open up numerous possibilities, these methods can often not be applied to new (relatively unknown) crop development, because of lack of basic information about the crops. Therefore, floriculturists still depend on the classical breeding methods for improvement in flower crops. As floriculture is a fashion driven industry, the aim and objective of improvement in flower crops varies with the growing environment, climatic conditions and consumer demand. Depending upon the need, breeders need to modify their objectives for crop improvement. In India, most of breeding work related to floriculture is for open field or exhibition purpose and now, time has come to go for varieties for protected cultivation. Certainly, aim is to improve the production of greenhouse grown cut flowers and combining some technological and agricultural qualities (such as appearance, colour, vase life, yielding ability, disease hardiness and low heating requirement) in the varieties.

Breeding Objectives of Commercial Flower Crops

Rose

Crop growth and yield of any crop are influenced by various factors like genotype, climate, soil, nutrition, various cultural manipulations, pest and diseases and interaction of all these factors. Among these factors, cultivars themselves contribute much to the performance of any flower crop. However, the performance of cultivars of any crop differs from one region to another region. When different cultivars are grown under identical conditions, it is the genetic potential that expresses the morphological differences. Hence, selection of cultivars is an important criterion for successful cultivation of any crop. For development of new cultivars in rose, various factors play important role like mother plant, donor plant, growing condition (open and greenhouse), compatibility of crosses, etc. Roses show high heterogeneity because of compulsory or preferential interspecific hybridization, but also show limited compatibility due to their complex ploidy levels (Rout *et al.*, 1999; Roberts *et al.*, 2003). Modern roses often produce hips that contain between 1 to 30 seeds, which germinate for less than 50%. Such discrepancy of seed numbers among different crosses will hinder the efficacy of rose breeding. In India, hundreds of cultivars were developed since the beginning of rose breeding. Most of the developed cultivars are suitable for open field and exhibition due to its floriferous flowering. Being a tropical country, Indian cultivars have faster rate of flower bud opening due to which these cultivars are not suitable for cut flower and cannot be grown under protected conditions. Very a few cultivars have good quality stem and flower bud which can be used for cut flower production. While breeding rose, character/objective changes with the growing environment and purpose. To develop suitable cultivars for protected cultivation, we need to have following objectives in mind:

- High yield (Flowers/m^2)
- Bud size and shape according to international standards
- Longer stem length
- Thornless stem and glossy leaves
- Disease resistance like powdery and downy mildew
- Colour – development of rainbow roses and blue roses
- Insect –pest resistance like mite and aphids
- Bent neck resistant
- Enhanced post harvest life
- Less sensitive to ethylene
- Abiotic stress resistant

The major supplier of rose cultivars to India is The Netherlands. Following cultivars and their performance data were based on cultivation in the Netherlands and grown on rockwool, recommended for cultivation in Indian conditions.

Variety/trademark	Flower size	Fl/m2	Vaselife
Red rose			
Amada	HT	180	12 days
Bordeaux	Medium	200	15 days
Devine	Medium	200	12 days
Grand amore	HT	160	14 days
Lucky star	Medium	180	12 days
Red ribbon	Medium	200	14 days
Respect	Medium	220	14 days
Red corvette	Medium	240	12 days
Red giant	Medium	200	10 days
Royal class	HT	220	12 days
Tiamo	Small	220	12 days
Yellow+Copper Roses			
Bella Donna	HT	140	14 days
Chantelle	HT	180	12 days
Charmant	Medium	180	14 days
Golden Gate	Small	220	10 days
Intense	Medium	160	10 days
Peach Avelanche	HT	200	12 days
Kerio	Medium	210	14 days
Maraia		180	16 days
Savitta	HT	190	12 days
Sirocco	HT	200	14 days
Svetlana	HT	220	12 days
Light Pink Roses			
Cadillac	HT	180	14 days
Cinderella	HT	180	12 days
Dekora	Medium	200	14 days
Duet	Medium	220	14 days
La Belle	HT	160	15 days
Sweet Avelanche	HT	200	12 days
Titanic	HT	180	12 days
Cadillac	HT	180	14 days
Dark Pink Roses			
Boogy	Medium	240	14 days
Bugatti	HT	180	12 days
Kingfisher	Medium	200	12 days
Marina	Medium	200	14 days
Opus	Medium	200	12 days
Pink Bordeaux	Medium	200	14 days
Orange Roses			
Kalahari	Medium	200	12 days
Fantasia	Medium	180	10 days
Picanto	Medium	160	10days

(Contd.)

Variety/trademark	Flower size	Fl/m2	Vaselife
Bi Color Roses			
N-Ensemble	Medium	180	12 days P/W
Rigoletto	HT	200	18 days O/P
Royal Circus	Medium	210	14 days R/Y
Wonderland	Medium	200	10 days O/C
White Roses			
Avalanche	HT	200	12 days
Sofie	HT	220	10 days
Spray Roses			
Jelena	creme	180	12 days
Fire Flash	strype y/o	140	10 days
Pink Flash	strype w/p	120	10 days
Snowflake	white	120	10 days

Gerbera

Gerbera is one of the most important cut-flowers, successfully grown under different conditions in several areas of the world and meeting the requirements of various markets. This success is primarily due to wide range in colour and shape of the flower. In the past, all Gerberas were grown from seed. This changed in the seventies, when techniques for multiplication with *in vitro* became available. These methods allowed a different approach in breeding and selection of Gerbera. Presently, Gerbera is a plant which is multiplied 100% *in vitro*, the most modern way of propagation in horticulture. Using this technique, it is possible to produce large amounts of new variety in a relatively short span of time and thus, intensifying the selection process. Breeding and selection are becoming increasingly more important, thus we need invest more in a new selection area with modern facilities. All these efforts resulted in a range of new varieties for gerbera growers with exceptional flower and stem quality and new colours. To undertake extensive improvement program in gerbera, we need to understand the market demand along with some basic requirements of gerbera growing like-

- More yield/ unit area
- Flower size
- Stem length and strength
- Tolerance to water stress
- Resistant to diseases like crown root, root rot
- Resistant to leaf minor
- Abiotic resistance
- More vase life/suitable for long distance transport

Most of the gerbera cultivars/hybrids were imported by the farmers. Some important commercial hybrids of gerbera marketed in India by Kumar Florist are -

Colour	Variety	Type
Red	Diablo,	Single and double with black Centre
	Ruby Red	Single with black centre
	Rubian	Single with black centre
	Sangria	Semi double with black centre
	Savannah	Single with black centre
	Testarossa	Single and double
Yellow	Cabana	Single and double with black centre
	Dana Ellen	Single and double
	Golden Gate	Double with black centre
	Piton	Single and double with black centre
	Solero	Single and double
	Tonneke	Single and double with black centre
	Thalassa	Single and double with black centre
White	Dalma	Single and double with black centre
	Himalayan	Single and double
	Junkfrau	Double
	Winter Queen	Single and double + Black Centre
Peach	Aida	Single and double + Black Centre
	Foske	Single with black centre
Pink	Charmander	Single and double
	Fleurance	Single and double
	Pink Elegance	Single and double
	Pink Sparklet	Single with black centre
	Primrose	Single and double
Orange	Careera	Single and double
	Gollath	Single and double
	Mystique	Single with black centre
	Oriella	Single and double
	Rodis	Single and double with black centre
	Sunset	Single and double with black centre
	Sunway	Single and double
Cream	Tiramisu	Single and double with black centre
	Vital	Single and double with black centre
Lilac	Magnum	Single and double with black centre
	Ringo	Single
Magenta	Aquilla	Single and double
	Tonga	Single

Carnation

Carnation breeding is directed for outcomes such as:

- A qualitatively better product;
- Improved productivity, more rapid flowering, better yield distribution; and
- New varieties to increase diversity and sustain market demand.

The breeding procedure typically consists of hybridization, self-pollination and selection (Holley & Baker 1992). If the desired trait is recessive, it may not be expressed in the F1 progeny. By self-pollinating the F1 and growing a large population of F2, selection of one or more individuals with desirable traits will be possible. The process of inbreeding (self-pollination) may, however, hinder the breeding objectives by generating recessive homozygotes expressing undesirable traits. A variation of the above breeding method has been described as pedigree breeding which entails selecting F2 individuals by continual soft-cut propagation (Galbally & Galbally 1997). While inbred parental lines are necessary to breed homogeneous F1 hybrid varieties, inbreeding detrimentally affects the inbred plants (Galbally & Galbally 1997). Inbreeding depression appears in the third selfed generation (S3) and therefore, it is almost impossible to produce S4 seeds (Sato *et al.* 2000). Hybrids between carnation and other Dianthus species can provide useful sources of genetic traits to achieve the above objectives (Segers 1987). For instance, interspecific hybrids are obtained through crossing *D. caryophyllus* and *D. capitatus*. These hybrids are highly resistant to bacterial wilt caused by *Pseudomonas caryophylli*. However, the flower quality is adversely affected and further improvement through backcrossing is necessary before commercial production (Onozaki *et al.* 1998). Interestingly, traditional techniques for carnation breeding were developed before the discovery of inheritance by Mendel. As a result, these traditional techniques do not include a self-pollination step. In absence of self-pollination, continuous hybridization has inadvertently resulted in highly heterozygous carnation varieties. The high level of heterozygosity promotes recombination mutations resulting in further new varieties (Holley & Baker 1992). Mutation breeding has been also employed to create new colour mutants. More recently, the development of doubled haploidy techniques has also permitted breeders to accelerate breeding and selection (Holley & Baker 1992). To develop hybrids/cultivars for protected cultivation, we need to have specific characters in mind like

- Higher yield
- Colour- Development of blue carnation and other novel colour
- Longer and sturdy stem
- Bigger bud size

- Development of Super carnation
- Resistant to calyx splitting and wilt

Standard type: White Sim, White perfection, Madonna, Arthur Sim, Dusty Sim, Shocking Pink

Spray type: White elegance, Red baron, Scarlet Ncelegae, Amalia, Annelies, Donora

Chrysanthemum

Chrysanthemum is a worldwide ornamental plant. The process of chrysanthemum breeding is slow. Successful cultivar production requires innumerable crosses and a large number of progeny to obtain the combination of desirable characters from a number of individuals. The apparent complex genetic heterozygosity makes the cultivated chrysanthemum an unlimited source of new flower forms and cultivars. Good quality flowers with early flowering cultivars are favourable to produce quality chrysanthemum cut flowers at lower cost of production. Flowering control should allow shorter breeding cycles. A possible strategy to speed up flowering in plants is genetic manipulation of key regulators of floral transition.

- Longer short day response group for flower bud development
- A few strap- shaped leaves near the terminal floret
- Stronger but more brittle stems
- Doubleness
- Novel colours
- Increased post harvest life
- Resistance to abiotic and biotic stresses

Lily

The genus Lilium comprises more than 90 species. Involvement of wild species is promising for the production of novel cultivars (Van Tuyl *et al.* 1986; Van Tuyl *et al.* 2000). The factors limiting interspeciûc crosses in Lilium can be separated into pre- and post fertilization barriers (Van Tuyl *et al.* 1991). To overcome pre-fertilization barriers, various pollination methods which tested: applying a mixture of pollens from several species, cut-style, grafted-style, placenta pollination and *in vitro* ovule pollination (Asano, Myodo 1977; Van Tuyl *et al.* 1982; Van Tuyl et al. 1991; Chi 2000). Embryo rescue, ovary slicing and ovule culture are used to overcome post-fertilization barriers (Chi 2002). Asiatic hybrid lilies (AH) are important today in the flower trade. However, they lack fragrance and are sensitive to Botrytis blight. The Oriental hybrids (OH) are ornamental but sensitive to virus. Lilium breeders are facing very deep challenge of fertilization barriers along with lack of suitable cultivars for resistance to various biotic and abiotic stresses. Before going for any breeding program, breeder should have following points in mind along with the fertilization barrier.

- Resistance to diseases like bulb rot, Botrytis and several viruses
- Development of new flower shapes, sturdy stem, colours and fragrance
- Tolerance to low light intensity, heat and leaf scorch
- Year round forcing ability
- Long term storage ability and rapid bulb growth
- Good bulb production (under plains)
- Tolerance/resistance to biotic stresses (insect and pests)

So far we don't have any variety of Lilium from our country. These are some of the exotic varieties which are popular among the farmers and commercial growers.

Asiatic Hybrid Lilies: Red Black Out, Monte Negro, Brunello, Lyon, Loreto, Pink Vivaldi, Toronto, Cannes, Farfalla, Navona ,Sorpressa, Vermeer, Renoir, Corrida, London, Gironde

Oriental Hybrid Lilies: Algarve, Brindisi, Dazzle, Glow, Fangio, Original Love, Litouwen, Timaru, Menorca, Salmon Classic, Madrid, Royal Trinity, Samur, Royal Sunset, Courier

Current Strategies and Future Prospects of Resistance Breeding in Ornamentals for Protected Cultivation

Current breeding strategies for resistance breeding are mostly intuitive and only in exceptional case based on a deeper knowledge about plant pathogen interactions. Therefore, most varieties are bred without special selection schemes and disease resistance is often considered only very late in the process of variety development. Many commercial breeding programs do not even consider disease resistance as a major goal at all. Thirty years ago around 500 publications about black spot of roses were reported (Baker and Linderman, 1979). Nothing of this knowledge has been utilized for developing new cultivars even 30 years later which is characteristic of lack of importance of resistance breeding in ornamental variety development. Carnation breeding is one of the a few exceptions where for example, disease testing is part of the selection process for Fusarium resistance. For this purpose, cuttings of advanced clones and in some cases, even seedlings are inoculated with defined spore suspensions of particular races and disease progress is monitored. These tests will be repeated in the field as it has been observed that interactions differ between greenhouse and field environments (Ben Yephet *et al.*, 1996). Less advanced is a strategy applied for Dahlia breeding for resistance to powdery mildew where artificial inoculations of excised leaves in the lab are combined with field observations. These field tests are manipulated for higher disease pressure in which highly susceptible genotypes are grown between candidate genotypes providing sufficient inoculum for the new genotypes to be tested.

Future Prospects for Resistance Breeding in Ornamentals for Protected Cultivation

Before predictions on the future of resistance breeding can be made, particular situation for each plant pathosystem has to be considered. This also includes market share and half life of varieties on the market of the particular ornamental crop as this will determine the possible investment in resistance breeding. Major factors influencing future strategies of resistance breeding in ornamentals will probably come from three major sources: the rapid technological progress e.g. new high throughput sequencing technologies fuelled by human genetics, the wealth of information from plant genome projects and increased knowledge about plant pathogen interactions obtained from a small number of model systems. This will have the following short term effect on ornamental plant breeding for resistance:

- Development of markers and their application to non model species will become much cheaper. This will also include diagnostic tools for fingerprinting host and pathogen genomes via DNA- or protein chips, PCR assays, etc.
- Results from sequenced genomes of plants will facilitate access to candidate genes for marker assisted selection in two ways: Understanding the regulatory networks responsible for disease resistance in model species will bring up a number of additional candidate genes for master switches of disease resistance responses.
- Isolation of orthologues of interesting candidate genes will be easier with every new related plant genome completely sequenced by homology based approaches.
- Once candidate genes are isolated, techniques like Tilling and EcoTilling will enable the evaluation of thousands of candidate sequences for useful genetic variation that can then be channeled into breeding programs.

Identification and isolation of resistance factors from host plants (e.g. transcription factors) will also lead to new candidates for transgenic approaches. Transgenic approaches will further shift away from genes coding for antifungal proteins to master switches for resistance, e.g. transcription factors. Induction of disease resistance will furthermore be limited to the main target tissues and only under conditions of pathogen challenge to reduce the metabolic load of resistance. This will be achieved by more specific promoters or even artificial promoters designed in the lab.

Candidate sequences from pathogens including virulence factors or indispensable housekeeping functions, might also lead to new targets for both transgenic approaches on the host side and plant protection by designer pesticides.

Greenhouse Breeding Problems

Indian flower crops possess limited gene pool. Many crosses that a breeder might like to make are actually incompatible. The outcome is random and unpredictable. Its difficult to breed for a specific trait, like fragrance for instance. Small, incremental improvements of a variety are not possible (suppose you wanted to improve the disease resistance). Average length of time it takes from the initial cross to getting the new variety to market is ten years. For example, each rose has approximately 100,000 genes and each variety has a different combination of all these genes. Infrastructural support (green houses, protected structures) to conduct the complete research work under fully controlled conditions is not competent to meet the desired results. Human resource engaged in floricultural improvement needs to be trained for capacity building as to expose them to sophisticated techniques to develop new varieties of flowers for greenhouse. Mission-mode approach is required to intensively document and evaluate the indigenous flora suitable for cut flower production and there is a dire need to develop indigenous cut flower varieties to insulate farmers from rising royalty. The research work in India has mainly focused on developing package of practices for a few crops like rose and gerbera India is lacking in organized research work to develop greenhouse cultivars.

Breeding Problems of Flower Crops in India: SWOT Analysis

The floriculture industry in India has registered considerable growth during the last one-and-a half decades and has emerged as a sunrise industry with apparently vast potential to grow further. It has contributed to the economy in several ways including generating employment opportunities and additional incomes in rural areas and fetching valuable foreign exchange. Floriculture provided a seemingly viable diversification option for the otherwise shrinking agricultural sector. Right from the early nineties, the central and state governments have introduced a number of policies and programs to promote export-oriented hi-tech floriculture and enormous government outlays have been made during the last three five-year plans in expectation that the sector would make a remarkable export breakthrough. Inspite of these efforts, India's share in the world's cut flower exports has hardly crossed even half of one percent. India continues to be a negligible player in the world's cut flower market, notwithstanding the massive area it has under flower cultivation. The reasons for lack of major strides in the export driven hi-tech floriculture are many. Lack of international quality plant materials and exotic varieties is one of the major drawbacks of export-oriented floriculture. The international varieties entail prohibitive patent royalties, while pirating attracts huge penalties. The only solution for this problem is to come up with the improved indigenous varieties that can match the international specifications. With limited resources, we have to go ahead. We need to assess and analyze our potential, through SWOT, we have throne some light on it.

SWOT associated with breeding new cultivars/hybrids of flower crops in India are as under

Strengths

- Favorable climatic conditions for breeding throughout the year (north and eastern hilly states)
- Good number of central and state agricultural universities engaged in breeding of flower crops
- Rich germplasm for breeding and product diversification

Weaknesses

- Lack of local germplasm suitable for protected cultivation for existing commercial crops
- Lesser-availability of quality planting materials of flower crops
- Extreme climatic variation unfavorable for hi-tech protected cultivation
- Poor facilities of infrastructure to conduct complete breeding program under protected conditions.

Opportunities

- Huge demand in domestic as well as in international markets
- Large number of private entrepreneurs are being involved in breeding pro cess which needs to be collaborated with public sector
- Local cultivars are rich in fragrance and resistant to a few of biotic and abiotic stresses which can be further utilized in improvement of flower crops.

Threats

- Competition from countries like China and African countries (Kenya, Ethiopia, etc.)
- Untrained manpower to handle hi-tech greenhouse structures to conduct complete breeding program.
- Very a few training centers available in India with full training and sophisticated equipments.
- Easy availability of imported planting material at door step of farmers.

Greenhouse Varieties Development: Future Strategies

To develop new varieties of cut flower for protected cultivation, we need to work on collective manner with multi-institutional approach. There is need to screen the potential of locally available varieties of various flowers for cut flower production under protected conditions. Various tools like Hybrid Breeding, Mutation Breeding, GMO Breeding and Marker Assisted Breeding should be used to improve flower quality and increase genetic base. New varieties suitable for protected cultivation need to be evolved through combination of various conventional and non-conventional methods of improvement. Florigene (The World's First Molecular Breeder), part of the Suntory Group, has used genetic modification technology to create valuable improvements to important flower species. "Applause" a blue rose cultivar bred first time through plant biotechnology in Japan, is a breakthrough in flower breeding. Also, we need to develop infrastructure for basic and applied research to develop indigenous cut flower cultivars. All research and development work to develop variety need to be carried out under fully controlled conditions. Young talent should be diverted towards the research and development by providing special incentives. So far, there are sporadic instances of its success for commercial cultivation in Maharashtra and Karnataka but its organized research need to be strengthened in suitable climatic regions to develop the cut flower cultivars. There is a huge domestic market in north and eastern India for high value cut flowers. Small farmers can greatly benefit from this market if they are provided with right mix of protected cultivation technology along with the royalty free indigenous varieties of high value flower crops.

References

Asano Y. and Myodo H. 1977, Studies on crosses between distantly related species of lilies. I. For the intrastylar pollination technique. *J. Japan. Soc. Hort. Sci.,* **46**: 59–65.

Baker, K. F. and R. G. Linderman. 1979, Unique features of the pathology of ornamental plants. *Annu. Rev. Phytopathol.,* **17**: 253-277.

Ben-Yephet, Y., M. Reuven, A. Zveibil, and Shtienberg, D. 1996, Effects of abiotic variables on the response of carnation cultivars to *Fusarium oxysporum fsp. dianthi. Plant Pathol.,* **45**:98–105.

Chandaragiri, R., Janakiram, T. and Srinivas, M., 2004, Performance of exotic Chrysanthemum varieties under greenhouse. *Proceedings of National Symposium of Recent Trends and Future strategies in Ornamental Horticulture.* University of Agricultural Sciences, Dharwad. pp 43-48

Chi H.S. 2000, Interspecific crosses of lily by in vitro pollinated ovules. *Bot. Bull. Acad. Sin.,* **41**: 143–149.

Chi H.S. 2002, The efficiencies of various embryo rescue methods in interspecific crosses of Lilium. *Bot. Bull. Acad. Sin.,* **43**: 139–146.

Gaikwad, A. M. and Dumbrepatil, S. S., 2001, Evaluation of chrysanthemum varieties under open and polyhouse conditions. *J. Orna. Hortic. New Series,* **4**(2) : 95-97.

Galbally, J. and Galbally, E. (1997). Carnation and pink for garden and greenhouse. Timber press, Portland, Oregon, USA.1-310.

Gupta, Y. C., Dein, Lequec, Dhiman, S. R. and Jain, R., 2004, Standardization of growing media under protected environment for gerbera in mid hill of Himachal Pradesh. *J. Orna. Hort.*, **7**(1) : 99-102.

Guttal, G. B. and Takte, R. L., 1992, Effect of partial controlled environment on growth of chrysanthemum in polyhouse. Nation. Conf. on Use of Plastics in Agriculture, 19-20th May, 1983, pp. 103-108.

Holley, W. D. and Baker, R. 1992, Breeding for better varieties. Chapter 3. In: WD Holley, R Baker, eds. *Carnation Production II.* Colorado State University, 21-30.

Khan, M. M., 1995, Relevance of green house – A report from training on construction and management of low cost greenhouse, 8-10th May, 1995, pp. 1-8.

Kumar, R and Sooch, M., 2003, Effect of shading and media on growth and flower production of Gerbera jamesonii variety priyadarshi. National symposium on Recent advances in Indian horticulture. Proceedings of Indian society of Ornamental Floriculture, Trichur,12-14 Nov., pp. 37-83.

Mahanta, P. and Paswan, L., 2003, Assessment of Comparative Performance of Some Gerber (Gerbera jamesonii Bolus) cultivars under open condition and plastic rain shelter in Assam condition. Nation. Symp. Recent, Adv. in Indian Flori. Trichur 12-14 Nov, Proc. Indian Soc. Orna. Hortic., pp. 154-165.

Malhotra, R. and Kumar, R., 2000, Effect of different treatments overhead shading and polythene covering on rose Cv. Raktagandha. Proceedings of National symposium on Recent Trends and Future Strategies in Ornamental Horticulture University of Agricultural Sciences, Dharwad.

Nagaraja, G. S., Narayanagowda, J. V. and Nagaraja, N. B., 1999, Influence of growingcondition and cultivars on growth, yield and quality parameters of exotic rosecultivars. *Mysore J. Agric. Sci.*, **33**:139-144.

Onozaki, T., Ikeda, H., Yamaguchi, T. and Himeno, M. 1998, Introduction of bacterial wilt *Pseudomonas caryophylli* resistance in *Dianthus* wild species to carnation. In ISHS Acta Horticulturae III: New Floricultural crops, Considine, J.eds, *Acta Horticultuirae*, Perth Western Australia, 127-132

Patil, B. R., Gohil, K. B. and Vaghasiya, P. M., 2003, Response of gladiator rose to fan-pad cooled greenhouse under humid climate of Navsari. All India Seminar on Potential and Prospects for Protective Cultivation, pp. 113-116.

Roberts, A.V., Gudin, S. and Debener, T. 2003. Biotechnologies for breeding and Breeding. p.5-30. In: Encyclopedia of rose science, Vol. 1, Elsevier, Oxford.

Rout, G.R., Samantaray, S., Mottley, J. and Das, P. 1999, Biotechnology of the rose: a review of recent progress. *Scientia Hort.,* **81**:201-228.

Sas Paszt L, 2010. Data-collection of existing data on protected crop systems (greenhouses and crops grown under cover) in Eastern European Member States in EU. Technical report to EFSA under procurement CT/EFSA/PPR/2008/07.

Sato, S., Katoh, N., Yoshida, H., Iwai, S. and Hagimori, M. 2000, Production of doubled haploid plants of carnation dianthus carnation L. by pseudofertilized ovule culture. *Scientia Horticulturae,* **83**:301-310

Segers, A.1987 the developemt of interspecific carnation hybrids. *Acta Horticulturae* **216**:373-375.

Sigrimis N, 2010. Data-collection of existing data on protected crop systems (greenhouses and crops grown under cover) in Southern European Member States in EU. Technical report to EFSA under procurement CT/EFSA/PPR/2008/06.

Sindhu, G.S. and Ramesh K., 2004, Performance of rose cut flower varieties under unheated polyhouse conditions. Proceeding of National Symposium on Recent Trends and Future Strategies in Ornamental Horticulture, University of Agricultural Sciences, Dharwad.

Singh, K. P. and Ramachandran, N., 2002, Comparison of greenhouses having natural ventilation and fan and pad evaporative cooling systems for gerbera production. *J. Orna. Hort.*, **5**(2):15-19.

Talukdar, M. C., Sarma, B., Das, S. and Mahanta, S., 2003, Evaluation of spray chrysanthemum cultivars under open and polyhouse conditions. Nation. Symp. Recent Adv. Indian Floric. Trichur, 12-14 November, Proc. Indian Soc. Orna. Hort., 12-14 Nov., pp. 166-171.

Van Tuyl J.M., Marcucci K.M. and Visser, T. 1982. Pollen and pollination experiments. Vll. The effect of pollen treatment and application method on incompatibility and incongruity in Lilium. *Euphytica,* **31**: 613–619.

Van Tuyl J.M., van Dien M.P., van Creij M.G.M., Kleinwee T.C.M., Franken J. and Bino R.J. 1991, Application of in vitro pollination, ovary culture, ovule culture and embryo rescue for overcoming incongruity barriers in interspecifi c Lilium crosses. *Plant Sci.*, **74**:115–126.

Van Tuyl J.M., van Dijken A., Chi H.S, Lim K.B., Villemoes S. and Van Kronenburg B. 2000, Breakthroughs in interspecific hybridization of lily. *Acta Hort.*, **508:** 83–90.

Van Tuyl, J.M., Franken, J., Jongerius, R.C., Lock, A.M., and Kwakkenbos, T.P.M., 1986, Interspeci®c hybridization in Lilium. *Acta Hort.,* **177**: 591-595.

Van Tuyl, J.M., Van DieÈn, M.P., Van Creij, M.G.M., Van Kleinwee, T.C.M., Franken, J. and Bino, R.J., 1991, Application of in vitro pollination, ovary culture, ovule culture, and embryo rescue for overcoming incongruity barriers in interspeciûc Lilium crosses. *Plant Sci.,* **74:** 115-126.

17

Structural Support vis-à-vis Quality Concerns of Protected Floriculture

Ranjan Srivastava, Satish Chand and D.K. Singh

Abstract

The tradition of growing of flowers is age old. The flowers have travelled from the bowl of prayers to gardens to commercial sector for revenue. The quality concerned world demands for round the year production and supply of quality produce which can only be possible under controlled environment production. A range of location specific and crop specific infrastructures are being commercially used for production of planting material and flowers. General and crop specific guidelines for trade of cut flowers are available which the producers and policy makers must know for successful promotion of export of cut flowers from the country. The producer can also use practical methods for vase life study of their produce which is of great relevance in cut flower trade.

Keywords: Protected floriculture, trade, crop specific guidelines, export, cut flower

Introduction

Indian floriculture industry is shifting from traditional flowers to cut flowers for export purpose. The liberalized economy has given an impetus to Indian entrepreneurs for establishing export oriented units (EOUs) under controlled climatic conditions. The wide variations in agro climatic conditions in India permit us to grow various types of tropical and subtropical flowers. Though such plants grow successfully in open conditions without cover, they fail to meet the quality requirements of being blemish free. High quality ornamental plants meant for export should be scientifically grown only in plastic, glass or fibre glass greenhouses or at least under partial cover. Though initial investment is high in protected cultivation, the products in them will obtain far more attractive prices.

Protected Cultivation

Protected cultivation is the cultivation of plants inside a structure which protects plants from wind, precipitation, excessive radiation, temperature extremes, insect and diseases. It is also of vital importance to create an ideal microclimate around the plants. Therefore, an ideal protective structure should have facilities for creating most acceptable growing conditions for a given plant species. The main aim is to boost productivity of the crop per unit area in addition to year round culture, improved quality and control of various environmental factors. This can be achieved in a protected environment. In India, flowers are being grow in 1, 90,000 hectare area (NHB, 2012). Out of this total area, 700 hectare is under protected cultivation and over 200 hectare in under export oriented unit (EOU). At present, about 250 companies got clearance for establishing floriculture units in India. Model units are established in fields wing academic departments.

1. Department of Horticulture, G.B. Pant University of Agriculture & Technology, Pantnagar
2. Department of Vegetable Science, G.B. Pant University of Agriculture & Technology, Pantnagar

Considerations Before Planting

a. **Bed or Bench Preparation:** Most of the flower crops are either grown on raised beds or in metallic benches. Orientation of beds should be along the length of greenhouse so that 45-60 percent of floor area can be utilized. In modern greenhouse, movable benches are used.

b. **Medium:** Soil is the natural growing medium for most of the greenhouses from economical consideration. But, it has drainage and pest disease problems. Nowadays soilless media like rockwool, perlite and cocopeat are common.

   A good medium must be stable and has optimal bulk density and pH range of 5.4-6.0 for soilless media and 6.2-6.8 for soil media.

c. **Irrigation and fertigation:** Automatic watering system through drips or overhead foggers are generally used depending upon the crop. Fertilizers are opened with irrigation water mostly. This avoids wastage of water and fertilizer and reduces cost.

d. **Choice by cultivar**: In India, major demand is for deep colour and it changes with season and occasion. Colour, productivity, disease resistance, vase life, growth rate and market demand must be taken into consideration.

e. **CO_2 Enrichment:** CO_2 enrichment accelerates vegetative growth, water use efficiency and photosynthetic capacity leading to high quality, blemish free blooms. For CO_2 enrichment, CO_2 generators and injectors are used.

Ideal CO_2 concentration for important flowers-

Crops	Ideal CO_2 conc. (ppm)	Toxic CO_2 conc. (ppm)
Rose	1000-1500	2500-3000
Carnation	1000-1500	2000-2500
Chrysanthemum	700-900	1500-3000
Gerbera	800-1200	1600-2000

General Quality Specifications for Cut Flowers

The General Quality Specifications for Cut Flowers apply to all types, varieties and cultivars that are traded as cut flowers via auctions affiliated with the Association of Dutch Flower Auctions (VBN). All products sold via the auctions are exclusively intended for decorative purposes and are not intended for internal use unless specifically otherwise indicated on the product. In the case of incorrect use, consumption, contact and/or allergy, the products may have harmful consequences for humans and/or animals.

The general specifications consist of requirements for quality, grading, packing and identification. Deviation of these requirements is only possible by explicit agreement between supplier and buyer, established beforehand via an intermediary office of a VBN auction .The requirements set forth in the General Specifications apply to all cut flowers unless otherwise stated in the product specifications.

Definition of Terms

Abnormal growth

Imperfections in the flower, leaf and/or stem parts resulting from disorders during growth/culture.

Acceptance Standard

Minimum standard of the Reference Test (R) with respect to cultivar-specific internal quality standards which must be met by the cultivar so that the product can be expected to have acceptable internal quality.

Active Ingredient Code

For the sake of clarity, pre-treatment agents are coded according to their active ingredients. The code is printed on the label.

Bent Necks

The uppermost part of the stem under the flower, drooping downward

Cluster Width

Size of the inflorescence of *Celosia argentea* Cristata Group (the cristate cluster), measured across the greatest width, expressed in cm. The cluster width can be indicated in the assortment code by means of characteristic code S23 for minimum flower diameter.

Colour Uniformity

The batch must be composed in such a way that the selling units are uniform in terms of colour composition.

Cut Flowers

The term "cut flowers" also includes ornamental green twigs and cut shrubs.

Damage

Damaged leaf- or flower parts resulting from jolting, bruising, pinching, cutting, etc.

Deficiencies

A shortage of nutrients leads to aberrant colouring in flower and/or leaf.

Do Not Trade

The product does not meet the requirements for trade. Depending on the situation, the supplier is given the opportunity to improve the product to such a degree that it can still be traded or the product is destroyed.

Fresh

State of the cut flower in which it looks fresh, showing healthy colour in flower and leaf, is not limp and shows no signs of ageing, as far as can be perceived by the eye.

Heel

The undermost thickened and dark-coloured part of the stem which cannot easily absorb water

Leaf Point

Dying-off of plant cells on the top of a leaf larger than 2 mm

Opening

The first stage of the flower opening up

Parasites

All organisms which are harmful to the cut flower, such as insects, mildew/ fungus, bacteria, viruses, etc. "100% free of animal and/or plant parasites" means that the product must be visually free of parasites. An exception to this rule applies to products which are traded with the designation "insect-free", whereby "100% free" truly means a zero tolerance policy.

Residue and/or Contaminants

Residue from crop protection agents and/or lime residue and/or other deposits on flower- and leaf parts

Shoots

The (generative) shoots which must be removed (disbudded) in some cases

Shrinker, Shrunk Flower

Ethylene damage to the flower, causing it to wilt at an early stage

Visible Remnants Of Main Flower

Remnants of the (final stage of the) main flower after removal, rendering a negative influence on the ornamental value.

Minimum Requirements for Trade

Cut flowers may be traded if they meet the requirements below. Products which do not meet the requirements are not traded and are destroyed, if necessary.

Internal Quality

Pre-treatment

- A quantity of cut flowers is required to be pre-treated. The product specifications indicate whether and which pre-treatment is required or recommended.

- Cut flowers without any further product specifications and supplied on water in a cut flower container must be pre-treated using pre-treatment agents containing the active ingredient chloride (code 9 on the label of the product), to be applied according tot the instructions.

Pre-treatment with STS

- Cut flowers standing in water with STS (STS solution) may not be brought in.
- Pre-treatment is verified by measuring the minimum concentration of STS in the flower bud in accordance with the VBN (Dutch Flower Auctions Association) protocol "Verification of the pre-treatment of cut flowers".
- In connection with the environmentally harmful characteristics of silver, used STS solutions may not be dumped in surface water. The solution may not be dumped until it has been neutralised with the supplied neutralising agents. The neutralised solution may only be dumped via the sewer system. The sediment (residue) must be brought to government designated sites (Small Chemical Waste Depot).

Pre-treatment with chlorine or aluminium sulfate

- Cut flowers pre-treated with agents containing the active ingredient chlorine or aluminium sulphate must be supplied standing in water with pre-treatment agent.
- Pre-treatment is verified by taking samples of the supply water or by determining the bacteria count in accordance with the VBN protocol

Pretreatment with quaternary ammonium compounds, gibberellins or sugars

- Cut flowers treated with agents containing the active ingredient quaternary ammonium compound, gibberellins or sugars must be supplied standing in water with pre-treatment agent
- Pre-treatment is verified by taking samples of the supply water in accordance with the VBN protocol
- Pre-treatment with quaternary ammonium compound is verified by determining the amount of active ingredient in the stem.

Penalty

If the above requirements are not met, the auction imposes a penalty.

External Quality

Minimum quality

- All cut flowers offered for sale must meet at least the requirements set for B1. value after trading and will retain this value for some time.
- All cut flowers must also be of such a quality that it can reasonably be expected that the flowers will still have ornamental value after trading and will retain this value for some time.
- Products that are offered for sale per bunch as unit of sale (instead of per stem) must comprise at least 3 branches/stem per bunch..

Ripeness

- The lowest minimum ripeness requirement for cut flowers is that the inflorescence must be well developed.
- The highest maximum ripeness requirement is that the flowers may not show any sign of ageing or have left off flowering. (See table below.)

min.ripeness ↓ max. ripeness ↓

	saleable					
not saleable	maturity stage 1; inspection code not ripe	maturity stage 2	maturity stage 3	maturity stage 4	Maturity stage 5 inspection code ripe	not saleable

Penalty Cut flowers which do not meet the above minimum requirements are not traded.

II. Quality and Grading Criteria

Quality and grading criteria per batch

Batches offered for auction must meet the requirements below (in order to be traded as A1).

- The batch must be of good internal quality.
- The batch must be fresh.
- The batch must be supplied in appropriate ripeness stage.
- The batch must be free of animal and/or plant parasites.
- The batch must be free of harmful effects and/or damage from animal and/ or plant parasites.
- The batch must be free of damage and/or deficiencies and/or deviations and/or contamination in– flower/inflorescence/bud;– branch/stem; – leaf/ needle/thorn.
- The batch must be of good form, composition and flower and leaf colour.
- The undermost 10 cm of the stems must be free of leaves.
- The stems must be straight and sturdy enough to bear the flower.
- The batch must be uniform in colour, thickness, sturdiness and bouquet volume.
- The batch must be graded in the proper manner.
- The batch must be properly packed.

Requirements and Quality Comments

If a batch shows a deviation from one or more of the mentioned requirements, the batch can be auctioned with a qualification (and pertaining inspection code). The seriousness of the deviation determines whether the batch is traded without a negative qualification, with a minor comment (an even inspection code) or with a serious comment (an odd inspection code). A comment is given on the basis of the tables below.

Table 1: Negative comments for deviations from the requirements, except for the deviations mentioned in Table 2

deviation / no. stems in batch	to a slight degree	to a heavy degree
less than 5 %	no comment	minor comment
5-25 %	minor comment	serious comment
more than 25 %	serious comment	do not trade

Table 2: Negative comments for deviations involving shrinkers, water stems, Botrytis leaf, miner leaf and freshness.

deviation no. stems inbatch	to a slight degree	to a heavy degree
less than 5 %	minor comment	serious comment
5 and up %	serious comment	serious comment

An exception to the above holds for deviations from the maturity requirement in section 1.1 when products must be graded on the basis of maturity and the maturity stage must be specified in the grade code. In that case no inspection codes will be assigned.

Supplemental Grading Requirements Per Batch

The following supplemental requirements apply to the grading of cut flowers if they are graded according to length, weight, maturity or another characteristic. The product specifications indicate which aspect serves as the basis for grading. Table 1 applies to these requirements. If the grade differ 1 class, a slight or moderate deviation is indicated. A serious deviation is indicated in grades differing 2 or more classes. These codes do not hold for products that must be sorted on the basis of maturity and the maturity stage must be specified in the grade code. If, in those cases, the maturity is incorrectly specified in the grading code, this will be corrected and the auction may impose sanctions.

Grading According to Length

- Flowers are graded in length classes in increments of 5 cm as shown in Table 3.
- The maximum allowed difference in length between the shortest and the longest stem in one bunch is 3.0 cm.
- The stem is measured from where it was cut to the uppermost flower (bud) and the measured length is rounded down.
- The length grade can be given during trading in the grade code by using characteristics code S20, according to the length of the shortest stem - rounded down - is indicated.

Table 3: Grading according to stem length

Minimum stem length	Characteristics code S20
25 cm	025
30 cm	030
35 cm	035
40 cm	040
45 cm	045
50 cm	050
55 cm	055
60 cm	060
And so on	And so on

Grading According to Weight

- Flowers are graded in weight classes of 1, 2, 5 or 10 grams, depending on the product.
- The weight grade can be mentioned during trading by in the grade code by using characteristics code S21.
- The average stem weight in the batch determines the grade to be indicated.

Grading According to Maturity

- Graded in classes, whereby a maximum of two successive stages of ripeness may occur in one batch.
- The grading according to maturity can be indicated in the grade code by using characteristics code S05.The spread within a batch is indicated by specifying the most immature stage occurring in the batch as the second digit (1, 2, 3, 4 or 5) and the most mature stage (1, 2, 3, 4, 5) as the third digit.

Grading on the Basis of other Characteristic

Depending on the product, a grading according to other aspects may occur, such as number of buds, spike length, flower diameter, number of clusters or sheath length.

Number of Flowers Buds

- Graded according to the number of flower buds in grading classes as shown in Table 4.
- If flowers are graded in a class with the designation "and up" (upgrading), an average 5 stems per bunch of 10 stems must then have more than the minimum number of buds.

- The number of flower buds can be mentioned during the trade in the grade code by using characteristics code S22. The code indicates the lowest number of buds per stem in the batch.

Table 4: Grading according to number of flower

Minimum buds/stem	Av. buds/ stem	Characteristics code S22
1	10	010
1 and up	15	015
2	20	020
2 and up	25	025
3	30	030
3 and up	35	035
4	40	040
4 and up	45	045
5	50	050
5 and up	55	055

Mixed Batch Requirements

Cut flowers may be supplied as "mixed" if they belong to product group level 3 and the batch consists of at least three clearly distinguishable colours, type or shades of flowers per bunch or packaging unit. A maximum of 50% of products in the batch may consist of one shape, type or colour. The mixed batch must be evenly uniform over the bunches and packaging units in the batch.

Quality Groups Based on Inspection Codes

Cut flowers can be traded in 3 quality groups: A1, A2 and B1, depending on the extent to which they meet the quality and grading criteria.

Inspection Codes

The degree to which the batch meets the quality and grading criteria is expressed in negative comments accompanying the batch. The comments are translated into a corresponding negative inspection code (characteristics codes K01 and K02) on the following basis:

Comments	Negative inspection code K01/K02
none	none
minor comment	even code
serious comment	odd code

Division into Quality Groups

Depending on the inspection codes, cut flowers are divided into quality groups as follows:

INSPECTION CODES

- No inspection code → A1
- 1 even inspection code → A2
- 2even inspection codes → A2
- 1 odd inspection code or 1 odd and 1 even inspection code → B1
  - With the exception of codes:
    - 149 (shrinkage)
    - 495 (water stems)
    - 613 (Botrytis in inflorescence)
    - 615 (Botrytis in leaf)
    - 645 (Miner leaf)
    - 905 (rejected)
    - 941 (not fresh)
    - 999 (no action)
- 2 odd inspection codes
  - With the exception of a minimum of 1 code beginning with 7 or 8 B1
- All other combinations of inspection codes → **do not trade**

The codes below must be interpreted as even codes:

117	171	447
819		
123	179	455
839		
125	181	463
867		
133	415	531
919		
139	417	763
939		
143	427	765
163	431	817

The codes below must be interpreted as odd codes:

148
494
612
614

Grade Code

Information about the grade is provided in the grade code by the use of characteristics codes. Indication of characteristics codes is mandatory or recommended, depending on the product. The Overview of Grade Codes for Cut Flowers (appendix to and part of these general specifications) shows per product which characteristics codes are to be indicated (compulsory or recommended) and on which position in the grade code. The grade code may contain a maximum of 20 characteristics codes. While trading via auction clock only the characteristics codes on the first 4 positions will be shown.

III. Packing Requirements

General

- Cut flowers are delivered for shipment in cut flower containers or in boxes.
- Cut flowers must be supplied in packaging materials approved by the VBN.
- For cut flowers which are supplied standing in water in a container, clean drinking water (with the prescribed pretreatment agents added to it) must be used. There must be sufficient water to prevent the flowers from becoming dry during the auction process. At the moment of inspection, a maximum of 5 cm water may be in the container. The water level is measured without the flowers in the container.
- A single-use box may not be used more than once.
- Products, including those with thorns, needles, hard branches and/or sharp leaf edges, and also products with (projecting) supporting materials must be packed so as to prevent the risk of any form of injury and/or damage to humans and/or the products' surroundings during sale
- Packaging must be chosen that properly protects the cut flowers during transport and sale in the rest of the chain.

Loading

- Optimal loading must be realised, whereby the volume of the bunches determines the number of supplied bouquets per container unit.
- There must be at least 5 cm free space between the product and the tray above it in the stacking cart.

Penalty

If cut flowers are not packed in accordance with the requirements, the auction reserves the right to impose penalty.

IV. Identification Requirements

Consignment note

Every stacking cart must be accompanied by a fully and correctly filled-in consignment note containing at least the following information provided by the supplier:

- Supplier information: company name, address, supplier's number
- Auction or selling date

- Number of container units
- Number of pieces per container unit/packing unit
- VBN container code
- VBN product code: the code must be chosen which describes the product as accurately as possible
- Grade code
- Accompanying product description
- Type of stacking cart: auction stacking cart or Danish container (if applicable)
- Auction or product group (if applicable)
- Positive quality comment (if applicable)
- Method of trade.

Product label

Every container must be provided with a label with product and supplier information.

Product information

Genetically modified products must be made identifiable as such by means of correct indication of their GM nature in the product information

Penalty

If the above requirements are not met, the auction reserves the right to impose a penalty.

V. Recommendations

- It is advised to specify the maturity grade of cut flowers in the grade code (by using characteristics code S05 and only then if the product specifications contain no other guidelines for the use of the positions concerned in the grade code).
- If sleeves are used, hot needle-perforated or porous sleeves or foil with a sleeve thickness of 30 μm (P30) are recommended.
- In the case of products offered for sale per bunch, it is advised to specify the minimum number of stems in the bunch in the grade code (by using characteristics code L11).
- It is advised to indicate the following text on the sleeves of cut flowers that

are wrapped in a sleeve per bunch or per bundle: "This product is exclusively intended for decorative purposes and is not intended for internal use. In the event of incorrect use, consumption, contact and/or allergy the product may have harmful consequences for humans and/or animals". The crop specific trade requirements are summarized below:

1. Rose

I. Minimum Requirements for Trade

Pre-treatment

Roses must be pre-treated and supplied in water with a pre-treatment agent. Products containing the active ingredient aluminium sulphate and spreading agent (code 3 on the label of the product) must be used and applied in accordance with instructions. • The minimum concentration of aluminium in the supply solution must be 64 ppm.

Minimum Quality

Roses, excluding spray Roses, which are sold per bunch (instead of per stem as unit of sale) are to be offered for sale in bunches of at most 20 stems.

Bacteria Content

- The bacteria content in the stem must be less than 1 million bacteria per gram of stem.
- The bacteria content is determined by means of stem samples.

Ripeness

- The minimum ripeness requirement for Rose is that a minimum of 95% of the number of stems in a batch must have a coloured bud
- Whose bracts must be completely separate from the petal.

II. Quality and Grading Criteria

- **Quality Requirements Per Batch:** The following additional requirements apply:
- The batch must be free of growth defects, including:
  - flat buds;
  - grass hearts;
  - crooked necks;

- The tolerances as shown in Overview 1 in the General Specifications for Cut Flowers apply.

Grading Requirements Per Batch

Rose must be graded according to

- Length, at which instead of an allowed 3 cm difference in length, all Roses must be bunched so that the stems in the bunch are even at the bottom. There is no tolerance in length difference at the bottom of the bunch.

Ripeness

- Number of bloomable buds;
- Height of flower bud, at which;
- Grading is done in classes according to the overview
- The grade can be mentioned in the grade code by using characteristics code S19
- The smallest height in the batch determines the code to be indicated.

Overview: Grading according to flower bud height

minimum flower bud height (in cm)	characteristics code S19
1	010
1.5	105
2	020
2.5	025
3	030
3.5	035
4	040
4.5	045
5	050
5.5	055
6	060
6.5	065
7	070
7.5	075
8	080

Packing Specifications

General Roses, except Rosa floribunda, must:

- Be supplied in bunches of 10 or 20 stems;
- Be packed in foil per container;
- Be bunched so that the flower buds are all at the same level or in two layers. When bunching with two layers, the separate layers may not touch each other.

Rosa floribunda must:

- Be supplied in bunches of 10 stems;
- Be packed in foil per container;
- Be bunched so that the highest placed flower buds are at the same level.
- Roses must be supplied in clean water (containing the prescribed pre-treatment agents).

Loading

- Rose must be supplied in packaging/ containers of type 533, 544 or 577, optionally with an additional rack. Rose flowers with stems shorter than 45 cm (length 035 and 040) must be supplied in packaging/containers of type 566. By way of exception, greater lengths may, if necessary, be supplied in packaging unit 566, provided permission is first obtained from the responsible auction inspector.
- Optimum loading is to be realised with the volume of the bunches determining number of bunches to be supplied per unit of packaging.
- To prevent the risk of formation of condensation during sale it is advised to:
- Wrap the roses in (entirely or partly) perforated film;
- Make sure the flowers are not too tightly encased by sleeves;
- Use well-fitting film that is not too wide.

2. Chrysanthemum

I. Minimum Requirements for Trade Minimum Quality

- All chrysanthemums must be visually free of living forms of leaf miner fly (*Liriomyza* and *Phytomyza*).

- All chrysanthemums must be 95% free of leaf miner.
- Cluster chrysanthemums and Santini must have a minimum of 3 bloomable buds.
- The maximum length grading class for Santini stems is 55 (code 055).
- The maximum flower diameter of Santini with full opening is 40 mm.

Ripeness

The lowest minimum ripeness for disbudded chrysanthemums and Santini is that 95% of the stems in the batch must have at least 3 half opened flower buds.

II. Quality and Grading Criteria

Quality standards per batch

The following additional requirements apply:

- The batch must be free of growth defects, including: – unevenly formed hearts or ray flowers; – split stems.
- The batch must be free of heels, which are the lowest, thickened and dark-coloured part of the stem (particularly outdoor chrysanthemums).
- The stems in the batch must be visually uniform.
- The stems must have a minimum of 6 flowers (buds). 4-5 buds mean a slight deviation; 3 buds mean a serious deviation.

Overview 1 in the General Specifications for Cut Flowers applies to the above requirements.

Grading standards per batch

Chrysanthemums must be graded according to:

- Length;
- Ripeness.

Disbudded chrysanthemums and cluster chrysanthemums from the greenhouse must also be graded according to stem weight, whereby:

- They are graded in weight grade classes in increments of 5 grams;the average stem weight in the batch is mentioned in the grade code by using characteristics code S21 (overview 1);
- All bunches in the batch must have a weight which equals to at least 85% and not more than 120% of the indicated average stem weight of the batch multiplied with the number of stems in the bunch;

- The content of all boxes in the batch must meet the average stem weight indicated for the batch.

Overview 1: Grading according to stem weight

average stem weight (grams)	characteristics code S21
25	025
30	030
35	035
40	040
45	045
and so on	and so on

Overview 2: Grading according to bundle weight

Min. bundle weight(grams)	characteristics code S35
500	500
550	550
600	600
650	650
and so on	And so on
950	959
1.000	100
1.050	105
1.100	110
1.150	115
and so on in increments of 50	and so on in increments of 5

- They are graded in weight grade classes in increments of 50 grams;
- The weight in the batch is mentioned in the grade code by using characteristics code S35 (overview 2);

III. Packing Specifications

General

Cluster chrysanthemums from the greenhouse must:

- Be supplied in bunches of 5 stems;
- Be packed one bunch in a sleeve.

Outdoor cluster chrysanthemums must be supplied in bunches of 5 stems. Santini must:

- Be supplied in bunches of 5 stems;
- Be tied in bundles of 5 bunches;
- Be packed per bundle in a sleeve. Disbudded chrysanthemums must be:
- Supplied in bunches of 10 stems; packed one bunch in a sleeve; It is required to supply outdoor cluster chrysanthemums and Santini in clean drinking water (possibly with the prescribed pretreatment agents) to prevent the flowers from becoming dry during the auction process; If chrysanthemums are traded per bunch (and not per piece), then the bunch must contain a minimum of 5 stems and all bunches in the batch must be of uniform volume. Chrysanthemums that are severely affected by Japanese rust (identified by means of symptom 6.33) must be packed in the boxes specially assigned for that purpose by the auction. If chrysanthemums are nevertheless supplied in reusable boxes, the auction will mark the boxes in question to prevent reuse.

IV. Loading

- The minimum quantity of stems of cluster chrysanthemums (with the exception of outdoor chrysanthemums) and disbudded chrysanthemums will be packed per container unit, depending on the average stem weight of the batch as mentioned in following tables:

Cluster chrysanthemum

Indicated average stem weight (grams)	No. stems per box (container code 520)
40 or less	120
45 and up	80

Disbudded chrysanthemum

Indicated average stem weight (grams)	No. stems per box(container code 520)
45 or less	80
50-85	60
90-150	40
155 or more	30

- At least the quantities of bunches of disbudded chrysanthemums below will be packed per container, depending on the average stem weight of the batch.

- For outdoor cluster chrysanthemums standing in water container 577 must be used and packed full.
- The minimum quantity of stems of Santini will be packed per container unit, depending on the average bundle weight as follows.

Indicated average bundle weight (grams)	No. stems per container with code 544
600 or less	100
601 or more	75

V. Recommendation

It is advised that the grade code concerning the length and weight be mentioned on the label with product information.

3. Gerbera

I. Minimum Requiremnts for Trade

Bacteria Count

- The maximum permitted bacteria content in the stem (bacteria count) is 1 million bacteria per gram stem.
- The bacteria content is determined by means of stem samples and by making a culture (in accordance with the inspection protocol).

Ripeness

The minimum ripeness requirement for Gerbera: a minimum of 95% of the number of stems in a batch must have a flower with a corolla with fully developed stamen.

II. Quality and Grading Ceriteria

Quality requirements per batch

The following additional requirements apply:

- Gerbera must be free of growth defects, including tears during growth and irregularly shaped hearts and flowers.
- Gerbera must be free of heels.
- The batch must have no limp flowers.

The tolerances as shown in Overview 1 in the General Specifications for Cut Flowers apply.

Grading criteria per batch

Gerbera must be graded according to:

- Ripeness.
- Flower diameter, whereby:

The flower is measured from the narrowest side of the flower and flat, meaning that by

laying a ruler on the flower, it will be flattened a little bit. The measured diameter is rounded down in accordance with the overviews;

- Contrary to the General Specifications for Cut Flowers, a tolerance of 0 does apply.
- For flowers with a diameter of less than 9.5 cm, the difference between the largest and the smallest flower may not be more than 1cm. For flowers with a diameter of 9.5 cm and more, the difference between the largest and the smallest flower may not be more than 1½ cm.
- During trade the flower diameter grade may be mentioned in the grade code by using characteristics code S23. The diameter of the smallest flower, rounded down, is indicated.

Grading according to flower diameter for large flowered, mini and other Gerbera

Minimum Large flower diameter code (in cm)	Characteristics code S23	Minimum Mini flower diameter code (in cm)	Characteristics code S23
9.5	095	7	007
10.5	105	8	008
11.5	115	9	009
12.5	125		

In addition Gerbera supplied standing in water is to be graded according to length.

III. Packing Requirements

General

- Gerbera with a minimum stem length of 40 cm are packed in interior boxes.
- Gerbera with a stem length between 35 and 40 cm may be packed in interior boxes, provided that the minimum length is mentioned on the auction note and on the box.
- Gerbera with a stem length shorter than 35 cm may not be packed in interior boxes; they must be supplied standing in water.

- The interior fittings must be stapled to the inside of the box. There must be no risk of the stems moving in the box; the auction may to this end demand the use of a strip of foam rubber.
- For delivery in boxes, Gerbera must first be pre-watered in clean water (with the recommended pre-treatment agents, if possible) and then packed dry.
- If Gerbera are delivered in bunches, then bunches of large-flowered Gerbera must consist of 10 stems and bunches of Gerbera mini of 25 stems, possibly with added Gerbera leaves.

IV. Loading

- For delivery in boxes, the boxes must be loaded upside-down (with the lid on the bottom) on the stacking cart.
- For delivery in boxes Gerbera will be packed in the quantities per box indicated below, depending on grade according to flower diameter.

Grade according to flower diameter (in cm)	Interior	Container code	No. of stems per box
7 and smaller	30- hole	614	60
8	35- hole	612	70
9.0 and 9.5	35-hole	612	70
	25-hole	612	50
10.5 and 11.5	25-hole	612	50
12.5 and larger	25-hole	612	50
	20-hole	612	40

For delivery in bunches, the minimum quantities to be loaded per container are indicated below:

Groups	No. of stems per bunch	No. of stems per container	Container code
Gerbera Mini	2525	100200	996577
Gerbera large flowered	10	100	577

V. Recommendations

Pre-treatment of Gerbera with chlorine lye or chlorine tablets is recommended.

4. Lilium

I. Minimum Requirements for Trade

Ripeness

The lowest minimum ripeness for Lilium is that at least 95% of the stems in the batch must have a flower bud showing colour.

Pretreatment

Lilium Asian Hybrids and Lilium LA Hybrids must be pre-treated with products containing the active ingredient STS (code 1 on the label), to be applied in accordance with instructions. Exceptions apply for 'Annabelle','Brindisi', 'Cannes', 'Chianti','Connecticut King','Festival','GranSasso','Hilde','Jolanda', 'Madras', 'Monte Rosa', 'Mont Blanc', 'Navona', 'Pronto', 'Sterling Star', 'Toscane', 'Vivaldi' & 'Wiener Blut'.

II. Qualty and Grading Criteria

Quality requirements per batch

The following additional requirements apply:

- The batch must be free of damage due to viruses.
- The batch must be free of root and bulb remnants.

Grading requirements per batch

Lilium must be graded according to:

- Length;
- Ripeness;
- Number of bloomable buds.

III. Packing Specifications

General

- Lilium must:
- be supplied in bunches of 10 stems;
- be packed in a sleeve per bunch;
- to be packed according to the survey. The supply of Lilium standing in water in containers of type 577 or 997 is allowed.

- *Lilium longiflorum*:
- when supplied in bunches of 10 stems, must be packed in a sleeve;
- when supplied loose, 20 stems must be rolled in paper (only for singles pips).
- It is required to strip the leaves from the bottom 10 cm of the stem. An exception to this requirement applies to *Lilium longiflorum* if it is rolled loose in paper.

IV. Loading

Products	weight per 10 stems (grams)	min. number stems per box/container	box/container (packaging code) depending on length
L. aziatische Group L. LA Group	< 600	100	519 for length < 85 cm525 for length > 85 cm
	601-900	80	
	901-1300	60	
	> 1300	40	
L. longiflorum Group	< 500	100	519 for length < 85 cm525 for length > 85 cm
	501-650	80	
	651-900	60	
	> 900	40	
L. oriental Group	< 600	80	519 for length < 85 cm525 for length > 85 cm
	601-850	60	
	851-1100	40	
	1101-1400	30	
	> 1400	20	527
other Lilium		full	

If Lilium is supplied standing in water in container 577 or 997, the container must be full.

V. Recommendations

In order to prevent the buds of Lilium Oriental Group from turning brown, the urgent recommendation is to avoid great drops in temperature (immediately) after harvesting. During the first hours after harvesting, the cold store temperature should not drop below 4 to 5 °Co Lilium Oriental Group should be harvested in time to prevent the product temperature from being too high (on warm days before 10:00 am). The product should then be transported from the greenhouse as quickly as possible. The product must have enough time to acclimatize (a few hours). It is also recommended that Lilium Oriental Group be pre-watered for a few hours (while acclimatizing).

5. Carnation

I. Minimum Requirements for Trade

Pretreatment

Dianthus must be pre-treated using pretreatment agents containing the active ingredient STS (code 1 on the label of the product), to be applied according to instructions.

Minimum quality

Excess shoots longer than 30 cm must be removed.

Ripeness

For **Dianthus**, a minimum of 95% of the number of stems in the batch must contain 1 coloured flower bud. For Dianthus standard, one bud must show a minimum of 1 cm coloured petal.

II. Quality and Grading Criteria

Quality requirements per batch

The following additional requirements apply:

- The batch must be free of shrunken flowers and/or "shrinkers".
- The batch must be free of woody stems.
- The batch must be free of crooked necks

The following additional requirements apply to spray Dianthus:

- The stems must have a minimum of 3 bloomable (mature) flower buds. 2 buds can show a slight deviation; 1 bud can show a serious deviation.
- The batch must be free of visible remnants of the main flower.
- The batch must be free of torn flowers.
- The batch must be free of excess shoots, whereby instead of Overview 1, the following applies:

Average no. of excess shoots per bunch	Quality comment
0-4	none
5-8	slight
9 or plus	serious

Dianthus standard

- The batch must be free of buds
- The batch must be free of stuck flowers.
- The batch must be free of torn flowers. The tolerances shown in Overview 1 in the General Specifications for Cut Flowers apply.

Grading requirements per batch

Dianthus must be graded according to:

- Length
- Ripeness
- Number of bloomable buds

In grading according to length, instead of an allowed length difference of 3 cm, all **Dianthus standard** must be bunched so that the stems at the bottom of the bunch are even. No tolerance is allowed in length difference at the bottom of the bunch. At Aalsmeer Flower Auction (VBA) the above requirement applies to all *Dianthus.*

III. Packing Specifications

General

Spray Dianthus must:

- Be supplied in bunches of 10 stems;
- Be packed per bunch in a hot needle perforated sleeve.

Dianthus standard must:

- be supplied in bunches of 20 stems;
- Must be bunched in two layers, whereby 12 flowers form the upper and 8 flowers the lower layer;
- Packed per bunch in a hot needle-perforated sleeve.

Other Dianthus must:

- Be supplied in bunches of 10 stems;
- Be supplied in bundles of 5 bunches.

IV. Loading

- **Spray Dianthus** must be packed in the following quantities per container, depending on the weight of the bunch:

Weight per bunch of 10 Spray Dianthus stems (grams)	No. of pieces (bunches) per container container code 577	container code 577
150-175		240 (24)
175-200	240 (24)	200 (20)
200-250	220 (22)	180(18)
250-300	220 (22)	160 (16)
300-375	200 (20)	140(14)
375-500	160 (16)	120 (12)
500-700	140 (14)	100 (10)
700-800	120 (12)	80 (8)
800 and up	120 (12)	60 (6)

- Dianthus with a stem length shorter than 45 (grade code 040) must be supplied in a small container (container code 566).

V. Recommendation

For **Dianthus standard**, extra florist's material can be used to support the flower buds, if necessary

6. Freesia

I. Minimum Requirements for Trade

Minimum quality

- For **single-flowered Freesia** each comb must have a minimum of 4 buds.
- For **double-flowered Freesia** each comb must have a minimum of 3 buds.

Ripeness

Minimum ripeness requirement for **Freesia**: A minimum of 95% of the stems in the batch must have at least 1 coloured flower bud. Freesia 04/11/UK 4

II. Quality and Grading Criteria

Quality requirements per batch

The following additional requirements apply:

- The batch must be free of growth defects, including:
  - deviant/ deformed combs
  - growth tears
  - thumbs
- The batch must be free of damage, including plucked flowers.
- The batch must be free of bent stems.

Overview 1 in the General Specifications for Cut Flowers applies to the above requirements.

- The batch must be free of visible damage from virus(es).
- The batch must be free of old parts of stems.

Overview 2 in the General Specifications for Cut Flowers applies to the above requirements.

Grading requirements per batch

Freesia must be graded according to:

- Length;
- Weight:
- Ripeness.

When grading according to weight, the following requirements apply in addition to the General Specifications for Cut Flowers:

- **Freesias** are graded in the following classes:
  - up to 20 grams in increments of 1 gram;
  - 20 grams and up in increments of 2 grams;
  - the average stem weight in the lightest bundle, rounded down.
- All stems in the batch must have a weight which is equal to at least 80% and no more than 120% of the indicated average stems weight of the batch.

Grading according to stem weight

Average stem weight (grams)	grade code
6	006
7	007
8	008
and so on in increments of 1 gram	and so on in increments of 1
20	020
22	022
24	024
and so on in increments of 2 grams	and so on in increments of 2

III. Packing Specifications

General

Freesia

- Must be supplied in buncShes of 10 stems;
- Must be supplied in bundles of 5 bunches;
- Each bundle must be packed in a sleeve.

IV. Loading

The container must be full.

V. Recommendations

It is recommended that **Freesia** be pre-treated with products containing the active ingredient STS (code 1 on the label) according to instructions.

7. Tulip

I. Minimum Requirements for Trade

Ripeness

At least 95% of the number of Tulipa stems in the batch must have a coloured flower bud.

II. Quality and Grading Criteria

Quality requirements per batch

The following additional requirements apply:

- The batch must be free of growth defects, including
  - water stems;
  - noses;
  - flowers with colour deviations.
- The batch must be free of bulb remnants or bases.

The flowers (buds) must have grown far enough out of the leaf. The tolerances listed in Overview 1 and the General Specifications for Cut flowers apply.

Grading requirements per batch

Tulipa must be graded according to:

- Length, in which respect, contrary to the General Specifications for Cut Flowers:
  - the flowers are graded in classes of 1 cm in accordance with overvie 1
  - the maximum allowed difference in length between the shortest and longest stem in a bunch is 3 cm;
  - the length is determined by measuring the stem from the cut up to the top of the flower(bud);
  - the grading is specified in the grade code in accordance with overview 1.
- Weight, whereby:
  - Tulipa is graded in weight grade classes of 1 gram (see overview 2);
  - all stems in the batch must have a weight at least equal to 80% of the indicated stem weight of the batch;
  - The grading is specified in the grade code in overview 2.

Ripeness

Overview 1: Grading according to length		Overview 2: Grading according to weight		
Min. stem length (cm)	Grade code	Minimum bundle weight per batch (g)	Average stem weight (g)	Grade code
16	016	600	12	012
17	017	650	13	013
18	018	700	14	014
19	019	750	15	015
20	020	800	16	016
21	021	850	17	017
22	022	900	18	018
and so on	and so on	950	19	019
		1000	20	020
		and so on	and so on	and so on

III. Packing specifications

General

- Tulipa must be:
  - bunched so that the flowers (buds) are all on the same level;
  - supplied in bunches of 10 stems;
  - supplied in bundles of 5 bunches;
  - rolled in paper per bundle (rolled in plastic foil is permitted per bunch).
- It is required that Tulipa be supplied standing dry in a container. The products must be properly pre-watered so that they do not droop during the auction process.

IV. Loading

- Loading must be optimal, i.e. the container must be full without damaging the quality of the product; the volume of the bunches determines the quantity of the supplied bunches per container.
- Tulipa must be supplied in container 533 or 560.
- If Tulipa is supplied in container 533, the container must contain at least 200 stems (4 bundles).
- If Tulipa is supplied in container 560, the container should be fully packed.

A Practical Guide to Conduct Keeping Quality Tests

Shelf or Vase life is an important aspect of the quality of plants and flowers. It provides an understanding of the life of plants and flowers among consumers and is therefore essential to such things as quality marks and guarantee programs. Shelf/vase life tests provide important information on such aspects as the length of transport and conditions under which the plants and flowers are transported. They also provide an insight into issues such as botrytis, bloom length, flower opening, ripeness etc.

The instructions will help to correctly prepare and carry out keeping quality tests. The guide will help to gear tests to specific situation and requirements. It is not intended as a standard because different situations and requirements will call for different keeping quality tests.

1. Aim of the tests

Keeping quality tests may be conducted for many different reasons. One aim could for example be to gain insight into your product's keeping quality once it has reached the consumer. But you may also want to conduct keeping quality tests in order to find out more about the effects of different cultivation conditions or operations. The setup, extent and execution of the tests and the requirements relating to the material to be used will depend on the aim of tests. Start the tests; formulate the aim intended to achieve.

2. Methods and materials

Sales simulation

To obtain realistic insight into the keeping quality of your product once, it has reached the consumer, you must try to simulate its route from the greenhouse to the consumer as accurately as possible. A sales simulation may comprise several parts, for example:

- grower-auction simulation phase
- auction-retail trader simulation phase
- retail trader simulation phase

Consider what duration and conditions hold most in each phase for the intended sales channel. Important conditions during the sales simulation that may affect a product's keeping quality are

- temperature, relative humidity, light, presence/absence of wrapping, repackaging, availability of water, use of pretreatment agent

Area in which the sales simulation is to be conducted

The area in which you are to conduct your sales simulation will depend on the simulation you have chosen. Aspects you should consider are temperature, relative humidity, light and the absence of ethylene (the area must be free of exhaust fumes, smoke and vegetables/fruit). When you repeat tests you should ensure that the conditions are as constant as possible at any time of the year. Record the conditions during the sales simulation, for example with the aid of a data logger for measuring temperature and relative humidity.

Consumer phase

Consider the duration of the consumer phase. What do you intend to test? The period up to the end of flowering? Up to the end of the guarantee period? etc.

Area in which the consumer phase simulation is to be conducted

The conditions under which the flowers or plants are to be tested (for example in a canteen or an office) must be as constant as possible. Guidelines for keeping quality tests are:

- A temperature of 20°C ±1°C
- A relative humidity of between 60 and 70%
- A light intensity of 1000 lux (table height) for 12 hours (and 12 hours of darkness)
- No direct sunlight
- No droughts
- Regular ventilation
- Absence of exhaust fumes, smoke and vegetables/fruit.

While repeating the tests it should be ensured that the conditions are as constant as possible at any time of the year. Record the conditions during the test, for example with the aid of a datalogger for measuring temperature and relative humidity.

Materials

Plant materials

Number of stems/plants: The greater the number of flowers or plants you use for your test, the more accurate your impression of your product's keeping quality at the test moment will be. Preferably harvest your cut flowers or plants from different parts of your cultivation area. A product's keeping quality may vary considerably throughout a season. So you will have to repeat the test at different moments to obtain a reliable impression of the keeping quality throughout the season. The VBN recommends the following numbers: *Cut flowers:* at least 1 vase per test. The number of stems per vase will depend on the product's practical application: at most 10. Quality of the plant material: Use plant material that is representative of your range: your most common length/grade and quality. Use products that are ready for auctioning. Describe the product in as much detail as possible, considering the following aspects:

- Cultivar/name; Date of harvesting; Period of cooling; Date of processing; Maturity stage; Length; Pretreatment; Quality group; Pot size;
- Other grading characteristics.

This detailed product description is necessary to test comparable products in subsequent repeat simulations to ensure reliable comparison of the test results.

Assessment cards

The VBN has developed special cards for assessing cut flowers and plants. There are general assessment cards for cut flowers and plants specifying the aspects you should consider in assessing your products. The cards also give guidelines for conducting the tests. For a number of products, there are specific assessment cards that cover in somewhat greater detail the aspects of the products in question that are important with respect to keeping quality.

Assessment forms

Prepare a diagram covering all the aspects to be considered in assessing your products. In preparing the assessment form, consider the frequency of the assessments:

- If you envisage a keeping quality of at most 7 days, assess the products every day until they lose their decorative value and the flowers or plants are to be rejected;
- If you envisage a keeping quality of between 7 and 21 days, assess the products at least once every three days;
- If you envisage a keeping quality of 3 weeks or longer, assess the products at least once a week.

Testing of Cut Flowers Vases

- Always use clean vases. After the test, wash the vases thoroughly with detergent and rinse them with hot water (optionally clean them further with a chlorine solution for two hours).

- Use vases with a height of at least one third of the stem length. Support flowers in lower vases with for example a bucket ring.
- Use tap water, with or without cut flower nutrients, for keeping quality tests.
- If using nutrients, make sure to use the right brand and the right dose. The brand and dose will influence the keeping quality in the consumers' home. If using nutrients, stir well
- To add extra water to the vases during the test, use tap water without nutrients. Labels: Products will have to be labelled if you subject them to different treatments or start testing them on different days.
- When testing cut flowers, you must cut the stems at the beginning of the consumer phase using a pair of sharp secateurs or a sharp knife. Try to avoid damage to the stems.
- Before cutting the stems, clean the secateurs or knife with alcohol or methylated spirits.
- Optionally a camera to record or illustrate the test.

3. Test Procedure

Use the materials selected to subject your products to the sales simulation chosen for the test and use the assessment cards to assess the results. Record the test results in the form of meticulous notes and optionally photos.

4. Processing the results

Upon completing the test, you can start processing your results. Use the example assessment form to record the average keeping quality and other relevant results. These results and the information provided by the dataloggers will give you a complete survey of your test for your own administration. Draw up a short report specifying:

- The day on which you started the test
- Description of the product (cultivar, grade, number of stems/plants)
- Duration and conditions of the sales simulation
- Duration and conditions of the consumer phase

These results (for example average keeping quality, own comments on flowering or quality of the foliage, photos of the test) may be communicated to the customers for better understanding of the products.

References

Bhattacharjee, S.K. Advances in Ornamental Horticulture. Pointer Publishers, Jaipur, pp19.

Bose, T.K. Maiti, R.G. Dhua, R.S. and Das, P. 1999. Floriculture and Landscaping, Naya Prokash, Calcutta, 757p

NHB, 2012. Indian Horticulture Database 2011. National Horticulture Board, Gurgaon, pp 279.

Singh, H.P. Zabeltitz, C.von, Schiva, Tito, Dadlani, N.K. and Gill, A.P.S.2001. An integrated approach for small scale greenhose floriculture in India. FAO Position Paper, FAO, 73p

Singh, M.C. and Singh, K.P. 2001. Protected cultivation of cut flowers. In: S.S. Sindhu and M.L. Choudhary (Eds) Commercial Flower Production. IARI, New Delhi. 87-96/

www.vbn.nl A practical guide to conduct keeping quality test, May 17, 2013.

www.vbn.nl. General specifications for trade of cut flowers, May 15, 2013.

Index

C

D

E

F

G

H

I

J

K

L

M

N

O

P

Q

R

S

T

U

V

W

X

Y

Z

FSC
www.fsc.org
MIX
Papier | Fördert gute Waldnutzung
FSC® C083411

Kolibri 360 GmbH
Ferdinand-Jühlke-Straße 7
99095 Erfurt, Deutschland
produktsicherheit@kolibri360.de